Holography for the New Millennium

Springer

New York
Berlin
Heidelberg
Hong Kong
London
Milan
Paris
Tokyo

Jacques Ludman   H. John Caulfield
Juanita Riccobono

Editors

Holography for the
New Millennium

With 156 Illustrations

Springer

Jacques Ludman
Northeast Photosciences, Inc.
18 Flagg Road
Hollis, NH 03049
USA
jludman@charter.net

H. John Caulfield
Department of Physics
Fisk University
1000 Seventeenth Avenue, N.
Nashville, TN 37208
USA
hjc@dubois.fisk.edu

Juanita Riccobono
Microphotonics Group
Optical Switch
   Corporation
65 Wiggins Avenue
Bedford, MA 01730
USA
jricco@attglobal.net

Library of Congress Cataloging-in-Publication Data
Holography for the new millennium / editors, Jacques Ludman, H. John Caulfield,
Juanita Riccobono.
    p.   cm.
  Includes bibliographical references and index.
  ISBN 0-387-95334-5 (alk. paper)
  1. Holography.   I. Caulfield, H.J. (Henry John), 1936–   II. Ludman, Jacques E.
III. Riccobono, Juanita.
TA1540 .H663   2002
621.36′75—dc21   2001042961

ISBN 0-387-95334-5   Printed on acid-free paper.

Printed in the United States of America.

9 8 7 6 5 4 3 2 1   SPIN 10844781

www.springer-ny.com

Springer-Verlag   New York Berlin Heidelberg
A member of BertelsmannSpringer Science+Business Media GmbH

Many of the ideas for work reported here and the idea for this book itself came out of the Peirmont Place Summer Conferences (1979–present). We gratefully dedicate this book to their sponsor, Jennie Hiller Ludman.

Preface

A universal law of some sort operates in technology that defies Moore's law and other laws of fast progress. It appears that it takes almost half a century between the invention of a revolutionary field and its widespread application. That was true for automobiles, airplanes, computers, and television. It appears to be true also as regards holography. Invented in 1947, holography has passed through several waves of optimism and pessimism to find itself at the beginning of a new millennium poised for the kinds of success those other fields eventually achieved. One sign of that is that holography has distinct roles to play in each of those older fields. In automobiles, versions of holograms are going into next-generation taillights and headlights for white LEDs. In aircraft, it provides the basis for head-up displays. In computers, holograms are being used to direct optical interconnections in massively parallel processor systems. In television, holographic screens are allowing brilliant autostereographic displays. But these are just the beginning. At long last, we know how to make acoustic holograms that will give live 3D images of internal parts of humans and machines without the danger of x-rays and the expense of tomographic systems. Holographic optical components are proving both more flexible and cheaper than are conventional components, and optics is the key field in information—gathering it, storing it, transmitting it, manipulating it, and displaying it. Holography occurs not just in space, but also in spectrum and in time. This has led to such useful concepts as coherence tomography (invented in holography in 1967, but only now being widely used by people who don't know it is holographic) to optical pulse shaping and time-domain optical memory. In short, holography appears to have served its 50 years of obscurity and to be ready to become a ubiquitous and even unnoticed tool.

This seems an ideal time to have some of the world's greatest holographers give their viewpoints on their specialties—where they have been, where they are, and where they are going. What better tool could we ask as holography enters the new millennium?

In this preface, we want to say something about the authors and their essays. These are, without exception, people who have earned the right to speak with authority on the futures of their fields.

Between its invention by Dennis Gabor in 1947 and its parallel reinvention in the United States and Russia 15 years later, many of the giants of optics saw the potential for the field but were unable to overcome the technical drawbacks of Gabor's original method. It remained for brilliant young scientists working in secret on optical processing for military radar in Russia and the United States to invent independent ways of overcoming those problems. Today, those inventors—Emmett Leith and Juris Upatnieks in the United Sates and Yuri Denisyuk in Russia—are good friends and colleagues. Leith and Denisyuk are similar people not only in their career paths, but also in their brilliance and humanity. They are much loved by the whole optics community. We are very fortunate that Emmett Leith agreed to write the first and featured chapter of this book. He speaks with an authority only he and Yuri Denisyuk have earned for the future of the field they, along with Gabor, created.

Of course, the excitement about holograms has always been about the wonderful 3D images they can produce. We have two chapters on the frontiers of holographic imaging. Many people have worked on full-color holography over the last 35 years, but the one who has made it work so well and beautifully is Hans Bjelkhagen, who has written a chapter on that field. Early on, people hoped to achieve holographic television. Full, realistic 3D images would leap from the "screen" (hologram). Gradually, the massive difficulties of that endeavor started to become apparent. We are aware of people trying now with much better computers, much better communication systems, and much better electronic hologram screens (called Spatial Light Modulators) to achieve this dream sometime in the moderately near future. But holography can be used as a screen to allow viewers to see stereographic movies without needing special glasses—a concept originally proposed and worked on by Gabor. Tin Aye, Andrew Kostrzewski, Gajendra Savant, and Tomasz Jannson describe here the history, current state, and future of this dramatically beautiful kind of display.

In the early days after the Leith and Denisyuk work, it seemed to many of us that everything that could be known about how to make a hologram was already known. Our job was to find new ways to use the already established methods. Of course, we were wrong. The next three chapters describe hologram formation methods on the current forefront—methods undreamed of back in the mid-1960s. Michael Metz describes the making and use of edge-lit holograms. The illumination comes neither from the front nor the back of the hologram but though its edge. Next, Joseph Mait, Dennis Prather, and Neal Gallagher describe a special kind of high-efficiency, very versatile, on-axis hologram. So far as we know, Neal made the first of these out of a copper plate and used it to focus microwave energy into a desired pattern. The optical version has created a significant new industry. The ability of two beams of light to form a stable interference pattern is called their mutual coherence. With low-coherence sources, we can cause light from one narrow slab of a 3D object to record a hologram, whereas light from elsewhere does not. Christopher Lawson and Denise Brown-Anderson describe the huge progress in that field from its invention in the mid-1960s. Incidentally, Emmett Leith is deeply involved in this field of research now. Designing a hologram by computer can be difficult. But even after you design it, you must make it. Suppose you could combine those steps and simply tune the hologram to have any physically allowable property you choose. Joseph Shamir shows how this is done.

Of course, all of these nice advances are "academic," in the bad sense of that word, unless holography is good for something. It is, in fact, good for many things. We offer a sampling of the applications likely to be important in the 21st century. There are many more. Nita Riccobono and Jacques Ludman discuss a technique just now being commercialized for using holograms to improve the efficiency of solar photovoltaic energy converters. Leith's group was among many in the mid- and late 1960s to invent ways of using holograms to do optical nondestructive testing with holograms. Contributing many things to this field and monitoring it from the beginning was the author of the chapter on this subject, Pramod Rastogi. Early on, the idea of making holographic diffusers was invented, but it remained for Gajendra Savant and the Jannsons (Joanna and Tomasz) to perfect this concept, commercialize it, and aim it toward the 21st century. That tale is told in their chapter. Key to almost all past, current, and future holography is the laser. In principle, it is the almost-perfect light source for this purpose. But, in fact, it is imperfect. Not all of the light emerging from a laser is of the nice, clean variety. We need to filter out the unwanted light, leaving only the "good" light. Ludman and Riccobono describe a way to use holograms to do that filtering when the standard methods fail.

Another application of very thick holograms is the storage and rapid random access to pages of data. Philip Hemmer, Selim Shahriar, Jacques Ludman, and John Caulfield describe how page-oriented holographic memory works and show some applications. At the opposite end of the thickness spectrum, very thin holograms can be used as a very powerful means to make antireflection coatings. This is described in the chapter by Jacques Ludman, Timothy Upton, David Watt, and John Caulfield.

One of the first applications of holography was the imaging of small particles in a gas or liquid. Unlike the other application we have discussed, this one worked well from the beginning. It works even better now, thanks in no small part to the efforts of the author of the chapter on that subject, Chandra Vikram.

But holography has some fundamental aspects as well as the practical ones just noted. A wonderful teacher and lecturer, as well as an innovative researcher, Nils Abramson has spent decades discovering and explaining what holography can teach us about relativity, and conversely; his chapter focuses on this ongoing work. We like to say that "Optics is Quantum Mechanics Writ Large." We opticists were doing quantum mechanics before we knew those words. Holography can be described by quantum mechanics as shown by Alex Granik and George Chapline in the final chapter.

Hollis, NH, USA   JACQUES LUDMAN
Cornersville, TN, USA   H. JOHN CAULFIELD
Carlisle, MA, USA   JUANITA RICCOBONO

Contents

Contributors

Nils Abramson
Industrial Metrology and Optics / IIP
Royal Institute or Technology
S-100 44 Stockholm
Sweden
nilsa@iip.kth.se

Denise Anderson
Research Analyst
Center of Naval Analyses
4825 Mark Center Drive
Alexandria, VA 22311-1850 USA
Andersob@cna.org

T. Aye
Physical Optics Corporation
20600 Gramercy Pl. Bldg 100
Torrance CA 90501-1821 USA
tinmaye@aol.com

Dr. Hans I. Bjeklhagen
De Montfort University
Centre for Modern Optics
Hawthorn Building
The Gateway
Leicester LE1 9BH, UK
hansholo@aol.com

H. John Caulfield
Distinguished Research Scientist
Fisk University
1000 17th Ave., N.
Nashville, TN 37208 USA
hjc@dubois.fisk.edu

G. Chapline
Physics Dept.
Stanford University
Stanford, CA
chapline@lanl.gov

Neal Gallagher
Dean
College of Engineering
Colorado State University
Fort Collins, CO 80523-1301 USA
nealg@engr.colostate.edu

Alex T. Granik
Associate Professor
Physics Department
University of the Pacific
Stockton, CA 95211 USA
agranik1@attbi.com

Philip Hemmer
Department of Electrical
Engineering, Rm 216H
Texas A&M University
3128 TAMU
College Station, TX 77843-1064 USA
prhemmer@ee.tamu.edu

Joanna Jannson
Physical Optics Corporation
20600 Gramercy Pl. Bldg 100
Torrance CA 90501-1821 USA
jjannson@aol.com

Tomasz Jannson
Physical Optics Corporation
20600 Gramercy Pl. Bldg 100
Torrance CA 90501-1821 USA
tomjannson@aol.com

A. Kostrzewski
Physical Optics Corporation
20600 Gramercy Pl. Bldg 100
Torrance CA 90501-1821 USA
akostrzewski@poc.com

Chris M. Lawson
Co-Director, Alabama EPSCoR Program
Director, Laser and Photonics Research
Center
Professor of Physics
University of Alabama at Birmingham
310 Campbell Hall, 1300 University Blvd.
Birmingham, AL 35294 USA
lawson@uab.edu

Emmett Leith
Professor
Dept. EECS
University of Michigan
Ann Arbor, MI 40109-2122 USA
leith@engin.umich.edu

Jacques Ludman
Northeast Photosciences, Inc.
18 Flagg Road
Hollis, NH 03049 USA
jludman@charter.net

Dr. Joseph N. Mait
U.S. Army Research Laboratory
AMSRL-SE-EM
2800 Powder Mill Road
Adelphi, MD 20783-1197 USA
jmait@arl.army.mil

Michael Metz
ImEdge Technology, Inc.
2123 Fountain Court
Yorktown Heights, NY 10598 USA
mmetz@imedge.com

Rafael Piestun
Assistant Professor
Department of Electrical and Computer
Engineering
University of Colorado at Boulder
Campus Box 425
Buolder, CO 80309-0425 USA
piestun@colorado.edu

Dennis W. Prather
Associate Professor
Department of Electrical and Computer
Engineering
University of Delaware
140 Evans Hall
Newark, DE 19716 USA
dprather@ee.udel.edu
www.ee.udel.edu/~dprather

Pramod Rastogi
IMAC–DGC
EPFL
Swiss Federal Institute of Technology
Lausanne
CH-1015 Lausanne
Switzerland
pramod.rastogi@epfl.ch

Juanita Riccobono
Microphotonics Group
Optical Switch Corp.
65 Wiggins Avenue
Bedford, MA 01730 USA
jricco@attglobal.net

Gajendra Savant
Physical Optics Corporation
20600 Gramercy Pl., Bldg 100
Torrance CA 90501-1821 USA
gsavant@aol.com

Selim Shahriar
Associate Professor
Northwestern University
Department of Electrical and Computer
Engineering
2125 Sheridan Road
Evanston, IL 60208-3118 USA
smshahri@mit.edu

Joseph Shamir
Professor
Dept. Electr. Eng.
Technion-Israel Institute of Technology
Haifa 32000, Israel
jsh@ee.technion.ac.il

Tim Upton
Northeast Photosciences, Inc.
18 Flagg Road
Hollis NH 03049 USA
tim@borneo2.unh.edu

Chandra S. Vikram
Center for Applied Optics
400-C Optics Building
The University of Alabama in Huntsville
Huntsville, AL 35899 USA
vikramc@email.uah.edu

David Watt
Associate Professor
Mechanical Engineering Dept.
University of New Hamshire
Durham, NH 03824 USA
dww@christa.unh.edu

Part I

Overview

Where Holography Is Going

E. Leith

Introduction

Holography has had an erratic past, but today it exists as a modest although significant industry and is growing at a respectable rate. Estimates by The Renaissance group estimates its dollar volume worldwide to be about one billion per year with a 30% annual growth rate. Whether these impressive figures can be maintained is uncertain, because the greatest part of the activity is in security devices and product verification, along with the rainbow type hologram for magazine covers, wrapping paper, and so on. It would of course be more comforting to see the commercial activity spread across a wider spectrum of application, and indeed, this seems to be happening, albeit at a slower growth rate. Here, I review the various major areas of holography and make some prognostications as to where they are going.

Display Holography

Display holography, in all of its various forms, is one of the most interesting, perhaps enigmatic, areas of holography. In 1964, when we first showed 3D holographic imagery of arbitrary, reflecting objects, the astonished viewers predicted that display holography would surely become one of the great application areas for holography. However, there were two major blocks to this realization. The first was a need for a cheaper and more convenient light source than a laser for viewing the hologram. The Denisyuk white light viewable hologram was the answer. Second, holograms had to be less expensive. The Benton rainbow hologram, combined with the hot stamping process,hologram!hot stamping process was the answer. Today, most of the world's holograms are of this kind.

Great as these advances were, and greatly as they enlarged commercial display holography, there is yet room for significant enlargement of the field, provided other limitations can be overcome. The Denisyuk hologram produces incredible realism and, as the years have passed, has become the standard for holographic realism. The image quality has steadily advanced. The holograms have become brighter, and the scatter has decreased, giving holograms of marvelous clarity. Such holograms, frequently seen at galleries, continually become better with the passage of even a few years. Visitors to these displays marvel at the realism of the image. In recent years, color holograms instead of monochrome have become commonplace. Yet, holograms of such superb quality are confined almost entirely to gallery exhibitions, because such holograms are expensive to make as opposed to hot stamped holograms and require special lighting, i.e., lighting from sources of moderate spatial coherence, in which the radiating element is only a few millimeters in its dimensions, and the lighting arrangement requires precise specifications. More convenient illumination methods are required before such holograms can be routinely viewed outside of galleries. The edge lit hologram is one avenue of improvement. Also, the bold efforts of Denisyuk to produce holograms that can be viewed with ordinary light instead of light of relatively high spatial coherence would be another great advance, although the difficulties in this direction are formidable. Such holograms, if developed, would possibly have low efficiency and fairly high scatter. Display holography has come far since 1964, but there is yet a further way to go.

There have been and are today a host of ways to produce 3D images; none has attained widespread use. All have serious limitations, either being complicated to implement or limited in realism or both. In general, 3D imaging carries a price, either monetary or in convenience, that precludes its widespread use. There is no 3D technique that, to my knowledge, comes even close to holography for its sheer realism, but also none that is more complicated and expensive to implement than is holography in its most dramatic forms.

Three-dimensional imaging (non holographic) may be at last on the verge of assuming a major importance, because of its alliance with the computer. Strong 3D perception is available without the stereo, or binocular disparity, that is often considered essential to 3D perception. Parallax without binocular disparity can give a strong perception of three-dimensionality, and it is readily available on a computer monitor by simply having the image rotate. Binocular disparity can be added with a certain amount of inconvenience, as with special glasses. Such 3D imaging can readily be dynamic and interactive; i.e., the image can change with time, and the observer can interact with it.

For holography to compete with this, I envision a real-time medium on which to form the hologram, say, a liquid crystal display, addressable with computer-generated data through a wideband data link. Such technology will almost certainly develop, and it could put holography in a dominant position over all competing 3D techniques, but this is far down the road.

Phase Conjugation

Optical phase conjugation, a process in which a wavefront is converted into its conjugate, can be used for imaging through irregular media. In 1962, we first proposed the use of the phase conjugation process to image through irregular media. The proposed process was in the context of holography. The conjugate image produced by Gabor's process of wavefront reconstruction produced the well-known twin image, and considerable effort was expended in those early days in conceiving methods for its elimination or reduction. We introduced a twist by proposing that this twin image could in fact have a usefulness, which is now the well-known process of conjugating the image and passing it back through the same irregular medium, whereupon the wavefront is "healed," and an image of excellent quality can thereby be obtained. Shortly afterward, in 1965, both we and Kogelnik independently reported a demonstration of this process.

The technique lay fallow for several years until more practical ways of carrying out the process became available. Among these is the use of four wave mixing in nonlinear crystals, which bears a strong similarity to holography, except that it is carried out in what could be called real time. Other ways of producing the process, such as Brillioun scattering, bear little or no relation to holography. The wavefront healing method, although originating in holography, has evolved to be distinct from holography, carried out by researchers remote from holography. Phase conjugation indeed qualifies as a field, and its historical connection to holography is nearly forgotten.

There is a theme here that recurs in holography. Holography spawns a new area that eventually grows beyond the boundaries of holography. Holography then may remain a discernible part of the new area; other times, the distinction as to where holography ends may become blurred to the extent that the drawing of a sharp boundary between holography and what may be called transholography becomes impossible.

Diffractive optical elements

Early in the history of holography it was found that holograms could function as optical elements, such as lenses, aberration correction plates, beam splitters and combiners, and so on. Somewhat later, in the 1960s, the full potential of holograms as optical elements began to be realized. The disadvantages of holograms, the color dispersion, and their less than 100% diffraction efficiency was countered with their potential advantages—lightness, the arbitrariness of their form—comparable to a refractive element of an arbitrary figure. Thus, it was thought that holograms could serve as an important kind of optical element, operating on diffraction principles instead of refractive and complementing the widely used refractive element. Much investigation was required before genuine applications were uncovered, as opposed to the many ill-conceived, nonviable applications that arose from many quarters.

The holographic optical element, the HOE, eventually became a success. Various worthwhile applications were found, including gratings, phase corrector plates, and beam combiners, for example. More specifically, holographic diffusing screens with controlled angular reflection patterns, the ERIM gunsight, and holographic knotched filters have been successful products.

As the success of HOEs expanded, there came the realization that all that diffracts is not necessarily a hologram. New classes of diffractive elements came into being that were similar to but not really holograms. Thus was born the DOE, the diffractive element, of which the HOE is a subclass. Of course, DOE in a limited way had been around for a long time, e.g., diffraction gratings. Today, diffractive optics has become a major branch of optical technology, with many applications, and many companies commercializing this technology. Some of the resulting DOEs are clearly holograms. Some are decidedly not. But the distinctions have become blurred, and in many cases, there may be genuine disagreement as to when DOEs are really HOEs.

Figure 0.1 is a diagram that places the various DOEs in their proper relation; within the entire realm of diffractive optical elements are placed various important subtypes. The hologram occupies a significant portion; how large its circle should be is a matter on which opinions differ. It may be as large as I depict it, or significantly smaller. A subclass of the HOE is the phase-only hologram (the Ph circle), in which only the phase part of a complex function is used, with the amplitude part being discarded. Also within the DOE circle is the kinoform, which its proponents insist is not a hologram, although it is a close relative. Also, we show the binary optical element, which is defined not so much in terms of its structure, but in how it is constructed, using the advanced technology of mask-making for computer chips. Finally, we show the Fresnel lens, which acts like a diffractive element, but its structure is so coarse that diffraction is negligible; the operations it performs on a transmitted beam are of a refractive nature; so it is properly placed outside the DOE circle. As the diagram suggests, there are yet other types of DOEs, one of which is a phase structure produced by diamond stylus cutting methods, a very traditional technique for ruling diffraction gratings, now extended to construction of more arbitrary phase structures.

Hologram interferometry

In 1965, a new kind of interferometry was born, hologram interferometry. In this technique, a hologram is made of a moving object, and the reconstructed image has superimposed on it a fringe pattern that indicates how the object had moved during the exposure time. It has three basic forms: time average, double exposure, and real time. In the first case, the object vibrates during the hologram exposure time; in the second, the object make a jump during the exposure time; and in the third, the object does not move at all during the exposure time; but the holographic conjugate image is interfered with the object and any movement of the object during this observation shows up as a fringe pattern. In most general terms, the object moves in any manner during the exposure time and the resulting fringe pattern on the holographic image gives an indication of the total motion.

BINARY OPTICS

Some related terms that are often confused:

Diffractive optics
Holographic optical elements
Phase-only holograms (especially phase-only holographic optical elements)
Binary optics
Kinoform
(Fresnel lens)

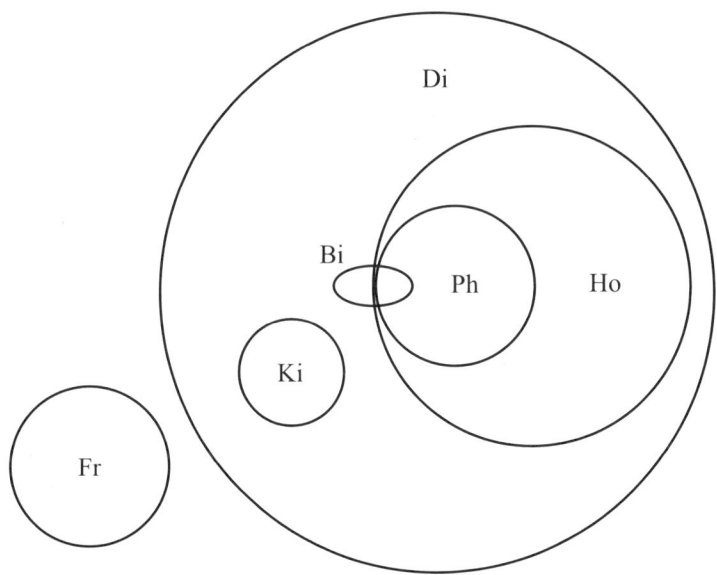

Di: diffractive optical element
Ho: holographic optical element
Ph: Phase-only holographic optical element
Bi: Binary optical element
Fr: Fresnel lens
Ki: Kinoform

FIGURE 0.1.

The technique was heralded as the first genuine application of holography, and hologram interferometry became a significant area of modern optics. Books were written on this topic, and entire sessions of technical conferences were devoted to it. As hologram interferometry grew in importance, related areas sprang up. If the basic process of hologram interferometry was carried out without the reference beam, the phase was not recorded and the process ceased to be holography. However, the highly coherent light produced, on reflection from the object, a speckle pattern and the contrast of the speckles depended on the object stability. Thus, a vibrating object would be imaged with a speckle pattern superimposed on it, with the contrast of the speckles being a measure of the amount of vibrational motion. Spatial filters could then convert the variable contrast speckles into a fringe pattern, thus, superimposing on the image a fringe pattern analogous to that of hologram interferometry. The process was less sensitive than was hologram interferometry, but much easier to implement; the speckle technique became an adjunct to holo-gram interferometry. Speckle interferometry is, I think, not holography, but it is close, and could be likened to the modulation of the object beam into a noise-like carrier, as opposed to the off-axis carrier of conventional holography. Then, there is also the related process of shearography, again similar to holography, but yet different. All three methods are useful, and they can be combined into a single instrument that employs a charge-coupled display (CCD) camera as a recording device, with extensive computer processing. One then chooses the mode preferred for the application at hand. When combined with electronic cameras and com-puters, it is inevitable that these closely related technologies will have a brilliant future.

These three techniques, diffractive optics, phase conjugation, and hologram interferomtry, together support the theme stated earlier. Holography, as it develops, has a tendency to meld with related techniques to form a broader technology in which the holographic portion becomes blurred. The broader technique may continue to grow and may coalesce even further with yet unknown technologies.

In each of these cases, a holographic technique was established, then expanded, becoming something more general than holography, and in the process became a major part of optical technology. In phase conjugation, the origin in holography is nearly forgotten. In diffractive optics, the connection to holography is well recognized and significant, but the field has been enlarged so that holography, although a significant part, is nonetheless a minor part. For laser metrology, the story is similar.

Some other technologies

Holography preserves phase by a spatial heterodyning process. A temporal signal, with a phase term with spatial dependence, $a(x, y) \exp[j2\pi vt + \phi(x, y)]$ is mixed with a temporal signal with a spatial phase term varying linearly, $\exp j(2\pi vt + \phi(x, y))$, giving a beat frequency $a(x, y) \cos[2\pi\alpha x - \phi(x, y)]$, or if we consider only one sideband, the complex equivalent. It is akin to a conventional mixing process, in which all variables are time functions $a(t) \exp(j2\pi\alpha t 4 + \phi(t))$, giving rise to the mixing term $a(t) \exp[j2\pi\alpha t + \phi(t)]$. The conventional mixing process is commonplace today, as it has been since long before the origin of holography. It is the spatial terms αx and $\phi(x, y)$ that are peculiar to holography and distinguish holography from a conventional hetrodyning process. With the spatial output being placed on a transparency and the subsequent reconstruction of the spatial data onto a beam of light, the holographic process is complete. And in the transformation of object information into the hard-copy hologram and then on readout, the conversion of the information back onto a light beam comes all of the astonishing properties of the hologram. With the removal of the spatial aspect of the mixing process, the holographic nature becomes blurred, or perhaps disappears, even though the same information is captured and processed in a mathematically similar way to produce the same final result.

Synthetic aperture radar (SAR) offers an excellent example. It is really not holography, but it has striking similarities in some of its embodiments, but can also be different. Here, an airplane carries an antenna along its flight path; the antenna emits a beam that scans the terrain. The signal returned to the aircraft is purely temporal; its temporal nature results from the scanning process, in which a spatial signal, the terrain reflectivity, is mapped into a temporal signal. In the very early days of synthetic aperture radar, the signal was processed in various ways to produce the final result, an image of the terrain. For example, the signal in one embodiment was placed into a recirculating delay line. Subsequent pulses of radiation were integrated as the aircraft progressed along the flight path, and the result was the image. Only the slightest resemblance to holography existed here.

Later in the mid-1950s, much more sophisticated systems were developed. The data were stored on photographic film, because that was the best storage medium in existence at that time. The processing to be done to convert the raw data into an intelligible image was then done with coherent optics, because its capability for the task far exceeded what the digital computers of that time could do. The recorded data were in fact the radar waves reflected back to the aircraft flight path. Illumination of the photographic record with a beam of coherent light produced a regeneration of the field on the flight path. These waves, optical replicas of the original radar waves, then proceeded to propagate and form an image. The synthetic aperture process in this embodiment was a very close analog of holography, and all of the basic physical and geometrical optical properties of the holographic process became manifest. The SAR process, in combination with coherent optical processing, was indeed generally described in holographic terms.

Now, many years later, the preferred and almost universal recording medium is no longer photographic film, but magnetic tape. The processing is done by digital computer. The basic physics of the process is unchanged, and the end result is again an image of the terrain being scanned, but the holographic nature of the process has essentially evaporated. The SAR process is more one of data collection and processing by computer, and the practitioners of this process generally do not think of it as holography.

We take a second example. We image through highly scattering media, such as biological tissue, by recording the light emitted from the exiting surface of the medium. Using the coherence gating properties of the holographic process and using light of short coherence length, the reference beam is adjusted to interfere only with the light that takes the shortest path through the scattering medium; this light, because it is scattered least, forms the best image of an embedded absorber within the medium. The hologram is formed on the surface of a CCD camera. The hologram exists for only a few seconds, just long enough for the hologram signal to be read into a computer, in which the reconstruction process is carried out digitally, and the image may be stored in memory, for display on a monitor when required.

The process described is clearly one of holography, although of a form far removed from mainline holography, with the hologram being unseen, ephemeral, and without an optical readout. Here is a form of holography, developed over a period of many years to a highly sophisticated level, using modern electronic imaging techniques and closely tied to the digital computer. It is a form of holography that will undergo tremendous advances in the 21st century, because of its ties to the rapidly advancing fields of electronic imaging and digital computers.

Now we carry the process one step further. The illumination emitted from a point on the exiting surface of the scattering medium is mixed with a second beam of light that is frequency shifted by a small amount, perhaps a few megahertz. The mixing process produces, after detection, an electrical signal in the megahertz range. The correlation process sorts out the first arriving light, and the detected signal then forms one image point of an image. Many points on the exiting surface can be simultaneously detected and recorded, or a scanning process can be used, with a point detector sequentially detecting the signal from different points of the exiting surface. The end result, an image, is essentially the same as with the holographic process, but now no hologram is formed at all, and the holographic aspect of the process has disappeared. Yet, the basic signal detection and processing process is essentially the same, except for being carried out in the time domain. Or, should one think of this alternative process as being time domain holography? I think probably not, although this is ultimately a semantic issue.

On the basis of the observations made here, I envision the following course of holography. Holography in the 21st century will continue to flourish. Its growth will result in large part from the advancement of the technologies on which holography depends: the computer, the electronic camera, the advances in real-time recording and display media, the advances in development of mask-making, and others. At the same time, as holography grows and as new forms develop, the boundary between holography and nonholography will become more indistinct.

Part II

Display Holography

1
Color Holograms

Hans I. Bjelkhagen

1.1   Introduction

Even after almost 40 years since the appearance of the first laser-recorded monochromatic holograms, the possibilities of recording full-color high-quality holograms are still limited. Although various special techniques of today allow for the production of holograms exhibiting several different colors, in most cases, the colors displayed in these holograms are not the true, original colors of the holographed object. These holograms are often referred to as *pseudocolor* or *multicolor holograms*. Applying Benton's rainbow technique, it has been possible to record transmission holograms as well as mass-produce embossed holograms with "natural" colors. However, a rainbow hologram has a very restricted viewing position from which a correct color image can be obtained.

It is common among artists to make multiple-exposed color reflection holograms using a single-wavelength laser in which the emulsion thickness has been changed in between the recordings of special objects. In this way, many beautiful pseudocolor holograms have been made.

Sometimes lifelike holographic images have been referred to as full color, natural color, or true color holograms. The most logical name for these holograms that comes to mind by analogy with color photography, color movies, and color television would be *color holograms*. However, this term is sometimes objected to on the grounds that some colors of objects we normally see around are impossible to record holographically because holograms can only reproduce colors of objects created by scattered laser light. Colors we see are often the result of fluorescence, which cannot be recorded in a hologram. For example, some dyed and plastic objects achieve their bright, saturated colors by fluorescence. This limitation in color holography does not seem to be, however, very dramatic.

This chapter will acquaint the reader with the topic of color holography by presenting a review of the history, as well as the current status and the prospects of this discipline.

1.2   The History of Color Holography

The first methods for recording color holograms were established in the early 1960s. Leith and Upatnieks proposed multicolor wavefront reconstruction in one of their early papers.[1] Mandel[2] pointed out that it may be possible to record color holograms in a more straightforward way, using a polychromatic laser and an off-axis setup. Lohmann[3] introduced polarization as an extension of the suggested technique. These first methods concerned mainly transmission holograms recorded with three different wavelengths from a laser or lasers, combined with different reference directions to avoid cross-talk. The color hologram was then reconstructed by using the original laser wavelengths from the corresponding reference directions. Color holograms of a reasonably high quality could be made this way, but the complicated and expensive reconstruction setup prevented this technique from becoming popular. The first transmission two-color hologram was made by Pennington and Lin.[4] The authors used the 15-μm thick Kodak 649-F (Rochester, NY) emulsion with a spectral bandwidth of about 10 nm. This narrow bandwidth eliminated, in principle, cross-talk between the two colors (633 nm and 488 nm) at the reconstruction.

The Lippmann color technique is very suitable for color recordings applying the technique of *reflection* holography. Lin et al.[5] made the first two-color reflection hologram that could be reconstructed in white light. They recorded a reflection hologram of a color slide illuminated with two wavelengths (633 nm and 488 nm). The material used here was the Kodak 649-F plate, which was processed without fixing to avoid shrinkage. This technique, using reflection holography and the white-light reconstruction technique, seems to be the most promising one as regards the actual recording of color holograms and will be further discussed later on. However, the three-beam transmission technique may eventually become equally applicable, provided inexpensive multicolor semiconductor lasers appear on the market in the future.

Relatively few improvements in color holography were made during the 1960s, and only a few papers were published.[6–12] Some publications on color holography appeared during the 1970s.[13–23] At the end of the 1970s, high-quality color holograms appeared thanks to the work by Kubota and Ose.[23]

During the 1980s, several new and improved techniques were introduced.[24–38] At the beginning of the 1980s, publications on color holography started to appear in the former Soviet Union.[25–27] A 1983 review of various transmission and reflection techniques for color holography can be found in a publication by Hariharan.[28] The possibilities of natural color stereograms are discussed in two papers.[29, 30] With regard to the development of color reflection holography, the work by Kubota[34] in 1986 must be mentioned.

From 1990 and until today, many high-quality color holograms have been recorded mainly due to the introduction of new and improved panchromatic recording materials.[39−71] Main contributions have been made by Kubota and coworkers,[39−48] Hubel and Solymar,[49] and Bjelkhagen and coworkers.[57−60]

Color reflection holography presents no problems as regards the geometry of the recording setup, but the final result is highly dependent on the recording material used and the processing techniques applied. The single-beam Denisyuk recording scheme has produced the best results so far. Color holograms have been recorded in single-layer silver-halide emulsions, or in two separate silver- halide emulsions in a sandwich. Dichromated gelatindichromated gelatin (DCG) emulsions or a DCG emulsion in combination with a silver-halide emulsion have been used. Panchromatic photopolymer recording materials are an alternative to silver-halide materials. In particular, DuPont's (Wilmington, DE) new color photopolymer film is suitable for mass production of color holograms.[67−70] Watanabe *et al.*[71] reported on the production of color holograms at DAI Nippon in Japan, recorded on DuPont's new color photopolymer material.

The following are some problems associated with the recording of color reflection holograms in silver-halide emulsions:

- Scattering occurring in the blue part of the spectrum found in Western silver-halide emulsions makes them unsuitable for the recording of color holograms.
- Multiple storage of interference patterns in a single emulsion reduces the diffraction efficiency of each recording. The diffraction efficiency of a three-color recording in a single-layer emulsion is lower than is a single wavelength recording in the same emulsion.
- During processing, emulsion shrinkage frequently occurs, which causes a wavelength shift. White-light–illuminated reflection holograms normally show an increased bandwidth on reconstruction, thus, affecting the color rendition.
- The fourth problem, related to some extent to the recording material, is the selection of appropriate laser wavelengths and their combination in order to obtain the best possible color rendition of the object.

Some of the problems mentioned above have been discussed in the paper by Lin and LoBianco.[12] Noguchi[16] tried to make a quantitative colorimetric comparison between images in recorded color holograms and the actual colors of test targets. He used the standard 649-F emulsion and four primary wavelengths for the recording. However, the reproduction of blue color was difficult. Kubota and Ose[23] demonstrated that a good color reflection hologram could be recorded in a dichromated gelatin emulsion with high efficiency and good blue reconstruction. Hariharan[24] introduced the sandwich recording technique to improve image luminance compared with the earlier triple-exposed 649-F emulsions. He used Agfa 8E75 emulsion for the red (633 nm) recording and the 8E56 emulsion for the green (515 nm) and blue (488 nm) recordings.

FIGURE 1.1. Sandwich color hologram recorded by Kubota in 1986, size 200 mm × 250 mm. *See color insert.*

The sandwich technique has been further used by Sobolev and Serov[25] for the recording of color holograms. The most successful sandwich recording technique has been demonstrated by Kubota,[34] who used a dichromated gelatin plate for the green (515 nm) and the blue (488 nm) components, and an Agfa 8E75 plate for the red (633 nm) component of the image. Because the DCG plate is completely transparent in red light, the silver-halide plate (containing the red image) was mounted behind the DCG-plate in relation to the observer. Kubota's sandwich color hologram of a Japanese doll is reproduced in Fig. 1.1.

Hubel and Solymar[49] used Ilford silver-halide materials for the recording of color reflection holograms. The sandwich technique (SP 672T for blue and green and SP 673T for red) was used with the recording illuminations at 458, 528, and 647 nm. Although Ilford blue/green material worked better with regard to light-scattering noise than did Agfa 8E56 HD material, nevertheless holograms produced on Ilford materials suffer from the blue light-scattering noise. The color rendition is, however, good, and Hubel and Solymar achieved at that time the best results on Western commercial silver-halide materials.

Very important is the Russian work by Usanov and Shevtsov[51] based on the formation of a microcavity (MC) structure in gelatin by applying a special processing technique to silver-halide emulsions. Their color holograms, recorded with only two wavelengths, have a high diffraction efficiency and exhibit a typical DCG hologram quality.

The recording of high-quality large-format reflection color holograms in a single-layer ultrahigh-resolution silver-halide emulsion was published by Bjelkhagen et al.[58] Markov and coworkers[61-64] have investigated recording problems and the characteristics of color reflection gratings and holograms in single-layer emulsions. The problem of multiple storage of interference patterns in a single layer and its influence on diffraction efficiency was treated by Chomát[72] and Shevtsov.[73]

1.3   Recording of Color Holograms

Currently, most transmission color holograms are of the rainbow type. Large-format holographic stereograms made from color-separated movie or video recordings have been produced. There are some examples of embossed holograms in which a correct color image is reproduced. In order to generate a true color rainbow hologram, a special setup is required in which the direction of the reference beam can be changed in-between the recordings of the color-separated primary images. However, as already mentioned, holographic color images of the rainbow-type can reconstruct a correct color image only along a horizontal line in front of the film or plate and are therefore of less interest for serious color holography applications.

To be able to record high-quality color reflection holograms, it is necessary to use extremely low light-scattering recording materials. This means, for example, the use of ultrahigh-resolution silver-halide emulsions.[74] Such materials have the advantage of being more sensitive compared with photopolymer or DCG, which are alternative materials for color holography. Ultrahigh-resolution silver-halide emulsions (grain size 10 to 20 nm) for monochrome holography have been manufactured in Russia since the 1960s, and these types of emulsions have now been panchromatically sensitized. Currently, holographic color plates (PFG-03C) are produced by the Micron branch of the Slavich photographic company located outside Moscow.[75, 76] The standard sizes range from 63-mm × 63-mm format up to 300-mm × 406-mm glass plates. Slavich has started to produce film, mainly monochromatic emulsions, but the PFG-03C color emulsion is also available on film on special order. Because there are great variations between each batch of this material, it is difficult to make a detailed characterization of the emulsion. The silver-halide grain size is, however, the most important parameter of this material, playing the most important role for the final quality of the holographic image. Some characteristics of the Slavich material are presented in Table 1.1.

In order to manufacture such emulsions, it is important to slow grain growth during emulsification by increasing the number of growth centers and introducing special growth inhibitors. The best emulsion ever made is probably the one obtained by Kirillov et al.[77, 78] in Russia. In this case, grain growth was hampered by the fact that a highly diluted solution was used in the emulsification process and that

TABLE 1.1. Characteristics of the Slavich color emulsion.

Silver halide material	PFG-03C
Emulsion thickness	7 μm
Grain size	12–20 nm
Resolution	∼ 10000 lp/mm
Blue sensitivity	∼ 1.0–1.5×10^{-3} J/cm^2
Green sensitivity	∼ 1.2–1.6×10^{-3} J/cm^2
Red sensitivity	∼ 0.8–1.2×10^{-3} J/cm^2
Color sensitivity peaked at:	633 nm, and 530 nm

the emulsion's concentration was increased by applying the method of gradual freezing and thawing.

Ultrahigh-resolution holographic emulsions are normally processed in solution-physical developers, creating colloidal silver. The most common procedure used for holograms of the Russian type is based on diluted developers of the solution-physical or semiphysical type, e.g., the Russian GP-2 developer, in such a way that silver particles of an appropriate size are formed (colloidal silver). However, although this processing technique works extremely well for monochromatic recordings, it is less suitable for color holography. The colloidal silver accreted in the emulsion introduces a light-red or brownish stain in the processed emulsion. This affects, in turn, color rendition, which is why the staining must be avoided in holographic color recordings. Although developers based on pyrogallol provide sufficient emulsion hardening, they are not suitable because of the brown stain they leave in the emulsion. By using suitable processing chemistry and processing baths, it has been possible to obtain high-quality stain-free color holograms, previously reported by Bjelkhagen et al.[58] These holograms are recorded in a single-layer emulsion, which greatly simplifies the recording process compared with all the earlier techniques. The aforesaid single-layer technique, including some improvements, is described in the following sections.

1.3.1 Laser Wavelengths for Color Holograms

The problem of choosing optimal primary laser wavelengths for color holography is illustrated in the 1976 CIE (Commission Internationale de l'Eclairage) chromaticity diagram (Fig. 1.2), which indicates suitable laser wavelengths for color holograms. It may seem that the main aim of choosing the recording wavelengths for color holograms would be to cover as large an area of the chromaticity diagram as possible. However, many other considerations must be taken into account when choosing the wavelengths for color holograms. One of these important considerations is the question of whether three wavelengths are really sufficient for color holography.

Hubel and Solymar[49] gave an exact quantitative definition of color recording in holography: "A holographic technique is said to reproduce 'true' colors if the average vector length of a standard set of colored surfaces is less than 0.015 chro-

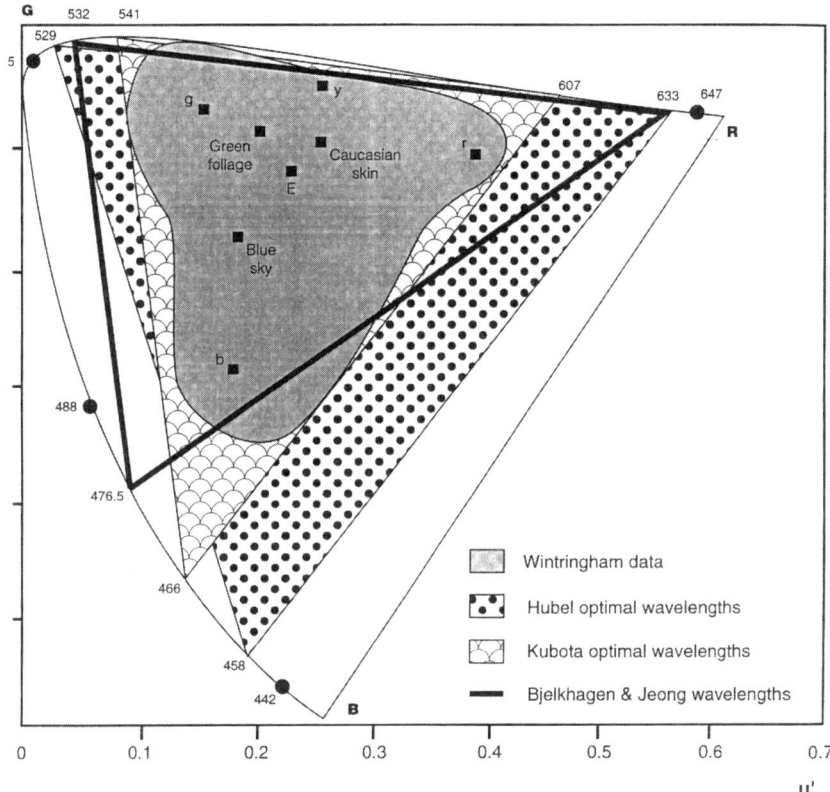

FIGURE 1.2. The 1976 CIE uniform scales chromaticity diagram shows the gamut of surface colors and positions of common laser wavelengths. Optimal color-recording laser wavelengths are also indicated.

maticity coordinate units, and the gamut area obtained by these surfaces is within 40% of the reference gamut. Average vector length and gamut area should both be computed using a suitable white light standard reference illuminant." The wavelength selection problem for color holography has been treated by, for example, Kubota and Nishimura,[40, 41] and Peercy and Hesselink.[53]

Most often, in early color holography experiments, the recording wavelengths were located at 476.5, 514.5, and 632.8 nm. These wavelengths cover the Wintringham data[79] sufficiently well. An important factor to bear in mind when working with silver-halide materials is that a slightly longer blue wavelength may give higher quality holograms because of reduced Rayleigh scattering during the recording. However, if the blue light scattering problem is disregarded, Hubel and Solymar[49] have made an important observation. They explained that, for example, only half of the relative efficiency is needed if a shorter (450-nm wavelength) is used as the blue primary (compared with the 480-nm wavelength) to obtain good white balance with the other primary wavelengths. This observation is very important for multiple exposures of holographic single-layer silver-halide emulsions. Using a short blue wavelength, the diffraction efficiency of this recording can be low, compared with the red and green recordings, and still produce a good color hologram. What is needed is a low light-scattering recording material that will make it possible to use the optimal short blue wavelength.

Another important factor to consider is the reflectivity of the object at primary spectral wavelengths. Thornton[80] has shown that the reflectivity of an object at three wavelength bands, peaked at 450, 540, and 610 nm, has an important bearing on color reconstruction. These wavelengths can also be considered optimal for the recording of color holograms.

The popular set of wavelengths (476.5, 514.5, and 632.8 nm) used in early color holography experiments covers the surface colors defined by Wintringham but causes a distortion in or lack of yellow. A triangle larger than necessary is normally used to compensate for color desaturation (color shifting toward white) that takes place when reconstructing color reflection holograms in white light. Hubel and Solymar[49] investigated the optimal wavelengths, both theoretically and experimentally. According to their color rendering analysis, these wavelengths are 464, 527, and 606 nm for the sandwich silver-halide recording technique. If the calculations are performed to maximize the gamut area instead, the following set of wavelengths is obtained: 456, 532, and 624 nm. The authors suggest that 458, 529, and 633 nm is the optimal wavelength combination in practical color holography. Kubota and Nishimura[40, 41] approached the wavelength problem from a slightly different angle. These authors calculated the optimal trio of wavelengths based on the reconstructing light source of $3400°$ K, a 6-μm thick emulsion with a refractive index of 1.63 and an angle of $30°$ between the object and the reference beam. Kubota and Nishimura obtained the following wavelengths: 466.0, 540.9, and 606.6 nm. These wavelengths have accurately reproduced the colors of the Macbeth ColorChecker (Munsell Color, GretagMacbeth, New Windsor, NY) chart in a hologram, according to the paper.

Peercy and Hesselink[53] discussed wavelength selection by investigating the sampling nature of the holographic process. During the recording of a color hologram, the chosen wavelengths point- sample the surface-reflectance functions of the object. This sampling on color perception can be investigated by the tristimulus value of points in the reconstructed hologram, which is mathematically equivalent to integral approximations for the tristimulus integrals. Peercy and Hesselink used both Gaussian quadrature and Riemann summation for the approximation of the tristimulus integrals. In the first case, they found the wavelengths to be 437, 547, and 665 nm. In the second case, the wavelengths were 475, 550, and 625 nm. According to Peercy and Hesselink, the sampling approach indicates that three monochromatic sources are almost always insufficient to preserve all of the object's spectral information accurately. They claim that four or five laser wavelengths are required.

Only further experiments will show how many wavelengths are necessary and which combination is the best for practical purposes. Another factor that may influence the choice of the recording wavelengths is the availability of cw lasers currently in use in holographic recording, e.g., argon ion, krypton ion, diode-

TABLE 1.2. Wavelengths from cw lasers.

Wavelength [nm]	Laser type	Single line power [mW]
442	Helium cadmium	< 100
457	DPSS blue	< 500
458	Argon ion	< 500
468	Krypton ion	< 250
476	Krypton ion	< 500
477	Argon ion	< 500
488	Argon ion	< 2000
497	Argon ion	< 500
502	Argon ion	< 400
514	Argon ion	< 5000
521	Krypton ion	< 100
529	Argon ion	< 600
531	Krypton ion	< 250
532	DPSS green	< 3000
543	Green neon	< 10
568	Krypton ion	< 100
633	Helium neon	< 80
647	Krypton ion	< 2000
656	DPSS red	< 1000

pumped solid state (DPSS) frequency-doubled Nd:YAG, helium neon, and helium cadmium lasers (Table 1.2). The recent progress in DPSS laser technology has brought to the market both red and blue DPSS lasers. These lasers are aircooled, small, and each laser requires less than a hundred watts of electric power to operate. Most likely, a set of three DPSS lasers will be the best choice of lasers for cw color holography in the future.

1.3.2   Setup for Recording Color Holograms

A typical reflection color recording setup is illustrated in Fig. 1.3. The different laser beams necessary for the exposure of the object pass through the same beam expander and spatial filter. A single-beam Denisyuk arrangement is used; i.e., the object is illuminated through a holographic plate. The light reflected from the object constitutes the object beam of the hologram. The reference beam is formed by the three expanded laser beams. This "white" laser beam illuminates both the holographic plate and the object through the plate. Each of the three primary laser wavelengths forms its individual interference pattern in the emulsion, all of which are recorded simultaneously during the exposure. In this way, three holographic images (a red, a green, and a blue image) are superimposed on one another in the emulsion.

The Denisyuk setup is the most demanding one as regards the material's resolving power and light scattering. Only materials with the lowest possible light

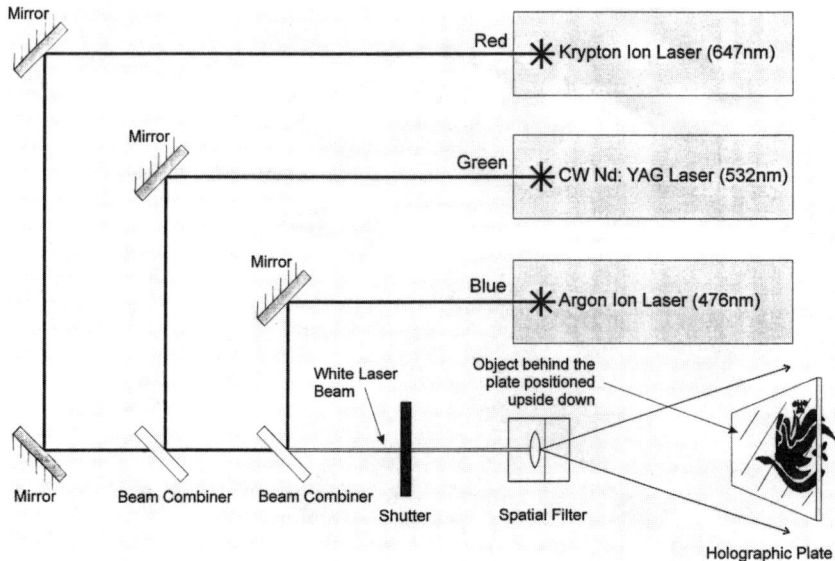

FIGURE 1.3. The setup for recording color reflection holograms.

scattering in the blue part of the spectrum can be considered for usage. Traditionally, only DCG and photopolymers have low blue light scattering, which is why they have been regarded as most suitable for color holography. Western silver-halide materials for holography have never been intended for color holography, and only ultrahigh-resolution silver-halide materials can be used in this area.

Three laser wavelengths are employed for the recording: 476 nm, provided by an argon ion laser; 532 nm, provided by a cw frequency-doubled Nd:YAG laser; and 647 nm, provided by a krypton laser. Two dichroic filters are used for the combining of the three laser beams. The "white" laser beam goes through a spatial filter, illuminating the object through the holographic plate.

By using the dichroic filter beam combination technique, it is possible to perform simultaneous exposure recording, which makes it possible to control independently the RGB ratio and the overall exposure energy in the emulsion. The RGB ratio can be varied by individually changing the output power of the lasers, whereas the overall exposure energy is controlled solely by the exposure time.

The hologram recording setup is arranged on an optical table. The lasers are installed on an independent vibration-free system isolated from the table surface on which the Denisyuk hologram recording setup is installed.

In the initial experiments performed by the author in 1993, a specially designed test object consisting of the 1931 CIE chromaticity diagram, a rainbow ribbon cable, pure yellow dots, and a cloisonné elephant, was used for the color balance adjustments and exposure tests. Later, another test target was employed, the Macbeth ColorChecker chart, which was used for color reproduction tests.

Currently, the Slavich PFG-03C emulsion is the most suitable for the recording of color holograms. The RGB sensitivity values of the recording plate are determined experimentally. Using the simultaneous exposure approach, the overall energy density for exposure is about 4 mJ/cm^2.

The temperature and relative humidity in the laboratory must be kept stable. A temperature of 20°C and 50%RH should be maintained in the laboratory. If the recording takes place in a warm room with, e.g., high humidity, the emulsion absorbs moisture and will increase in thickness. After processing, the emulsion shrinks, which will affect the image colors. The unexposed emulsion is very sensitive to humidity changes. During processing, the emulsion hardens, and when the hologram is finished, it is much less sensitive to variations in humidity and temperature.

1.3.3   Processing of Color Holograms

The processing of the plates is a critical stage. The Slavich emulsion is soft, and it is important to harden the emulsion before the development and bleaching takes place. Emulsion shrinkage and other emulsion distortions caused by the active solutions used for the processing must be avoided. In particular, when recording master color holograms intended for photopolymer replication, shrinkage control is extremely important. The processing steps are summarized in Table 1.3.

The following bath is used for the first processing step:

Distilled water	750 ml
Formaldehyde 37% (Formalin)	10 ml (10.2 g)
Potassium bromide	2 g
Sodium carbonate (anhydrous)	5 g
Add distilled water to make	1 l

The time in this solution is 6 minutes.

The developer used is the holographic CWC2 developer:

Distilled water	750 ml
Catechol	10 g
Ascorbic acid	5 g
Sodium sulfite (anhydrous)	5 g
Urea	50 g
Sodium carbonate (anhydrous)	30 g
Add distilled water to make	1 l

The developing time is 3 minutes at 20° C.

The catechol-based CW-C2 developer[81] is a popular developer for processing monochrome reflection holograms recorded with cw lasers. The use of urea serves to increase the developer's penetration into the emulsion, which is important for uniform development of the recorded layers within the emulsion depth. Catechol has also a tanning effect on the emulsion but with less staining effect compared with pyrogallol. This developer can therefore be considered suitable for processing color holograms.

TABLE 1.3. Color holography processing steps.

1. Tanning in a Formaldehyde solution	6 min
2. Short rinse	5 sec
3. Development in the CWC2 developer	3 min
4. Wash	5 min
5. Bleaching in the PBU-amidol bleach	~ 5 min
6. Wash	10 min
7. Soaking in acetic acid bath	1 min
8. Short rinse	1 min
9. Washing in distilled water with wetting agent added	1 min
10. Air drying the holograms	

Using the correct bleach bath to convert the developed silver hologram into a phase hologram is very important. The bleach must create an almost stain-free clear emulsion so as not to affect the color image. In addition, *no emulsion shrinkage* can be permitted, as it would change the colors of the image. A special rehalo-genating bleach (PBU-amidol bleach)[82] can be used here. This bleach has very good performance concerning both high efficiency and low noise, and in addition, there is no emulsion shrinkage. The bleach is mixed in the following way:

Distilled water	750 ml
Cupric bromide	1 g
Potassium persulfate	10 g
Citric acid	50 g
Potassium bromide	20 g
Add distilled water to make	1 l

After the above-mentioned chemicals have been mixed, 1-g amidol [$(NH_2)_2C_6H_3OH.2HCl$, 2,4- diaminophenol dihydrocloride] is added. The bleach can be used after a few minutes of having been mixed. Sufficient oxidation of the developing agent amidol must take place. One-part stock solution must be diluted in one-part distilled water for use. Bleaching takes normally about five minutes. The process must continue until the plate is completely clear. The plate is then washed for at least ten minutes, and after that, it is soaked in water with 20-ml/l acetic acid (glacial) added. This is done to prevent printout of the finished hologram. Washing and drying must be done so that no shrinkage occurs. Let the plate slowly dry in the air of room temperature. Hot air may introduce emulsion shrinkage. The best technique is to dry the plates in a high-humidity drying chamber, but then the drying takes a long time. Finally, to prevent any potential emulsion thickness variation by humidity variations, the emulsion needs to be protected by a glass plate being sealed onto the hologram plate.

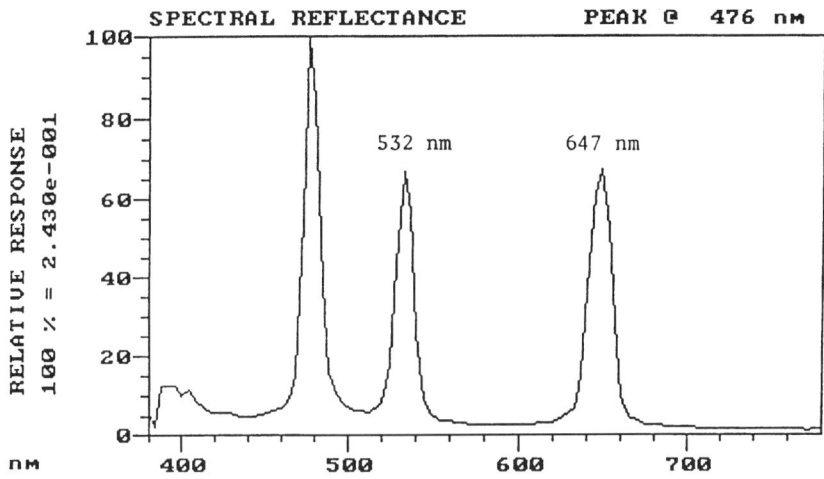

FIGURE 1.4. Normalized spectrum from a white area of a color test target hologram.

1.4   Evaluation of Color Holograms

The recorded color holograms of the two test targets have been evaluated using a PR-650 Photo Research SpectraScan SpectraCalorimeter. The illuminating spotlight to reconstruct the color holograms was a 12-V 50-W Phillips 6438 GBJ (Philips Lighting, Somerset, NJ) halogen lamp. This type of spotlight is suitable for displaying color holograms. The selection of a suitable lamp for the reconstruction of color holograms is much more important than is the selection of lamps for monochrome hologram display. This is why the color balance during the recording of a color hologram must be adjusted with what type of spotlight that is going to be used for the display of the finished hologram in mind. Figure 1.4 shows a typical normalized spectrum obtained from a white area of the color test target color hologram. This means that the diffraction efficiency of each color component is obtained assuming a flat spectrum of the illuminating source. One should note the high diffraction efficiency in blue needed to compensate for the low blue light emission of the halogen spotlight. The noise level, mainly in the blue part of the spectrum, is visible and low. The measured hologram is processed in such a way that no shrinkage occurs. The three peaks are exactly at the recording wavelengths, i.e., 647 nm, 532 nm, and 476 nm.

In Table 1.4, some results of the Macbeth ColorChecker investigation are presented. The 1931 C.I.E. x-and y-coordinates are measured at both the actual target and the holographic image of the target. The measured fields are indicated in the table by color and the corresponding field number.

Color reproductions of recorded color holograms by the author are presented in Figs. 1.5–1.8. The photographs of the reconstructed color holograms were recorded using the halogen spotlight described above. The photograph of the holographic

TABLE 1.4. Chromaticity coordinates from color hologram recording tests using the Macbeth ColorChecker.

Object	White #19	Blue #13	Green #14	Red #15	Yellow #16	Magenta #17	Cyan #18
CIE x y	x/y	x/y	x/y	x/y	x/y	x/y	x/y
Target	.435/.405	.295/.260	.389/.514	.615/.335	.517/.450	.524/.322	.285/.380
Image	.354/.419	.335/.362	.337/.449	.476/.357	.416/.437	.448/.338	.295/.366

FIGURE 1.5. Hologram of the CIE test object recorded with 476-, 532-, and 647-nm laser wavelengths. Notice the rainbow ribbon cable, the pure yellow dots, the full range of colors, and a balanced white in the center of the CIE color test target. *See color insert.*

image of the CIE test target object is shown in Fig. 1.5. A 100-mm × 120-mm display color hologram is featured in Fig. 1.6, showing a recording of two watches. Prints of two large-format color holograms are presented in Fig. 1.7, a 200-mm × 250-mm hologram of a Chrysler Viper (Daimler Chrysler Corp., Auburn Hills,

FIGURE 1.6. A hologram of watches, size 100 mm × 120 mm.

FIGURE 1.7. A color hologram of a Chrysler Viper model car, size 200 mm × 250 mm. *See color insert.*

FIGURE 1.8. A color hologram of an old Chinese vase, size 300 mm × 400 mm. *See color insert.*

MI) model car, and in Fig. 1.8, a 300-mm × 400-mm recording of an old Chinese vase.

1.5   Color Holography in the 21st Century

The recording of large-format color reflection holograms in single-layer silver-halide emulsions has been demonstrated. In spite of the common opinion that silver-halide materials are inferior to grainless recording materials, it was found that the performance of the ultrahigh-resolution emulsions is very good. With a correct choice of the three recording laser lines, good color rendition can be achieved on such materials. For large-format color holography, the following laser wavelengths are recommended: the blue 476 nm, the green 532 nm, and the red 647 nm. The processing of such holograms requires special attention to avoid shrinkage or other emulsion distortions. The diffraction efficiency of color holograms produced this way is about 25 to 30%, but it can be increased. Although good color rendition can be obtained, problems connected with color desaturation still need to be solved. The development process can be further improved to avoid color desaturation caused by nonuniform development. The problem of emulsion shrinkage and the resulting wavelength shift, as well as the color desaturation problems, make holographic color reproduction difficult. The white-light reconstruction of a color hologram shows a lower signal-to-noise ratio and a larger bandwidth, compared with the wavelengths used at the recordings. Desaturation is caused primarily by noise, but partly also by the increased bandwidth. Phillips discussed holographic color saturation and presented a potential solution to the problem.[66]

As already mentioned, other limitations concerning holographic color recording include the fact that some colors we see are the result of fluorescence, which cannot be recorded in a hologram. There are some differences in the recorded colors of the Macbeth ColorChecker test chart as shown in this investigation. However, color rendition is a very subjective matter. Different ways of rendition may be preferable for different applications, and different people may have different color preferences.

At the present time, research on processing color reflection holograms recorded on ultrahigh-resolution materials is still in progress. Work is carried out on techniques that could increase diffraction efficiency by *SHSG processing*. SHSG means Silver-Halide Sensitized Gelatin, and represents a method for converting a silver-halide hologram into a DCG-type hologram. Such holograms are also known as MC structure holograms. For best results, SHSG processing requires emulsions with very fine silver-halide grains to produce low-noise reflection recordings.

For producing large quantities of color holograms, the holographic photopolymer materials from E.I. Du Pont de Nemours & Co. are necessary for the progress.[67−70] The monochrome materials (OmniDex) have been on the market for some time, and the panchromatic materials have recently been introduced for commercial applications. Photopolymer film can become a very suitable recording material for mass replication by contact-copying color holograms and color HOEs from SHSG-processed silver-halide masters.

Another important field of research is the development of a three-wavelength pulsed laser. Employing such a laser, dynamic events, as well as portraits, can be recorded in a holographic form. In 1972, a multiwavelength 2.5-kW pulsed xenon laser was used in Russia for recording pulsed color holograms: (LGI-37 with seven wavelengths: 495.4 nm, 500.8 nm, 526 nm, 535.3 nm, 539.4 nm, and 595.5 nm).[83] The pulse length was 300 ns. Pulsed color holograms were recorded at the Leningrad Institute of Nuclear Physics on TEA-sensitized PE-1 plates. The plates were processed in a pyrogallol developer, fixed and reswelled in TEA. Brown staining was one problem, and the lack of a deeper red wavelength in the spectrum was another obstacle to obtaining good color holograms. Since then, only a few papers have discussed pulsed color holography.[84−88] Currently, research work is being done at the French German Research Institute, ISL, in Saint Louis, France, as reported by Lutz et al.[86] and Albe et al.[87]

The virtual color image in a Denisyuk holographic plate represents the most realistic image of an object that can be obtained today. The extensive field of view adds to the illusion of beholding a real object rather than an image of it. Such an image can be regarded as real virtual reality. The wavefront reconstruction process recreates accurately the three-wavelength light scattered off the object during the recording of the color hologram. This imaging technique has many obvious applications, in particular, in displaying unique and expensive artifacts. There are also many potential commercial applications of this new feature of holographic imaging. Display holography may well become this century's highly recognized accurate imaging technology. To reach this goal, special display devices for holography have to be designed and manufactured. Development of integrated lighting for holograms is important, for example, edge-lit technology.

Today, it is technologically possible to record and replay *acoustical waves* with very high fidelity. Hopefully, holographic techniques will be able to offer the same possibility in the field of *optical waves*, wavefront storage, and reconstruction. Computer-generated holographic images of this type would make it possible to display extremely realistic full-parallax 3D color images of nonexisting objects, which could be applied in various spheres; e.g., in product prototyping, as well as in other applications in 3D visualization and for 3D art. The progress in this field has been demonstrated by Zebra Imaging in Texas, where very large full-parallax color reflection computer-generated holographic images have been made. The generation of a 3D hardcopy of digitized images or computer graphics models is based on a technique described by Klug et al.[89] The 3D image is made in 60-cm × 60-cm tiled increments. So far, the largest image created was of Ford's P2000 Prodigy (Detroit, MI) concept car, in which ten such hologram tiles make up one very large color reflection hologram. This technique has opened the door to real 3D computer graphics.

This century may also see the possibility to generate true-holographic 3D color images in real time, provided that computers become faster and possess greater storage capacity. However, the most important issue is the development of very high-resolution electronic display devices that are needed for electronic holographic real-time imaging.

Acknowledgments

The author's initial work on color holography was performed in cooperation with Dalibor Vukičević, Université Louis Pasteur, Strasbourg, France. Work in the United States has been carried out both at the HOLOS Corp., New Hampshire, in cooperation with Qiang Huang, and at Lake Forest College, Illinois, with Tung H. Jeong.

References

[1] E.N. Leith and J. Upatnieks, "Wavefront reconstruction with diffused illumination and three-dimensional objects," *J. Opt. Soc. Am.* **54**, 1295–1301 (1964).
[2] L. Mandel, "Color imagery by wavefront reconstruction," *J. Opt. Soc. Am.* **55**, 1697–1698 (1965).
[3] A.W. Lohmann, "Reconstruction of vectorial wavefronts," *Appl. Opt.* **4**, 1667–1668 (1965).
[4] K.S. Pennington and L.H. Lin, "Multicolor wavefront reconstruction," *Appl. Phys. Lett.* **7**, 56–57 (1965).
[5] L.H. Lin, K.S. Pennington, G.W. Stroke, and A.E. Labeyrie, "Multicolor holographic image reconstruction with white-light illumination," *Bell Syst. Tech. J.* **45**, 659–661 (1966).
[6] A.A. Friesem and R.J. Fedorowicz, "Recent advances in multicolor wavefront reconstruction," *Appl. Opt.* **5**, 1085–1086 (1966).
[7] J. Upatnieks, J. Marks, and R. Fedorowicz, "Color holograms for white light reconstruction," *Appl. Phys. Lett.* **8**, 286–287 (1966).
[8] G.W. Stroke and R.G. Zech, "White-light reconstruction of color images from black-and-white volume holograms recorded on sheet film," *Appl. Phys. Lett.* **9**, 215–217 (1966).
[9] E. Marom, "Color imagery by wavefront reconstruction," *J. Opt. Soc. Am.* **57**, 101–102 (1967).
[10] A.A. Friesem and R.J. Fedorowicz, "Multicolor wavefront reconstruction," *Appl. Opt.* **6**, 529–536 (1967).
[11] R.J. Collier and K.S. Pennington, "Multicolor imaging from holograms formed on two-dimensional media," *Appl. Opt.* **6**, 1091–1095 (1967).
[12] L.H. Lin and C.V. LoBianco, "Experimental techniques in making multicolor white light reconstructed holograms," *Appl. Opt.* **6**, 1255–1258 (1967).
[13] A.A. Friesem and J.L. Walker, "Thick absorption recording media in holography," *Appl. Opt.* **9**, 201–214 (1970).
[14] E.T. Kurtzner and K.A. Haines, "Multicolor images with volume photopolymer holograms," *Appl. Opt.* **10**, 2194–2195 (1971).
[15] S. Tatuoka, "Color image reconstruction by image plane holography," *Jpn. J. Appl. Phys.* **10**, 1742–1743 (1971).

[16] M. Noguchi, "Color reproduction by multicolor holograms with white-light reconstruction," *Appl. Opt.* **12**, 496–499 (1973).

[17] R.A. Lessard, S.C. Som, and A. Boivin, "New technique of color holography," *Appl. Opt.* **12**, 2009–2011 (1973).

[18] C.P. Grover, and M. May, "Multicolor wave-front reconstruction of partially diffusing plane objects," *J. Opt. Soc. Am.* **63**, 533–537 (1973).

[19] R.A. Lessard, P. Langlois, and A. Boivin, "Orthoscopic color holography of 3D objects," *Appl. Opt.* **14**, 565–566 (1975).

[20] J. Růžek and J. Mužik, "Some problems of colour holography," *Tesla Electron.* **9**, 60–61 (1976).

[21] P. Hariharan, W.H. Steel, and Z.S. Hegedus, "Multicolor holographic imaging with a white-light source," *Opt. Lett.* **1**, 8–9 (1977).

[22] H. Chen, A. Tai, and F.T.S. Yu, "Generation of color images with one-step rainbow holograms," *Appl. Opt.* **17**, 1490–1491 (1978).

[23] T. Kubota and T. Ose, "Lippmann color holograms recorded in methylene-blue-sensitized dichromated gelatin," *Opt. Lett.* **4**, 289–291 (1979).

[24] P. Hariharan, "Improved techniques for multicolor reflection holograms," *J. Opt. (Paris)* **11**, 53–55 (1980).

[25] G.A. Sobolev and O.B. Serov, "Recording color reflection holograms," *Sov. Tech. Phys. Lett.* **6**, 314–315 (1980).

[26] G. Ya. Buimistryuk and A. Ya. Dmitriev, "Selection of laser emission wavelengths to obtain color holographic images [in Russian]," *Izv. VUZ Priborostr. (USSR)* **25**, 79–82 (1982).

[27] Yu.N. Denisyuk, S.V. Artemev, Z.A. Zagorskaya, A.M. Kursakova, M.K. Shevtsov, and T.V. Shedrunova, "Color reflection holograms from bleached PE-2 photographic plates," *Sov. Tech. Phys. Lett.* **8**, 259–260 (1982).

[28] P. Hariharan, "Colour holography," in *Progress in Optics* **20**, 263–324 (North-Holland, Amsterdam, 1983).

[29] L. Huff and R.L. Fusek, "Color holographic stereograms," *Opt. Eng.* **19**, 691–695 (1980).

[30] W.J. Molteni, "Natural color holographic stereograms by superimposing three rainbow holograms," in *Optics in Entertainment II*, C. Outwater, ed., *Proc. SPIE* **462**, 14–19 (1984).

[31] F.T.S. Yu, J.A. Tome, and F.K. Hsu, "Dual-beam encoding for color holographic construction," *Opt. Commun.* **46**, 274–277 (1983).

[32] F.T.S. Yu, and F.K. Hsu, "White-light Fourier holography," *Opt. Commun.* **52**, 384–389 (1985).

[33] F.T.S. Yu and G. Gerhart, "White light transmission color holography, a review," *Opt. Eng.* **24**, 812–819 (1985).

[34] T. Kubota, "Recording of high quality color holograms," *Appl. Opt.* **25**, 4141–4145 (1986).

[35] V.P. Smaev, V.Z. Bryskin, E.M. Znamenskaya, A.M. Kursakova, and I.B. Shakhova, "Features of the recording of holograms on a two-layer photographic material," *Sov. J. Opt. Technol.* **53**, 287–290 (1986).

[36] A.D. Galpern, B.K. Rozhkov, V.P. Smaev, and Yu.A. Vavilova, "Diffraction characteristics of color transmission holograms," *Opt. Spectrosc. (USSR)* **62**, 810–812 (1987).

[37] A.D. Galpern, B.K. Rozhkov, and V.P. Smaev, "Recording of rainbow holograms," *Opt. Spectrosc. (USSR)* **63**, 226–229 (1987).

[38] N.G. Vlasov, A.N. Zaborov, and A.V. Yanovskii, "Production of color specimens by rainbow holography," *Opt. Spectrosc.* (*USSR*) **67**, 243–245 (1989).

[39] T. Mizuno, T. Goto, M. Goto, K. Matsui, and T. Kubota, "High efficient multicolor holograms recorded in methylene blue sensitized dichromated gelatin," in *Practical Holography IV*, S. Benton, ed., *Proc. SPIE* **1212**, 40–45 (1990).

[40] T. Kubota and M. Nishimura, "Recording and demonstration of cultural assets by color holography (I) - Analysis for the optimum color reproduction," *J. Soc. Photogr. Sci. Tech. Jpn.* **53**, 291–296 (1990).

[41] T. Kubota and M. Nishimura, "Recording and demonstration of cultural assets by color holography (II) - Recording method of hologram for optimizing the color reproduction," *J. Soc. Photogr. Sci. Tech. Jpn.* **53**, 297–302 (1990).

[42] T. Kubota, "Lippmann color holography," *J. Opt.* (*Paris*) **22**, 267–273 (1991).

[43] T. Kubota, "Creating a more attractive hologram," *Leonardo* **25**, 503–506 (1992).

[44] S. Namba, K. Kurokawa, T. Fujita, T. Mizuno, and T. Kubota, "Improvement of the transmittance of methylene blue sensitized dichromated gelatin," in *Practical Holography VI*, S.A. Benton, ed., *Proc. SPIE* **1667**, 233–238 (1992).

[45] H. Ueda, K. Taima, and T. Kubota, "Edge-illuminated color holograms," in *Holographic Imaging and Materials*, T.H. Jeong, ed., *Proc. SPIE* **2043**, 278–286 (1993).

[46] M. Kawabata, A. Sato, I. Sumiyoshi, and T. Kubota, "Photopolymer system and its application to a color hologram," *Appl. Opt.* **33**, 2152–2156 (1994).

[47] M. Iwasaki, H. Shindo, T. Tanaka, and T. Kubota, "Spectral evaluation of laboratory-made silver halide emulsions for color holography," *J. Imaging Sci. Technol.* **41**, 457–467 (1997).

[48] M. Iwasaki and T. Kubota, "Ultra-fine-grain silver halide emulsions for color holography: preparation and spectral characterization," in *6th Int. Symp. on Display Holography*, T.H. Jeong and H.I. Bjelkhagen, ed., *Proc. SPIE* **3358**, 54–63 (1998).

[49] P.M. Hubel and L. Solymar, "Color reflection holography: theory and experiment," *Appl. Opt.* **30**, 4190–4203 (1991).

[50] P.M. Hubel and M.A. Klug, "Color holography using multiple layers of Du Pont photopolymer," in *Practical Holography VI*, S.A. Benton, ed., *Proc. SPIE* **1667**, 215–224 (1992).

[51] Yu.E. Usanov and M.K. Shevtsov, "The volume reflection SHG holograms: principles and mechanism of microcavity structure formation," in *Holographic Imaging and Materials*, T.H. Jeong, ed., *Proc. SPIE* **2043**, 52–56 (1994).

[52] C. Jiang, C. Fan, and L. Guo, "Color image generation of a three-dimensional object with rainbow holography and a one-wavelength laser," *Appl. Opt.* **33**, 2111–2114 (1994).

[53] M.S. Peercy and L. Hesselink, "Wavelength selection for true-color holography," *Appl. Opt.* **33**, 6811–6817 (1994).

[54] A.P. Yakimovich, "Recording of rainbow holograms of color images using white light," *Opt. Spectrosc.* **77**, 84–88 (1994).

[55] P. Zhu, X. Liu, and Z. Xu, "Color holography using the angular selectivity of volume recording media," *Appl. Opt.* **34**, 842–845 (1995).

[56] L.M. Murillo-Mora, K. Okada, T. Honda, and J. Tsujiuchi, "Color conical holographic stereogram," *Opt. Eng.* **34**, 814–818 (1995).

[57] H.I. Bjelkhagen and D. Vukićević, "Lippmann color holography in a single-layer silver-halide emulsion," in *5th Int. Symp. on Display Holography*, T.H. Jeong, ed., *Proc. SPIE* **2333**, 34–48 (1995).

[58] H.I. Bjelkhagen, T.H. Jeong, and D. Vukičević, "Color reflection holograms recorded in a panchromatic ultrahigh-resolution single-layer silver halide emulsion," *J. Imaging Sci. Technol.* **40**, 134–146 (1996).

[59] T.H. Jeong, H.I. Bjelkhagen, and L. Spoto, "Holographic interferometry with multiple wavelengths," *Appl. Opt.* **36**, 3686–3688 (1997).

[60] H.I. Bjelkhagen, Q. Huang, Q., and T.H. Jeong, "Progress in color reflection holography," in *6th Int. Symp. on Display Holography*, T.H. Jeong, ed., *Proc. SPIE* **3358**, 104–113 (1998).

[61] V.B. Markov, "Some charateristics of single-layer color hologram," in *Practical Holography IX*, S.A. Benton, ed., *Proc. SPIE* **2406**, 33–40 (1995).

[62] G. von Bally, F. Dreesen, V.B. Markov, A. Roskhop, and E.V. de Haller, "Recording of color holograms on PFG-03Ts," *Tech. Phys. Lett.* **21**, 667–668 (1995).

[63] V.B. Markov and A.I. Khizhnyak, "Selective charateristics of single layer color holograms," in *Practical Holography X*, S.A. Benton, ed., *Proc. SPIE* **2652**, 304–311 (1996).

[64] V. Markov, A. Timoshenko, G. von Bally, and F. Dreesen, "Single-layer color hologram recording," in *Holographic Materials IV*, T.J. Trout, ed., *Proc. SPIE* **3294**, 122–130 (1998).

[65] B. Wesskamp, A. Jendral, and O. Bryngdahl, "Hybrid color holograms," *Opt. Lett.* **21**, 1863–1865 (1996).

[66] N.J. Phillips, "Colour saturation issues in modern holography," in *Practical Holography XI and Holographic Materials III*, S.A. Benton and T.J. Trout, eds., *Proc. SPIE* **3011**, 216–223 (1997).

[67] W.J. Gambogi, W.K. Smothers, K.W. Steijn, S.H. Stevenson, and A.M. Weber, "Color holography using DuPont holographic recording film," in *Holographic Materials*, T.J. Trout, ed., *Proc. SPIE* **2405**, 62–73 (1995).

[68] T.J. Trout, W.J. Gambogi, and S.H. Stevenson, "Photopolymer materials for color holography," in *Applications of Optical Holography*, T. Honda, ed., *Proc. SPIE* **2577**, 94–105 (1995).

[69] K.W. Steijn, "Multicolor holographic recording in Dupont holographic recording film: determination of exposure conditions for color balance," in *Holographic Materials II*, T.J. Trout, ed., *Proc. SPIE* **2688**, 123–134 (1996).

[70] S.H. Stevenson, "DuPont multicolor holographic recording film," in *Practical Holography XI and Holographic Materials III*, S.A. Benton and T.J. Trout, ed., *Proc. SPIE* **3011**, 231–241 (1997).

[71] M. Watanabe, T. Matsuyama, D. Kodama, and T. Hotta, "Mass-produced color graphic arts holograms," in *Practical Holography XIII*, S.A. Benton, ed., *Proc. SPIE* **3637**, 204–212 (1999).

[72] M. Chomát, "Diffraction efficiency of multiple-exposure thick absorption holograms," *Opt. Commun.* **2**, 109–110 (1970).

[73] M.K. Shevtsov, "Diffraction efficiency of phase holograms for exposure superposition," *Sov. J. Opt. Technol.* **52**, 1–3 (1985).

[74] H.I. Bjelkhagen, *Silver Halide Recording Materials for Holography and Their Processing*, Springer Series in Optical Sciences, Vol. 66. (Springer-Verlag, Heidelberg, New York 1993).

[75] Y.A. Sazonov and P.I. Kumonko, "Holographic materials produced by the "Micron" plant at Slavich," in *6th Int. Symp. on Display Holography*, T.H. Jeong, ed., *Proc. SPIE* **3358**, 31–40 (1998).

[76] SLAVICH Joint Stock Co., Micron Branch Co., Pereslavl-Zalessky, Russia.

[77] N.I. Kirillov, N.V. Vasilieva, and V.L. Zielikman, "Preparation of concentrated pho-tographic emulsions by means of their successive freezing and thawing [in Russian]," *Zh. Nauchn. Prikl. Fotogr. Kinematogr.* **15**, 441–443 (1970).

[78] N.I. Kirillov, N.V. Vasilieva, and V.L. Zielikman, "A method for the concentration of the hard phase of the photographic emulsion by consecutive freezing and thawing [in Russian]," *Uspkhi Nauchno i Fotografii* **16**, 204–211 (1972).

[79] W.T. Wintringham, "Color television and colorimetry," *Proc. IRE* **39**, 1135–1172 (1951).

[80] W.A. Thornton, "Luminosity and color-rendering capability of white light," *J. Opt. Soc. Am.* **61**, 1155–1163 (1971).

[81] D.J. Cooke and A.A. Ward, "Reflection-hologram processing for high efficiency in silver-halide emulsions," *Appl. Opt.* **23**, 934–941 (1984).

[82] H.I. Bjelkhagen, N. Phillips, and W. Ce, "Chemical symmetry—developers that look like bleach agents for holography," in *Practical Holography V*, S.A. Benton, ed., *Proc. SPIE* **1461**, 321–328 (1991).

[83] V. Tolchin and B. Turukhano, "Color holography [in Russian]," in *Proc. 5th All-Union School of Holography*, G. Scrotsky, B. Turukhno, and N. Turukhno, ed., (Leningrad, Russia, 1973), Chap. 22.

[84] V.G. Bespalov, V.N. Krylov, and V.N. Sizov, "Pulsed laser system for recording large-scale colour hologram," in *Three-Dimensional Holography: Science, Culture, Education*, T.H. Jeong and V.B. Markov, ed., *Proc. SPIE* **1238**, 457–461 (1991).

[85] J.-Y. Son, V.G. Dmitriev, V.N. Mikhailov, and H.-W. Jeon, "Solid-state lasers for color holography," in *Practical Holography X*, S.A. Benton, ed., *Proc. SPIE* **2652**, 96–105 (1996).

[86] Y. Lutz, F. Albe, and J.-L. Tribillon, "Étude et réalisation d'un laser à solide pulsé émettant dans le bleu pour l'holographie couleur d'objects dynamiques," *C. R. Acad. Sci. Paris* **323**, Série IIb, 465–471 (1996).

[87] F. Albe, Y. Lutz, M. Bastide, and J.-L. Tribillon, "Pulsed color holography," in *6th Int. Symp. on Display Holography*, T.H. Jeong, ed., *Proc. SPIE* **3358**, 114–118 (1998).

[88] A.M. Rodin and A.S. Dement'ev, "The development of a 2–colour laser based on high-efficiency Raman amplification for multi-colour holography," in *6th Int. Symp. on Display Holography*, T.H. Jeong, ed., *Proc. SPIE* **3358**, 211–217 (1998).

[89] M.A. Klug, A. Klein, W. Plesniak, A. Kropp, and B. Chen, "Optics for full parallax holographic stereograms," in *Practical Holography XI and Holographic Materials III*, S.A. Benton and T.J. Trout, ed., *Proc. SPIE* **3011**, 78–88 (1997).

2

Real-Time Autostereoscopic 3D Displays

Tin M. Aye, Andrew Kostrzewski,
Gajendra Savant, Tomasz Jannson,
and Joanna Jannson

2.1 Introduction

As in the late 1960s when stereo replaced monaural audio, we are on the verge of a change in visual presentation in movies, computers, TV, video, telemedicine, and robotics. Current efforts to present visual information in true 3D format covers a wide range of concepts, products, and quasi-product prototypes. Three-dimensional displays must present images to the two eyes from slightly different angles. The method by which we transmit the correct image to the correct eye defines the technology used. For example, stereoscopic displays require viewers to wear glasses, helmets, or other devices.

Visual display of electronic media has been an area of intense development since the advent of the first cathode ray tubes. Increased resolution and color capability have added detail and realism. Liquid crystal devices are maturing rapidly and will ultimately allow images to be displayed in many new applications.[1] Some of the image content, however, is missing from these types of displays, because depth is conveyed only by the viewer's interpretation of the 2D depth cues. This is particularly true when the specific object being observed is unfamiliar, as on unmanned space missions when 2D pictures are interpreted to determine the 3D structure of, for example, Mars terrain. The lack of depth cues in the 2D images limits an observer's true visualization of the scene.

Stereo vision is based on the perception of depth. Stereo display systems are based on the most dominant depth cue, parallax, which is defined as the angular separation between corresponding points of left and right perspective images. In this respect, human 3D vision does not require "real" 3D information, because the human brain has contact only with two 2D retinal images, neither of which preserves z-axis (i.e., depth) information. This third dimension is reconstructed only in the brain. Therefore, 3D displays can be based on capturing and displaying a 2D left image and a 2D right image in order to replicate the corresponding left and right retinal images.

In many technical fields, 3D displays enable scientists and technicians to better interpret the physical parameters of an image. Past 3D display techniques[2–7] have been hampered by the need for the observer to wear external devices, the incapability for real-time display, a lack of full screen resolution, the need for heavy computation and special formatting, very high cost, or any combination of these. A truly practical device must interface with conventional 2D display systems to increase acceptance, support real-time interaction, and allow 3D look around without special glasses. Cost effective production must make the device commercially attractive.

As noted, stereoscopic displays give an observer parallax depth cues by presenting each eye with a view of the same object from a different perspective. When the difference between these viewpoints approximates normal interocular separation, the viewer has the impression of viewing an actual solid object. Conventionally, some form of viewing device such as glasses channels the view into each eye. This is effective in presenting a 3D impression, and requires only twice as much image data as a 2D view. The major drawback is the need to wear a device or to peer into eyepieces. This has led to many years of effort to create autostereoscopic images—directly viewable stereoscopic imaging without glasses.

Autostereoscopic displays[4–7] create a "window" through which an unaided observer can view what appears to be a solid object (see Fig. 2.1). This is a very natural and desirable situation from the standpoint of the observer, but it poses two large technical challenges. First, the light rays leaving the display must be directed so that each of the observer's eyes receives a complete but different image. Second, images of the object from a range of viewpoints must be presented simultaneously. There may be as few as two perspective views, but typically four or more are required for a range of viewing positions (scenes). This greatly increases the amount of image data to be managed.

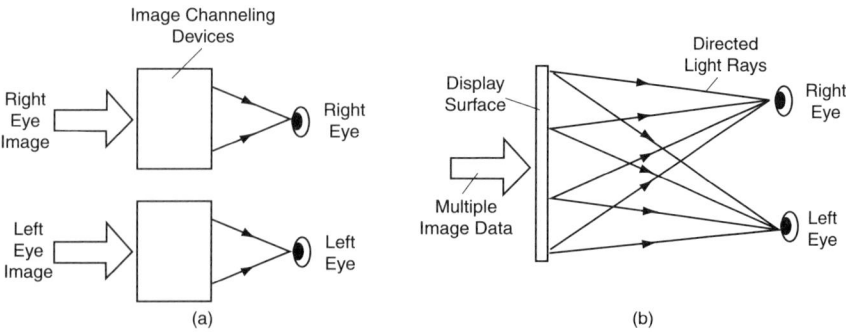

FIGURE 2.1. Comparison of (a) conventional stereoscopy and (b) autostereoscopic display.

Holographic displays and lenticular photographs are the most familiar examples of autostereoscopic images. A holographic display (or hologram) is an interference pattern of an object that can be reconstructed as an image by a collimated beam.

Recently, considerable interest has been directed to autostereoscopy based on the multiplex carrier because of its simplicity, practicality, and potential low cost. This creates a limited look-around 3D effect if more than two perspectives are projected into the viewing eyebox. If the number of perspectives projected in real time is large, the data processing demands can be substantial. The image projection area must be divided into two or more parts, which linearly reduces the image's spatial resolution. It requires sequential scanning of N perspective ($N > 1$) for each vertical scan line. The line feed rate R to the multiplexed monitor is related to the flicker-free line rate r by $R = N \times r$. Thus, the feed signal rate is two or more times the rate for a monocular monitor.

2.2   2D Views to 3D Perception

Although the human eye is an excellent receptor of images in 2D, the brain does not require real 3D information to visualize 2D images as 3D objects. Reconstruction of a third dimension is automatic in the brain. Because this is a natural process, it varies from person to person according to training of the eyes to see artificial or stereoscopic 3D. In fact, some people do not see 3D. The schematic picture in Fig. 2.2 illustrates a 2D retinal image that does not preserve 3D information about a scene; the brain has reconstructed 3D information.

In the case of a remote telerobotic system, two cameras positioned side-by-side and separated by a distance corresponding to the distance between an observer's eyes can be used to produce the left (L) and right (R) perspective views. From Figure 2.3 it can be observed that a single left or right 2D perspective image alone cannot show a true terrain profile. Stereo vision, on the other hand, immediately

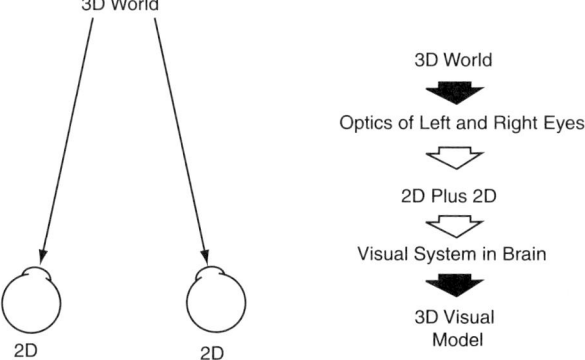

FIGURE 2.2. Retinal images do not preserve 3D information about a scene. Instead, the brain reconstructs the 3D information by inferring a 3D visual model.[8]

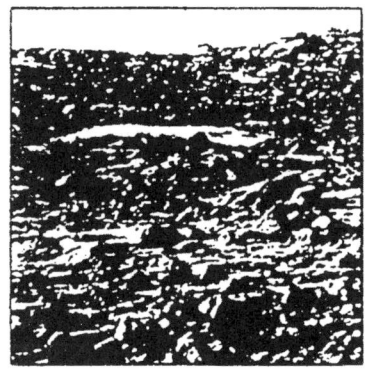 

FIGURE 2.3. Left (a) and right (b) parallax views of Mars terrain.[9]

shows the actual 3D scene of the Martian landscape. In order to increase or decrease the parallax, the camera separation distance can be made larger or smaller than the normal eye separation distance ($\sim$ 65 mm). For real-time applications, the output of these video cameras can be fed into an autostereoscopic 3D holographic display system to regenerate the 3D scenes as they would be perceived by an observer. This requires a camera separation equal to the observer's ocular separation distance. In general, the display geometry determines the camera setup geometry. Figure 2.4 illustrates the relationship between the observer's position from the display and the stereo camera convergence distance and focal length (i.e., field-of-view, FOV) of the camera lens. The ratio between the observed distance (D) and screen width (S) determines the focal length of the L and R camera lenses, according to the formula:[8]

$$\frac{D}{S} = \frac{\text{Focal Length}}{\text{CCD Chip Width}}. \tag{1}$$

This ensures that the display image FOV is the same as that of the cameras, and it avoids "telephoto" distortion.

These perceptual design considerations were extended to the stereo camera system design to produce the correct perspective (relative size, magnification, FOV, etc.) for inverted, or pseudoscopic, 3D images, which integrate smoothly with the normal, orthoscopic, 3D images when projected on the screen.

2.3   Autostereoscopic 3D Display System

Physical Optics Corporation's (POC's) autostereoscopic 3D display system has a unique capability for real-time autostereoscopic display with look-around. The principle of this 3D system is based on a generalization of Gabor's 3D picture projection.[10] POC's approach combines simple, conventional projection systems with a holographic diffusion screen (HDS),[11] building on multiplexed volume holographic optical element (HOE) technology.[12–15] The display screen consists

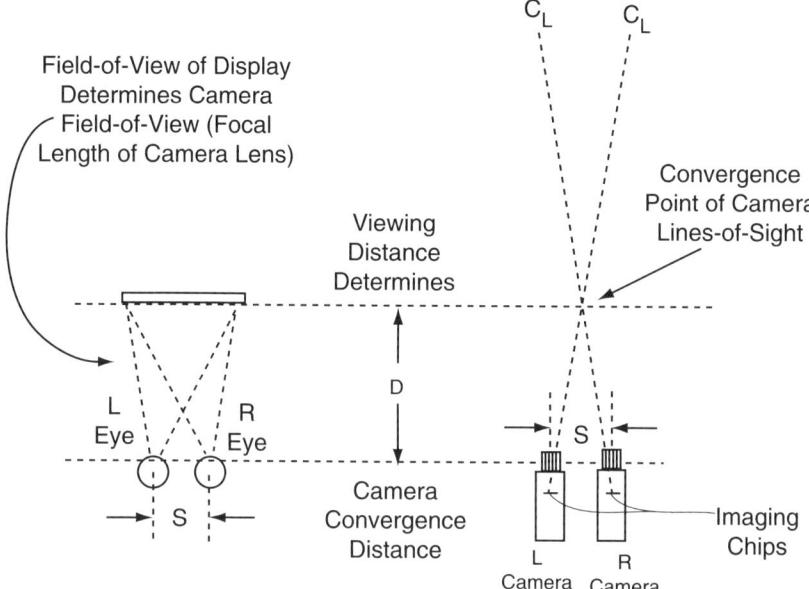

FIGURE 2.4. Two-camera system and geometry for generating left and right perspective view for teleoperator 3D display.[8]

of two or more volume holograms of diffuse perspective regions, one for each perspective viewing zone. These regions are multiplexed in a thin film volume of material, each recorded with a different reference beam angle. Projecting the corresponding perspectives of a scene simultaneously along these reference directions simultaneously reconstructs a spatial sequence of perspective views in these viewing zones, projecting a 3D scene to the viewer. The high angular Bragg selectivity of volume holograms makes it feasible to record many holograms with high efficiency and without cross-talk. Full color only requires either multiplexing the three primary colors (RGB) or dispersing the light from each hologram along the vertical axis.[13]

The unique features of POC's 3D autostereoscopic display system include:

- The volume holographic screen is stationary.
- There are no sequential switching elements and, thus, no flicker.
- Bragg selectivity makes the entire screen area available simultaneously for both the left and right perspectives.
- The system is easily scalable and compatible with large displays and currently available liquid crystal TV (LCTV) technologies.
- It interfaces easily with state-of-the-art liquid crystal display systems, including active matrix liquid crystal displays (AMLCD), and it does not require any special video formatting of data.
- Illumination uniformity is high (within 2%) as required for medical applications.

2.3.1   Design of POC's Autostereoscopic 3D Display

The design of the autostereoscopic 3D display includes both static and dynamic architectures. In designing these systems, the relationship between visual accommodation (focus) and convergence of the two eyes is extremely important, because the convergence depends on the amount of parallax in the views. Therefore, it is important to reduce parallax by increasing the distance between the screen and the observer.

2.3.2   Static 3D System

In the static 3D system design as of late 1999, the beam projected from a single projector is split in half using a simple mirror arrangement (see Fig. 2.5a); the actual breadboard demonstration model appears in Fig. 2.5b). A 10-in. × 10-in. holographic diffuser screen with a viewing distance of $h = 24$ in. to 30 in. and a viewing region of 8 in. × 8 in. creates a large 3D viewing zone. Ideally, a viewer's eyes should be in the full L and R view zones for 3D viewing. Outside of these zones, the viewer can still see 3D images, but the images are partially obscured. In the worst case, having both eyes in either the L or R viewing zone produces monocular (single-perspective) viewing.

Two holograms, one for the left view and one for the right, were multiplexed in a single holographic recording plate. The size of the hologram is limited only by the size of the optics (i.e., 12-in. diameter collimating mirrors). The reference beam angle, which corresponds to the projection beam angle, was chosen to be 25° to 30° to produce a wideband hologram, as is necessary for full-color display. For convenience, the reference beam was collimated rather than converging, which would require larger collimating mirrors. The 3D viewing range is a function of the size of the screen, viewing distance, and the size of the L and R viewing regions. The amount of chromatic dispersion in the vertical direction also determines the full-color viewing region. The efficiency of the HDS need not be high, as even 30% to 50% efficiency produces high light throughput.

2.3.3   Dynamic 3D System

Although the static system requires only two slides of a single object, the dynamic system accepts real-time input from a special tandem video camera to view scenes or images projected through a spatial light modulator. The dynamic system consists of a light source, a spatial light modulator, a projector, two cameras, and a holographic diffuser screen. POC's dynamic 3D stereoscopic system design is illustrated in Fig. 2.6, and the design layout in Fig. 2.7.

The dynamic stereoscopic 3D display system projects real-time, 3D high-resolution full-color imagery. It separates the visual information for the left and right eyes with low cross-talk, without the need for goggles. This dynamic system employs a liquid crystal display SLM with its associated driving electronics. For the first prototype, both the SLM and the driving electronics were taken from commercial

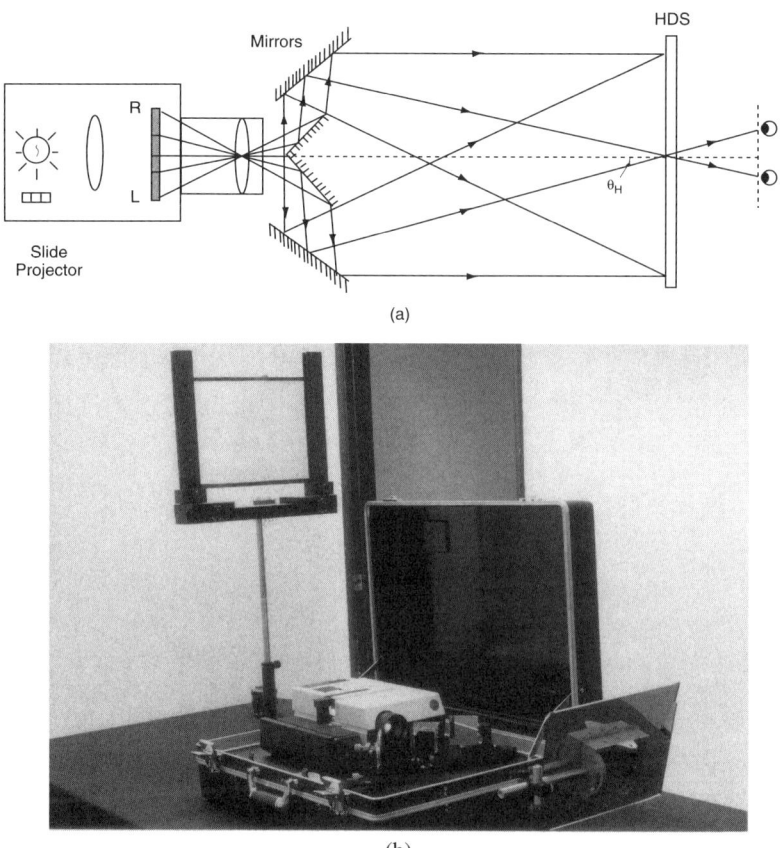

(a)

(b)

FIGURE 2.5. (a) Design of static slide projection system with mirror arrangement for stereo pair image projection on holographic display screen (HDS). Optimization includes large HDS screen and compact packaging. (b) POC's static 3D autostereoscopic display system.

Epson (Nagano-Ken, Japan) and Sony (Tokyo, Japan) liquid crystal televisions. Next-generation prototypes use high-resolution liquid crystal television or Digital Micromirror Displays (DMDs).

2.4   3D Holographic Technology

Simple HOEs can be used to project 2D images from plane to plane within a narrow field-of-view. To expand viewing to a wide angular field-of-view, the point source of the object beam is replaced with a suitable diffusing screen. This essentially amounts to coherently multiplexing an infinite number of point sources on the diffuser plane, simultaneously extending both angular and spatial fields-of-view. As illustrated in Fig. 2.8, the entire 2D image of the object can now be viewed

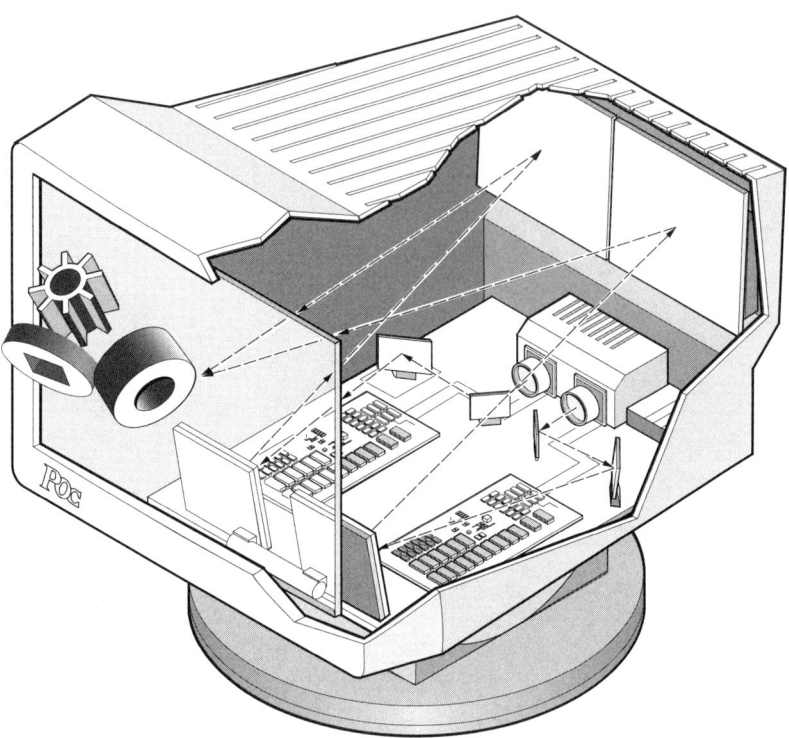

FIGURE 2.6. Compact and portable dynamic 3D autostereoscopic display system.

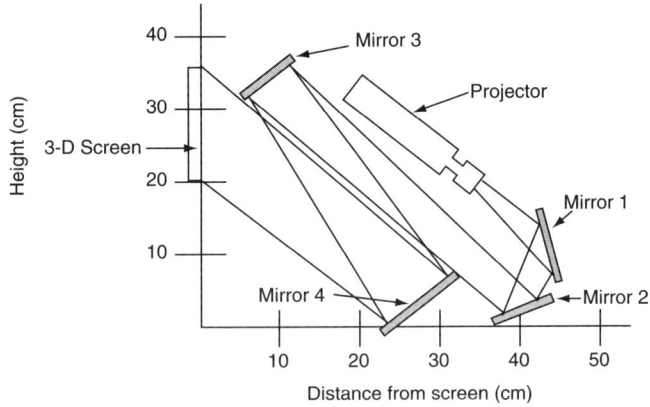

FIGURE 2.7. Optical layout of the dynamic 3D stereoscopic display system.

from anywhere within the viewing window, which is determined by the size of the diffuser.

Using white light to reconstruct a hologram produces spectral dispersion, focusing each color in a different direction. When the holographic diffuser screen is

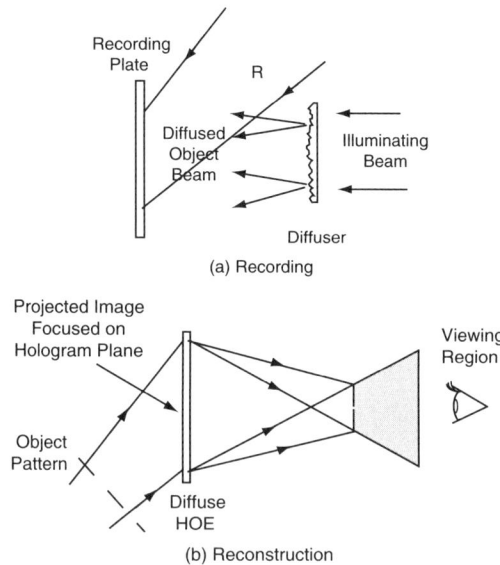

(a) Recording

(b) Reconstruction

FIGURE 2.8. Holographic diffuser screen.

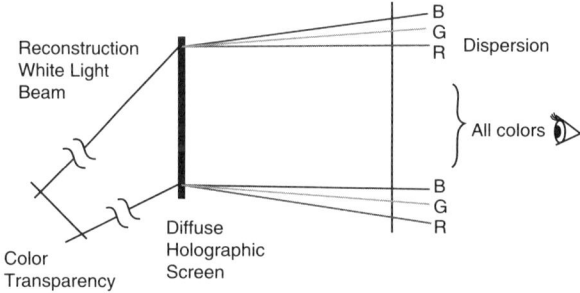

FIGURE 2.9. Full-color viewing, taking advantage of the dispersive property of a diffuse HOE.

illuminated with a white light reference projection beam (e.g., a full-color scene from a standard projector) focused on the hologram plane, the reconstructed images of the diffuse plane in the various colors are shifted, so that they overlap. The center overlap region, however, accommodates full-color viewing. In contrast to holographic imaging, no image smearing occurs because the projection images are focused onto the screen. This is illustrated in Fig. 2.9.

As of late 1999, POC has a projector that produces up to six projected view regions—three left and three right. We anticipate increasing this to ten projected view regions—five left and five right—as illustrated in Fig. 2.10. The left and right projectors reconstruct alternating view regions, spatially separated. The observer, with eyes positioned in any two adjacent regions, sees two perspective views of the 3D object.

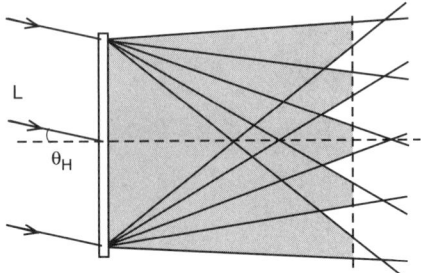

(a) Left perspective image projection to five view regtions.

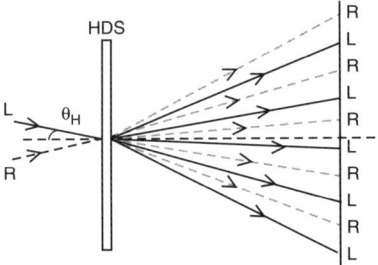

(b) Both left and right perspective image projection in parallel.

FIGURE 2.10. Multiple perspective view projection using left and right projection beams.

2.4.1   Holographic 3D Multiplexing Screen Development

The holographic 3D multiplexing screen is the most important component of a 3D display system. It separates left and right eye information and duplicates left and right viewing zones to enlarge the 3D field-of-view so as to operate as an angular image multiplexer. The development of this element required a combination of two critical holography technologies. The first is the volume holographic structure that performs efficient and angularly selective image multiplexing.[1-5] The second is the holographic non-Lambertian diffuser, whose small scattering angles blur the borders between viewing zones. Both technologies are based on POC proprietary holographic techniques, with optimized diffraction efficiency and chromatic aberration, using a special holographic material and processing, and optimizing the hologram recording geometry.

The structure of the holographic system for recording 3D multiplexing screens is shown in Fig. 2.11. Fabricating a 3D screen capable of producing multiple viewing zones requires a special multislit diffuser mask to create a set of left (L) and right (R) viewing zones. The slit center-to-center separation is 3 in., corresponding to the average separation between eyes. The multislit diffuser used in the recording process was based on POC's proprietary non-Lambertian diffuser components, which have scattering angles of only a few degrees. In contrast to conventional Lambertian diffusers, which scatter light in all directions, these diffusers have controlled scattering angles so that they deliver nearly all the light where it is needed.

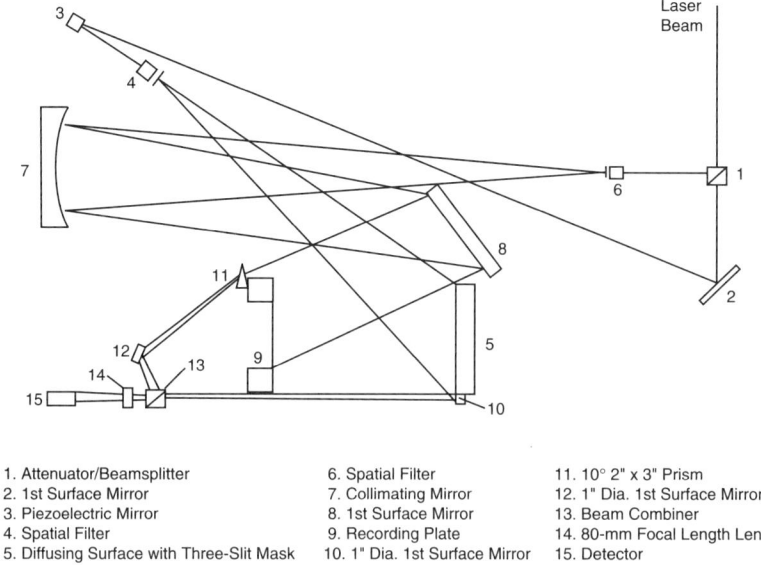

1. Attenuator/Beamsplitter
2. 1st Surface Mirror
3. Piezoelectric Mirror
4. Spatial Filter
5. Diffusing Surface with Three-Slit Mask

6. Spatial Filter
7. Collimating Mirror
8. 1st Surface Mirror
9. Recording Plate
10. 1" Dia. 1st Surface Mirror

11. 10° 2" x 3" Prism
12. 1" Dia. 1st Surface Mirror
13. Beam Combiner
14. 80-mm Focal Length Lens
15. Detector

FIGURE 2.11. Setup for recording volume holographic image multiplexing 3D screens.

The holograms recorded using the system diagramed in Fig. 2.11 can produce an image of the original slit array when illuminated by a white light beam from a single direction. However, if a hologram is illuminated from two slightly different directions, two diffused slit arrays are reconstructed (producing multiple viewing zones) as illustrated in Fig. 2.12. By adjusting the angle between the two illuminating beams, the viewing zones can be tiled together without gaps or overlaps, producing a continuous 3D viewing space.

Using these techniques, POC fabricated several sample holographic screens with up to six angularly separated fan-out channels (i.e., with up to three autostereoscopic 3D channels). The process was optimized for maximum diffraction efficiency, low noise, and high angular Bragg selectivity in order to produce clear 3D viewing in each zone. The tests performed on these screens demonstrated high brightness, contrast, and light uniformity. The intensity distribution of a sample 3D zone screen, plotted in Fig. 2.13, demonstrates that screen uniformity was maintained within 2%. These tests also demonstrate an effective system channel-to-channel separation as high as 20 dB (based on the peak intensity of 0.35 and noise level of 0.003). It is possible to extend the holographic screen technology to produce larger screens with over 30 viewing zones for full look-around capability.

2.4.2 Experimental Fabrication of 3D Multiplexing Screen

The holographic 3D multiplexing screen is designed to separate left and right eye information and duplicate left and right viewing zones to enlarge the 3D field-

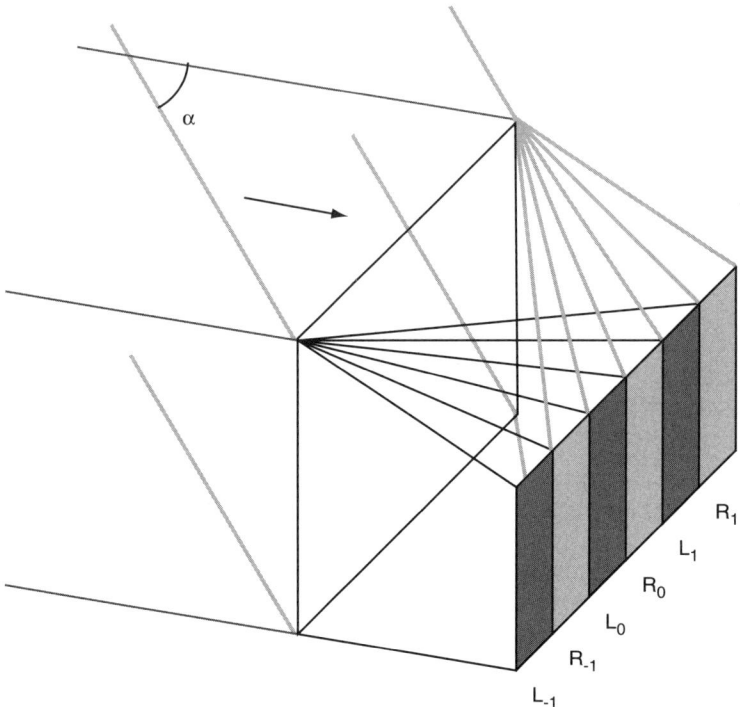

FIGURE 2.12. Reconstruction of 3D holographic screen using two beams intersecting at angle α. This screen is designed to produce three 3D zones (L_{-1}, R_{-1}), (L_0, R_0), (L_{+1}, R_{+1}).

of-view. The specifications of the 3D holographic screen demand high-efficiency holographic recording material. The authors evaluated three types of photosensitive holographic materials for fabrication of 3D screens: their own holographic material based on an emulsion from Russia, Agfa A.H.I. Millimask, and Agfa Holotest 8E56HD. Of the three materials, the first demonstrated the lowest absorption and the highest Bragg diffraction efficiency. The absorption characteristics for the three materials are compared in Fig. 2.14. POC's material also showed much lower overall absorption (the absorption peak at short wavelengths is related to UV absorption by the glass substrate).

Diffraction efficiency was calculated by measuring the maximum light intensity transmitted through the holographic sample I_{MAX} and the minimum light intensity I_{MIN} (maximum intensity of the diffracted beam).

The measured values for these three holograph materials are shown in Table 2.1. POC's emulsion efficiency is close to 33%, compared with 21% and 16% for the other materials.

In the near future this technology will be used to mass produce screens for applications such as 3D displays, head-mounted displays, cockpit displays, and dome displays.

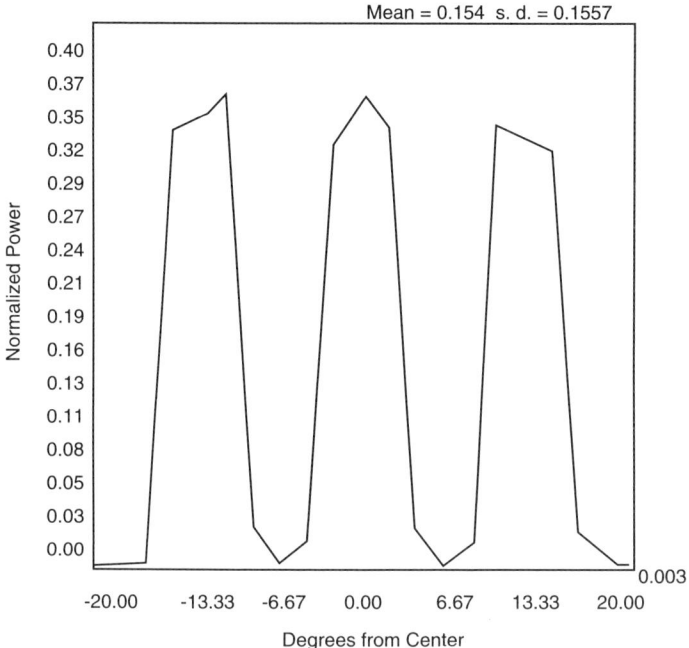

FIGURE 2.13. Intensity distribution plot for three-slit recording. The screen exhibits high diffraction efficiency and low noise.

TABLE 2.1. Diffraction efficiency for three holographic materials.

Parameter	POC emulsion	Agfa Millimask	Agfa Holotest
I_{MAX} (mW)	7.497	7.502	7.51
I_{MIN} (mW)	5.037	5.870	6.31
$\eta(\%)$	32.810	21.170	15.97

2.5   Development of Portable Monitor for 3D Video and Graphics

The major components of one such 3D system are electronic driving modules, optics (projection lens and mirror to minimize the system size by folding the optical path), projection modules (LCDs and light sources), and the holographic screen. Major optical components were optimized in terms of their optical properties and mechanical alignment (see Fig. 2.15).

Integration of a compact 3D video monitor and demonstration of 3D real-time motion video[16] required designing and fabricating driving electronics and system packaging.[17] Based on these designs, POC fabricated a dedicated electronic driver for the Sony LCX07AK liquid crystal display panel, supporting both high-resolution and variable aspect ratio video display. Specifications for a projection

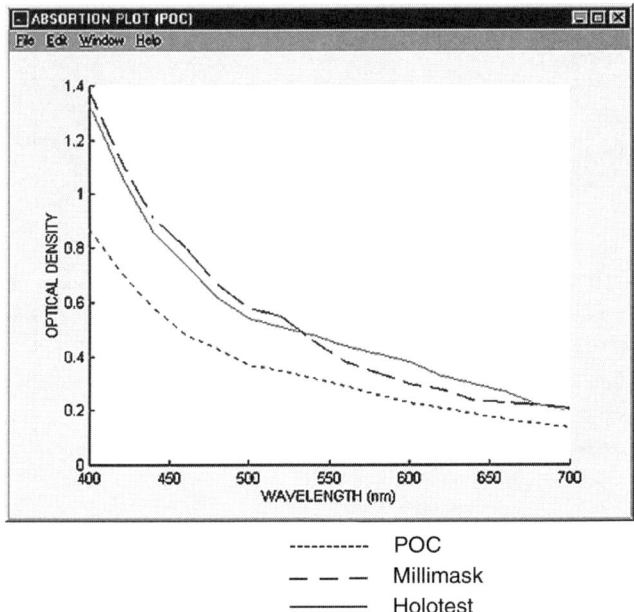

FIGURE 2.14. Absorption characteristics for three holographic coatings: POC's holographic material, Agfa A.H.I. Millimask, and Agfa Holotest 8E56HD. POC's material exhibits the lowest absorption across the entire visible spectral range.

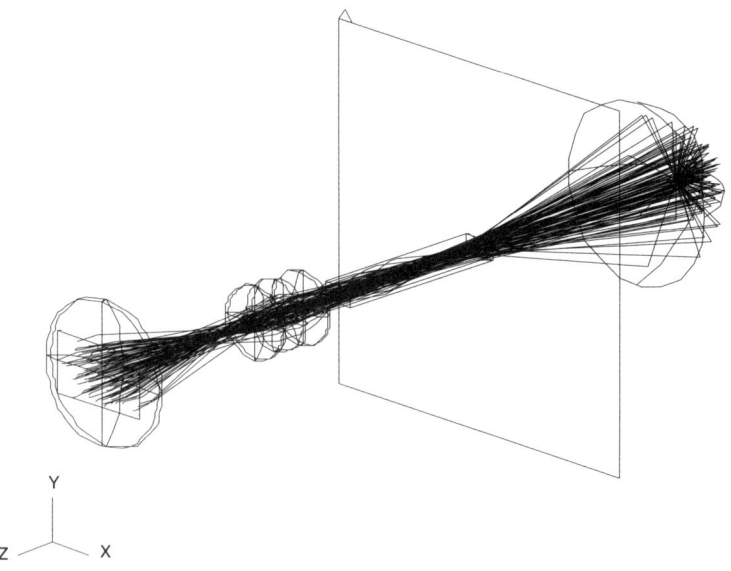

FIGURE 2.15. Projection system ray tracing. The major components of the system include the light source, integrating light tunnel, LCD panel, and projection lens.

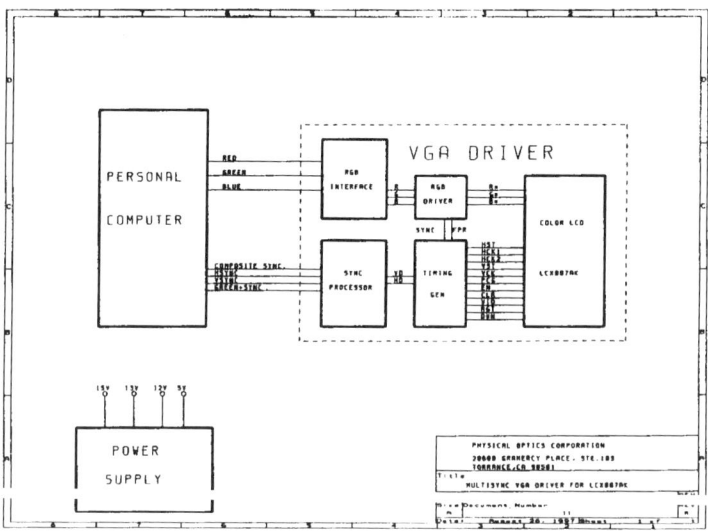

FIGURE 2.16. Electronic driving board design for dual channel VGA compatible projector.

TABLE 2.2. Specifications of 3D system.

Type	LCX07 AK
Display size	3.4 cm (1.35 in.)
Number of dots	16:9 (1024 × 480) or 4:3 (798 × 480)
Optical transmittance	3%
Contrast ratio	200:1
Horizontal resolution	600 TV lines (9:16) or 400 TV lines (3:4)
Vertical resolution	480 TV lines
Display mode	NTSC/NTSC-WIDE/HD (Bandwidth: 20 MHz)

system based on the LCX07AK panel are listed in Table 2.2. The top-level drawing for the electronic driver board is shown in Fig. 2.16. This driver is capable of automatically selecting the proper timing through a multisync input video module. It has multiple functions for displaying images: up, down, left, and right shift and inverse, with aspect ratios of 4:3 and 16:9.

Demonstration of True Three-Dimensional Monitor—The 3D holographic screen was integrated with the two LCD projectors based on Sony LCX07AK panels. Two-channel electronic drivers were designed, built, and optimized for use in this 3D display. Figure 2.17 shows the two-channel electronic driver.

Based on ray-tracing, the authors designed and fabricated a functional prototype demonstration system (see Figs. 2.18 and 2.19), using a Vario-Prolux f/2.8 lens with a 70–120-mm zoom. Antireflection coated mirrors fold the optical path to keep the projection subsystem compact. Figure 2.20 shows the internal structure of the system.

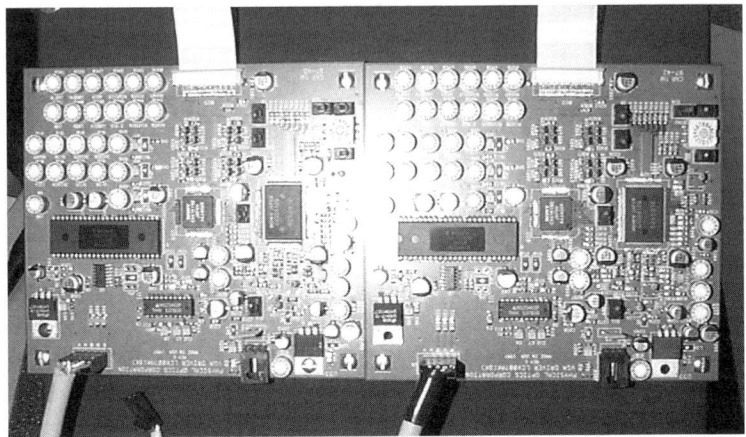

FIGURE 2.17. Two-channel driver electronics for monitor.

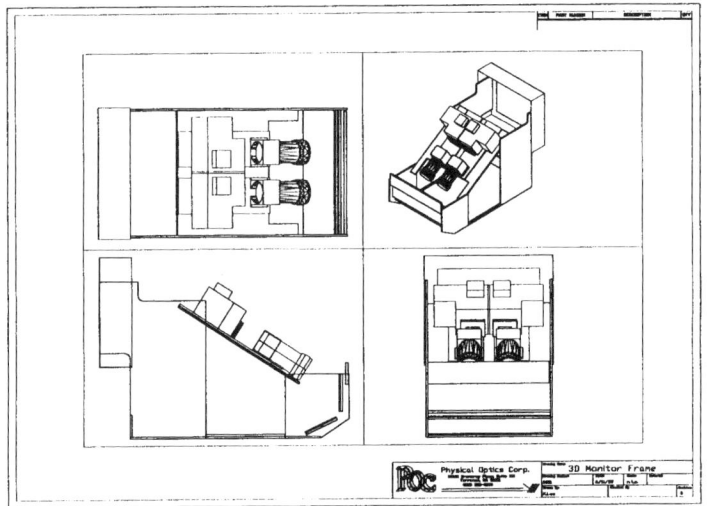

FIGURE 2.18. Three-dimensional monitor design.

The monitor has been built; its dimensions are 21 in. × 17 in. × 12 in. (L × H × W), with a 15 in. diagonal screen. Image quality can be judged from the photograph in Fig. 2.21.

2.6   Conclusions

POC's 3D monitor projects into multiple viewing zones for multiple viewers. It does not require any type of headgear, and it can accommodate flexible head

FIGURE 2.19. Portable monitor, with full-motion 3D video capability.

and body movement. This multichannel holographic screen projects into three adjacent autostereoscopic viewing zones. It displays real-time 3D motion video at a 60-Hz image refresh rate for real-time 3D visualization of an environment indistinguishable from natural viewing. The technology is being extended to large-screen reflective 3D projection.

Acknowledgements

The authors gratefully acknowledge the support of the National Science Foundation, National Institutes of Health, Department of Defense, Department of Energy, and National Aeronautics and Space Administration for their partial support.

References

[1] D.G. Hopper, "Real time holographic displays," *Proc. 11th Annu. IEEE/AESS Dayton Chapter Symp.*, 41–49 (1990).

[2] W.J.A.M. Hartmann and H.M.J. Hikspoors, "Three-dimensional TV with cordless FLC spectacles," *Inform. Display*, **3**(9), 15 (1987).

[3] J. Pollack, "Using both eyes for graphics and imaging," *Lasers Optron.*, May, 45 (1989).

[4] D.G. Hopper, "Dynamic holography for real-time 3D cockpit display," *Proc. NAECON 1986*, 166–172 (1986).

FIGURE 2.20. Internal structure of monitor.

[5] P. St. Hilaire, S.A. Benton, M. Lucente, M.L. Jepsen, J. Kollin, H. Yoshikawa, and J. Underkoffer, "Electronic display system for computational holography," *Proc. SPIE*, **1212**, 174 (1990).

[6] J.B. Eichenlaub, "An autostereoscopic display for use with a personal computer," Dimension Technologies Inc., Rochester, NY, in *Stereoscopic Displays and Applications*, *Proc. SPIE*, **1256**, 156–163 (1990).

[7] R.E. Holmes, R.M. Clodfelter, and G.K. Meacham, "Autostereoscopic video display for battle management systems," Electronic Image Systems, Inc., Xenia, OH, publication RADC-TR-89-386, Feb. 1990.

[8] J. Merritt and D. McAllister, "Stereoscopic display applications issues," *Short Notes, IS&T/SPIE Symp. Electronic Imaging Sci. Technol.* San Jose, CA (1992).

[9] B.K.P. Horn, *Robot Vision*, (MIT Press, Cambridge, MA, 1986).

[10] D. Gabor, U.S. Patent 3,479,111, (November 18, 1969).

[11] T. Jannson, D. Pelka, and T. Aye, "GRIN type diffuser based on volume holographic material," U.S. Patent 5,365,354 (November 15, 1994).

[12] T. Jannson and J. Jannson, "High-efficiency Bragg holograms in the IR, visible, UV and XUV spectral region," *Proc. SPIE* **833**, 84–92 (1988).

[13] T. Jannson, J. Jannson, and P. Yeung, "High channel density wavelength division multiplexer with defined diffracting mean positioning," U.S. Patent 4,926,412 (June 15, 1990).

[14] R.T. Chen, L. Sadovnik, T.M. Aye, and T. Jannson, "Submicron lithography using lensless high efficiency holographic systems," *Opt. Lett.*, **15**, 869–871 (1990).

[15] T. Jannson, T. Aye, G. Savant, and J. Hirsh, "Gabor/Bragg neural association in information storage system," presented at OSA Meeting (1991).

FIGURE 2.21. Photograph of 3D screen (one channel activated).

[16] T. Aye, T. Jannson, A. Kostrzewski, and G. Savant, "Autostereoscopic display system with fan-out multiplexer," U.S. Patent 5,886,675 (March 23, 1999).

[17] A. Kostrzewski, T. Aye, G. Savant, and D.H. Kim, "Direct view autostereoscopic monitor without goggles," in *Cockpit Display IV, Flat Panel Display Applications, Proc. SPIE* **3057**, 507–12 (1997).

Part III

Special Modalities and Methods

3
Edge-Lit Holograms

Michael Metz

3.1 Introduction

Edge-lit holograms offer great promise as compact optical systems, illuminators, and displays for various commercial and industrial applications. A diverse range of products, from self-illuminated, 3D, hang-on-the-wall pictures to thin electronic fingerprint sensors, can be realized through the use of edge-lit holograms.

Throughout its history, holography has had much more potential than actual financial success. Numerous display, commercial, and industrial applications have been suggested or tried; however, few have given entrepreneurial holographers the rewards they hoped for. Surface-relief holograms have enjoyed the most commercial success, due to their low cost and ease of manufacture, using adaptations of well-known printing methods, such as embossing, casting, and hot stamping. Their many fields of use include security devices on credit cards and packaging, trading cards, labels, magazine covers, stickers, and novelty items.

Volume display holograms, due to their realistic 3D imagery, have captivated the excitement and imagination of the public more than surface-relief holograms, but to date have remained more of a curiosity than a commodity. There are many reasons for this. Part of it lies in the public's misconception about what holograms are and are not and what holograms can and cannot do. For example, many people find it hard to recognize that looking at a virtual object in a deep, 3D volume hologram is more analogous to looking at the object through a window than looking at a 2D photograph or poster. Imaginative science fiction writers and well-intentioned, but at times misinformed, reporters, have given us lofty and exciting goals for what holography may be like in the future, but they have also fueled unrealistic expectations of what it is now. To many people, therefore, the practical realities and limitations of the current state-of-the-art of holography are disappointing. They want bright, clear, full-color, large-size, deep, 3D images, viewable with minimal effort in any lighting. Volume-reflection holograms have typically been dark, requiring a strong point light source several feet away to light them properly. A clear image of any depth cannot be reconstructed by a broad source such as a fluorescent lamp. This also limits the practical size of such holograms. The lighting issue means that the volume of space required for a hologram and its light source is significant. Thus, the appeal of edge-lit holograms, in which the light source can be contained in the frame around the hologram, allowing flat, hang-on-the-wall holograms with no externally needed lighting. An additional advantage for industrial applications is the ability of an edge-lit hologram to be used as a holographic optical element to produce compact, broad area illuminators.

A common problem with reflection holograms is that because the viewer and the reconstructing light source are located on the same side of the hologram, close inspection of the hologram is not possible because the viewer's head obstructs the light required for reconstruction. By using an edge-lit hologram, this problem can be eliminated.

This chapter will present an overview of edge-lit holograms. The reader is referred to the bibliography for references containing more depth on particular aspects, mathematics, and applications of edge-lit and waveguide holograms. Section 2 covers definitions and descriptions of a number of different types of holograms that fall into a general family with edge-lit holograms. It also contains a discussion of the historical background of edge-lit holograms and some of the key papers in the field. Some basic mathematics describing properties of edge-lit holograms are presented in Sections 3 and 4. Section 5 contains suggestions for additional issues to be solved, as well as a discussion of the future potential of the field. The references in Section 6 contain a thorough, but not exhaustive, list of sources of historical and technical interest.

3.2   Background

The term "edge-lit hologram" (ELH), has been used loosely. What is commonly referred to as an edge-lit hologram is really a system, consisting of a dielectric substrate, usually a glass or plastic slab, upon which is affixed a hologram. We'll call the large surfaces (usually two) of the substrate **faces**, and the smaller surfaces (usually four) **edges**. The hologram can be either a surface-relief type, in which the diffraction pattern is produced by undulations on the surface, or a volume hologram, in which changes in the refractive index within the bulk of the hologram-recording medium creates a diffractive structure, known as a fringe pattern. In volume holography, the fringe pattern of a standard transmission hologram aligns with the normal (perpendicular) to the face plane, and the fringe pattern of a standard reflection hologram aligns parallel to the face plane. The edge-lit holo-

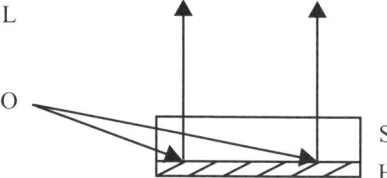
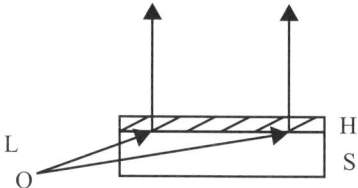

FIGURE 3.1. (a) Reflection edge-lit hologram and (b) Transmission edge-lit hologram.

gram generally embodies a diffractive fringe structure that is slanted with respect to the face plane of the hologram. Thus, it is a hybrid between standard transmission and reflection holograms. Figure 3.1a shows an edge-lit hologram used in reflection mode, and Fig. 3.1b shows an edge-lit hologram used in transmission mode, where L is the illuminating light source, S is the substrate, and H is the hologram.

Light for reconstructing the hologram is sent through the edge of the substrate, and then travels through the substrate at a "steep" angle (a large angle with respect to the normal to the face plane of the hologram, approaching grazing incidence to the face plane). The light then traverses the boundary between the substrate and the hologram, and it passes into the hologram. In some configurations, the light in the substrate may totally internally reflect (TIR), bouncing off the internal faces of the substrate before (and sometimes after as well) reaching the hologram. In other applications, the light is fed directly through the substrate to the hologram, without bouncing within the substrate. The fringe structure prerecorded in the hologram diffracts the light, forming the desired output reconstructed wavefront or holographic image. If the hologram is used in reflection mode, the light passes through the substrate again after being diffracted by the hologram. If the hologram is used in transmission mode, the light leaves the hologram from the side opposite where it entered.

Because the actual hologram is typically a very thin structure (e.g., 30 microns or less), it would be very difficult to actually illuminate the hologram through its edge. Therefore, the substrate is used to feed the light into the hologram. "Edge lit" commonly refers to how the light got into the substrate, through its edge. However, the light can enter the substrate in ways other than via its edge, such as through a prism or diffraction grating (e.g. another edge-lit or waveguide hologram) on the face of the substrate.[1, 2, 11, 12, 25, 73] These elements send the light into the substrate, allowing it to travel at the proper angle within the substrate, with or without TIR bounces, depending on the application, so that the light will cross the substrate-output hologram interface, and interact with the output hologram at the desired angle for reconstruction. Even though the light did not pass through the edge of the substrate, these structures are also sometimes called edge-lit holograms, because the output hologram does not know how the light got to it, through the substrate edge, or its face. Various authors have used other names for such substrate/hologram systems that are related in a general class, such as "substrate referenced hologram," "trapped beam hologram," "substrate mode hologram," "substrate guided-wave hologram," "slanted fringe hologram," and "grazing incidence hologram."

Much early work in the field was done on *evanescent wave holograms*.[8, 9, 37, 38, 50–53, 61, 74–76] In this case, the hologram was typically formed from the interference of an evanescent wave with a plane wave, or another evanescent wave. When a wave travels from a denser medium toward a rarer medium, at an angle (with respect to the normal to the interface) greater than the critical angle for the interface, total internal reflection occurs, and the wave bounces back into the first medium. However, depending on the angle of incidence, the relative indices of the media, and the wavelength of light being used, some light can penetrate into the second medium and travel along the interface between the two media. This type of wave is called an evanescent wave. Its penetration distance into the second medium, however, is typically small, on the order of a wavelength. Therefore, evanescent wave holograms are extremely thin, limiting their ability to yield high diffraction efficiency. Bryngdahl[8, 9] suggested evanescent wave holography would be useful in various studies of wave properties, in the recording of very high-resolution optical images, and in integrated optics for introduction and extraction of spatial information.

In another configuration, Stetson[62–64] defined *total internal reflection* holograms as being recorded through a prism in such a way that the object or reference beam suffers total internal reflection at the outer surface of the hologram-recording material. This method actually produces several sets of fringes, and it may thus limit the available bandwidth or signal-to-noise of the hologram. Stetson described the use of total internal reflection holograms as a means of allowing objects, such as transparencies, to be placed very close to the hologram-recording material without blocking the reference beam. Using this recording method allows holograms to be insensitive to conditions that degrade imagery of conventional holograms. Having an object very close reduces the coherency requirements for the recording and reconstruction, the third-order aberrations in the image, and the intermodulation noise commonly associated with phase holograms. Stetson brought a steep angle reference beam into the recording medium via a prism. The Bragg angle shift in a slanted fringe hologram due to shrinkage of the processed recording medium is also presented in his paper.

Laser-lit transmission volume holograms can reconstruct spectacularly deep, clear images. However, the use of a laser for reconstruction can be hazardous to the eyes of the viewer. In 1970, Lin[36] presented the *edge-illuminated hologram* as a way to overcome this and, as Stetson pointed out, the limitation placed on the minimum distance between the object and the hologram, which limits the maximum image resolution. In Lin's system, which would fall within the waveguide hologram category, light enters the edge of the substrate, and it undergoes total internal reflection at the substrate's two faces, trapping the reference or reconstructing beam totally within it, until it either emerges through the hologram or is absorbed at the opposite edge.

Following on the evanescent wave holography work of Bryngdahl and others, various researchers studied the development of thin film waveguide holograms for use in optical integrated circuits and optical communications: for grating input-output couplers that feed light into or out of a waveguide, for optoelectronic devices, as optical interconnects, for wavelength multiplexing and demultiplexing, and as filters and reflectors in distributed feedback and distributed Bragg reflector GaAlAs lasers.[1, 3, 21, 30, 32, 43, 44, 65, 77] A system in which the light undergoes one or more total internal reflection bounces within the substrate before passing into the hologram, is commonly referred to as a *waveguide hologram.* The terms "trapped beam hologram" and "substrate mode hologram" would also tend to imply that some total internal reflection has taken place within the substrate. Sometimes the hologram is used as a means of getting light into a substrate, rather than taking it out of the substrate. The hologram may actually be a pattern formed right into the substrate face. These holograms include the existence of twin conjugate primary images. One advantage of the waveguide hologram is that unlike the reconstruction of a conventional off-axis hologram, the reconstruction of a waveguide hologram is not limited by any unwanted waves (undiffracted and conjugate waves). Therefore, it is possible to reconstruct an image with an extremely wide field-of-view.

Beginning in the late 1980s, a new interest emerged among a number of researchers to develop edge-lit and waveguide holograms for displays and other commercial and industrial applications. Upatnieks did pioneering work on edge-lit holographic displays and the use of edge-lit holograms for compact gun sights.[69–72] He discussed applications such as holographic displays that hang on the wall like a painting, compact light-collimators, light collectors, and head-up displays (HUD) in vehicles, where sunlight on a standard HUD may form an objectionable bright diffraction pattern. An ELH solves this problem. Moss researched variations of edge-lit holograms for vehicle indicators and displays.[45–49]

Caulfield, Huang, and others studied waveguide holograms for displays and as holographic optical elements for illumination of, e.g., spatial light modulators[10, 13, 14, 24–28, 59] or in conjunction with diffusing capabilities for flat panel display backlighting.[66]

Benton and others researched the extension of rainbow techniques to edge-lit holograms.[5–7, 19, 22] A rainbow ELH allows a greater vertical viewing zone for rainbow holograms than for conventional holograms. It also allows white light illumination of the ELH, and a much more compact display than for standard rainbow holograms. Benton also described the use during recording of an immersion tank filled with index matching fluid to minimize spurious reflections.

In Japan, Kubota and others also studied edge-lit holograms for compact displays and illuminators.[31–35, 54, 55, 60, 67, 68]

Phillips and others took a fresh look at the ELH problem, using Dupont photopolymer rather than silver halide recording film.[15, 16, 56–58] They pointed out that the key to making successful edge-lit holograms is in careful setup and alignment of the reference beam. By using photopolymer, they achieved less shrinkage of the emulsion than is typical with silver halide recording materials. The slanted fringe structures of edge-lit holograms are particularly sensitive to emulsion shrinkage, in which a small change in emulsion thickness can result in slant angle changes so that the edge-induced input reference beam can no longer satisfy the Bragg angle.

This condition then reduces the available diffraction efficiency, or even the ability of the hologram to function at all.

Metz and others founded a company, ImEdge Technology, Inc., specifically devoted to edge-lit holograms. They researched ELHs for visual displays, as holographic optical elements for backlighting and frontlighting flat panel displays, and as compact areal illumination systems for industrial and biomedical applications.[39] In particular, they developed prototypes of compact electronic fingerprint sensors using edge-lit holograms to illuminate the finger.[40-42] Other researchers also investigated slanted fringe holograms for use in fingerprinting applications.[4, 7, 18, 20, 29]

Research and development in the field is continuing at various institutions around the world.

3.3   Slanted Fringes

Numerous approaches and notations have been taken in the literature to describe the propagation and interference of waves, particularly as they pertain to holograms. We will present in this section the basic mathematics comparing the fringe formation of reflection, transmission, and edge-lit or slanted fringe holograms. The reader is directed to the references for various articles that present related calculations in more depth.

For simplicity, we have chosen to limit most of the discussion to two dimensions, using the X-axis to the right, and the Y-axis upward. Ray directions, or the direction of propagation of the light waves will be denoted in a $360°$ system, starting at the positive X-axis and rotating counterclockwise. We will assume that the reference and object beams are plane-polarized light waves. Based on personal preference for visualization, we have chosen to keep the plane of the hologram (and substrate) along the X-axis and consider the formation of volume holograms. We apologize if this flies in the face of certain established conventions.

In two dimensions, following the notation of Hopkins,[23] if a plane-polarized wave is propagated along an arbitrary direction OP, where $P(x, y)$ is an arbitrary point in space, the magnitude of E, the electric vector, assumes the form

$$E(x, y, t) = A \cos[kn(px + qy) + \phi - \omega t], \qquad (1)$$

where p and q are the direction cosines of OP with respect to X and Y, respectively; A is the amplitude of the wave, which is an exponentially decreasing function for absorbing media; $t =$ time; λ is the wavelength of the light, in air; T is the period for one complete vibration; $k = 2\pi/\lambda$; $\omega = 2\pi/T$; ϕ is the phase angle; and n is the refractive index of the medium.

Inserting the direction cosines $p = \cos\theta$, $q = \sin\theta$, we get

$$E(x, y, t) = A \cos[kn(x \cos\theta + y \sin\theta) + \phi - \omega t]. \qquad (2)$$

As a mathematical convenience, we can put this in the form

$$E(x, y, t) = A \exp[i(kn(x \cos \theta + y \sin \theta) + \phi - \omega t)], \qquad (3)$$

where the real part of (3) represents the instantaneous value of E. We denote this as the complex amplitude. The time-averaged energy density of E is represented by

$$|E|^2 = EE^* = A^2, \qquad (4)$$

where $*$ denotes the complex conjugate.

Suppose we want to calculate the interference of two waves, E_1 and E_2. The total amplitude E_T of the two waves at some point $P(x, y)$ is given by

$$E_T = E_1 + E_2. \qquad (5)$$

The intensity I at point P is then

$$I = E_T E_T^* = (E_1 + E_2)(E_1 + E_2)^*. \qquad (6)$$

Now we will apply this to the standard reflection hologram, showing how the interference fringes align parallel to the plane of the hologram, to the standard transmission hologram, in which the fringes align perpendicular to the plane of the hologram, and to the edge-lit hologram in which the fringes are slanted with respect to the plane of the hologram (see Figure 3.2).

3.3.1 Standard Reflection Hologram

A standard reflection hologram is made by interfering two beams, an object beam and a reference beam, traveling in opposite directions (i.e., one beam enters the recording medium from one side, and the other beam enters from the other side of the recording medium). The resulting interference pattern is recorded in a holographic film material. For simplicity, assume that the object and reference beams are mutually coherent plane waves traveling in a nonabsorptive medium, and have equal phase, frequency, and amplitudes. In these examples, we are creating a simple grating; however, it is to be understood that the object beam can carry information, such as that of an object, and the fringe structures can get very complicated depending on the wavefront shapes of the object and the reference beams. Let the reference beam be traveling in the $+Y$ direction, and the object beam be traveling in the $-Y$ direction. The complex amplitude, E_R, of the reference beam is given by,

$$E_R = A_R \exp[i(kny + \phi - \omega t)], \qquad (7)$$

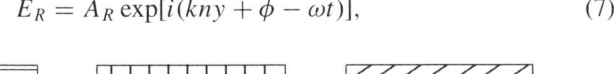

FIGURE 3.2. Interference fringe geometries for (a) the standard reflection hologram, (b) the standard transmission hologram, and (c) the edge-lit hologram, made from two interfering plane waves.

and the complex amplitude, E_O, of the object beam is given by

$$E_O = A_O \exp[i) - kny + \phi - \omega t)] \tag{8}$$

Following (6), the intensity, I, at a point $P(x, z)$ is given by,

$$I_{\text{refl}} = A_O^2 + A_R^2 + 2A_O A_R \cos[2kny]. \tag{9}$$

Note that the intensity variation is only a function of y, showing that the fringes are parallel to x, the hologram plane, or the bisector between the object and reference beam directions. We'll call this a fringe slant angle of $0°$. Because $k = 2\pi/\lambda$ the argument of the cos function becomes $4\pi ny/\lambda$. The fringe cycle repeats when this is equal to 2π or when $y = \lambda/(2n)$. Because the wavelength inside the medium $\lambda_m = \lambda/n$, the fringe cycle repeats for every $y = \lambda_m/2$.

Thus, for example, if $n = 1.5$, $\lambda = 647$ nm and, for normalization, $A_O = A_R = 0.5$, the fringes repeat every 647-nm/$(2 \times 1.5) = 216$ nm.

3.3.2   Standard Transmission Hologram

In a reflection hologram, the object and reference beams enter the holographic film from opposite sides. A transmission hologram is created when the object and reference beams enter the holographic film from the same side. Typically, for a standard transmission hologram, the object and reference beams enter the holographic film from equal but opposite angles about the normal to the film. For example, let the object beam (in the film) have an angle θ_0 (measured from the x-axis to the object ray) and the reference beam have an angle $\theta_R = \theta_0 + 2\alpha$ where $\alpha = 90 - \theta_0$, or $\theta_R = 180 - \theta_0$. The complex amplitude, E_O of the object beam is given by

$$E_O = A_O \exp[i(kn(x \cos(\theta_0) + y \sin(\theta_0) + \phi - \omega t)], \tag{10}$$

and the complex amplitude, E_R, of the reference beam is given by

$$E_O = A_O \exp[i(kn(x \cos(180 - \theta_0) + y \sin(180 - \theta_0)) + \phi - \omega t)]. \tag{11}$$

The intensity, I, at a point $P(x, y)$ is given by

$$I_{\text{trans}} = A_O^2 + A_R^2 + 2A_O A_R \cos[2knx \cos(\theta_0)]. \tag{12}$$

In this case, for the standard transmission hologram, the intensity variation is only a function of x, showing that the fringes are perpendicular to the hologram plane, or the bisector between the object and reference beam directions. We'll call this a fringe slant angle of $90°$. The fringe spacing is a function of the angle between object and reference beams. In this example, the fringe spacing is determined by setting

$$2knx \cos(\theta_0) = 2\pi \tag{13}$$

or

$$x = \lambda/(2n \cos \theta_0). \tag{14}$$

3.3.3 Slanted Fringes

If the object and reference beams enter the holographic film at asymmetrical angles about the normal for the transmission hologram, or about the plane of the hologram for the reflection hologram, slanted fringes will be produced. As the fringe slant angle approaches 45°, the hologram takes on a hybrid form, which has characteristics between those of the standard transmission and reflection holograms. This type of slanted fringe structured hologram is commonly referred to as an edge-lit hologram.

3.3.4 Limits on Fringe Slant Angle Defines Boundaries of Edge-Lit Hologram

For mechanical stability, flat edge-lit holograms typically consist of a recording medium (e.g., silver halide, dichromated gelatin (DCG), photopolymer, etc.), which is coated, laminated, or otherwise affixed to a rigid substrate. The substrate is commonly a transparent material, such as glass or plastic. For example, the substrate may be a square or rectangular slab of BK7 glass with perpendicular corners. The slab has four edge surfaces and two larger face surfaces. The light input edge must be polished to an optical finish (e.g., 60–40). The recording medium is affixed to one face.

Consider a transmission hologram, in which both beams enter the hologram-recording material directly, from the same side, rather than from the substrate side. We will show that using this geometry, the range of possible fringe slant angles inside the recording medium is limited. This can be seen by applying Snell's law to the substrate and film. For example, say the index of refraction of the hologram material is 1.5.

As shown in Fig. 3.3, for a transmission hologram, if we hold the object beam fixed at 90° (i.e., parallel to the y-axis, normal to the hologram plane), the fringes will become increasingly slanted as the reference beam approaches 0° (or 180° for a slant in the opposite direction). Because the object beam is along the normal to the substrate face, there will be no refraction. In the extreme limit, for this calculation, imagine if the reference beam could enter (the face, not the edge!) at

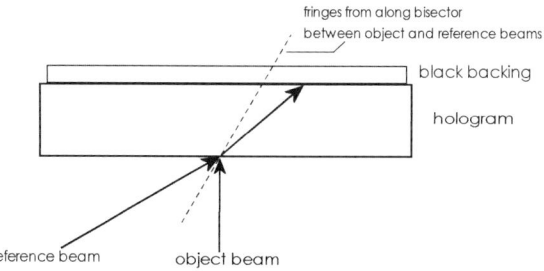

FIGURE 3.3. Diagram for Snell's law in transmission hologram: determining fringe slant-angle limitations.

an angle of 0°. Thus, the reference beam angle with respect to the normal to the surface is 90°. So from Snell's law, the angle of refraction, β' (with respect to the surface normal), is

$$n_1 \sin \beta = n_2 \sin \beta', \tag{15}$$

where n_1 is the index of refraction of the first medium, which in this case is air, and n_2 is the index of refraction of the holographic medium. Substituting $n_1 = 1$, $\beta = 90°$, $\sin \beta = 1$:

$$1 \times 1 = 1.5 \sin \beta', \tag{16}$$

$$\beta' = 42°. \tag{17}$$

Note that in our use of Snell's law here, we assume the angle of incidence and the angle of refraction are both positive quantities between 0° and 90°.

According to our sign convention for the hologram, now, we remeasure that angle from the x-axis; so the angle of propagation in the film is then the complement of β', or 48°. The slant angle would then be the bisector between the object and reference beam propagation angles, or $(90 + 48)/2 = 69°$. So for the transmission hologram in our example, the fringe slant angle is limited between about 68° and 90°, or about 90–112° for the opposite slant. (Note that the angle ranges will be different for different media and different geometries.)

Performing a similar calculation for a reflection hologram, in the limit as the propagation angle of the reference beam approaches 0°, (from the other side) and the object beam is at 90°, the refracted beam will propagate at an angle of −48°, yielding a fringe slant angle of $(90 - 48)/2 = 21°$. So the slant angle for a reflection hologram in which the object and reference beam enter through the face is limited between around 0° and 21° (or 159–180°).

How then do we achieve fringe slant angles between 21° and 69° (or between 112° and 159°)? We must send the reference beam through the substrate at a steeper angle than can be achieved by refraction through the face. One way to do this is to send the reference beam through the edge of the substrate. So, for example, say a reference beam strikes the edge of the substrate at a 10° propagation angle, or 10° to the normal to the edge. If we use, for example, a BK7 substrate and 488 nm laser light, by Snell's law,

$$1 \times \sin(10) = 1.522 \sin \beta', \tag{18}$$

$$\beta' = 6.55°, \tag{19}$$

where the index of refraction of BK7 at 488 nm is about 1.522.

For a transmission hologram with an object beam propagating at 90°, the fringe slant angle is now $(90 + 6.55)/2$ or about 48°. Therefore, we have breached this inaccessible zone of slant angles, and can achieve slant angles close to 45°. A similar argument pertains to the reflection hologram. Holograms with slant angles between around 20° and 70° (or 110° to 160°) are thus commonly referred to as edge-lit holograms, even if the reference beam did not originate through the edge of the substrate. So in this geometry, holograms with slant angles between around 20°

and 45° (or around 135° to 160°) are referred to as reflection edge-lit holograms, and holograms with slant angles between 45° and around 70° (or around 110° to 135°) are referred to as transmission edge-lit holograms.

As an example, let the object beam travel in the $-y$ direction, and the reference beam travel through the substrate at an angle θ_R within the substrate. The complex amplitude of the object beam is then

$$E_O = A_O \exp[i(-kny + \phi - \omega t)], \tag{20}$$

and the complex amplitude of the reference beam is given by

$$E_R = A_R \exp[i(kn(x\cos(\theta_R) + y\sin(\theta_R)) + \phi - \omega t)]. \tag{21}$$

When these two beams interfere, the resulting intensity is given by

$$I_{\text{ELH}} = A_O^2 + A_R^2 + 2A_O A_R \cos[kn(y + x\cos(\theta_R) + y\sin(\theta_R))]. \tag{22}$$

The fringe planes are defined by

$$kn(y + x\cos(\theta_R) + y\sin(\theta_R)) = \text{constant} = C. \tag{23}$$

Solving for y, we get

$$y = -(-C + knx\cos(\theta_R))/(kn(1 + \sin(\theta_R))). \tag{24}$$

The slope of the fringes is given by dy/dx. We set that equal to $\tan[\gamma]$ as follows:

$$\tan(\gamma) = -\cos(\theta_R)/(1 + \sin(\theta_R)). \tag{25}$$

For example, if the reference beam is at 85° to the normal to the face, or $\theta_R = 5°$, the fringe slant angle γ is $-42.5°$.

3.4 Fresnel Reflections

For light traveling from a medium of lower index of refraction to a medium of higher index of refraction, as the angle at which light strikes the boundary between the two media becomes more and more parallel to the surface (grazing incidence), more and more of the light will be reflected at the boundary, rather than transmitted. This is a very important point to consider when making edge-lit holograms. In the case in which the light is traveling from a medium of higher index of refraction to a medium of lower index of refraction, if the angle of travel is greater than the critical angle, the light will be totally internally reflected at the boundary. However, depending on the angle and the relative indices, a small amount of the light will penetrate a short distance into the medium (typically around a half a wavelength), travel for some distance, and then return to the first medium. This phenomenon is called evanescence, and it is discussed in detail in Bryngdahl's article.[9] We will show some of the basic mathematics for determining the amount of light that will be reflected within the substrate and transmitted from the substrate to the hologram. Knowing this information enables edge-lit holographers to select the most appropriate film and substrate for their applications, to properly match the intensities of the object and reference beams, and to determine the exposure times required for the hologram. For edge-lit holograms, because the beams interfere nearly perpendicular to each other, s-polarization (perpendicular to the plane of incidence) is usually used. Otherwise, if p-polarization is used (parallel to the plane of incidence), little interference will occur due to the mismatch of polarization vectors between the object beam and the reference beam. In other words, because the beams are nearly perpendicular to each other, the polarization component, for example, in the object beam direction, will be much larger in the object beam than will be the component in the reference beam parallel to the object beam direction.

We can predict the reflected and transmitted beam intensities using Fresnel's equations. For s polarization, let A_i be the amplitude of the light traveling in a substrate whose index of refraction is n_s. The light travels through the substrate at an angle with respect to the normal of ϕ_s and is incident on the boundary between the substrate and the holographic-recording medium. The index of refraction of the holographic-recording medium is n_m. Let A_t be the amplitude of the refracted light that is transmitted from the substrate into the recording medium, and let ϕ_m be the angle of refraction with respect to the normal to the boundary. A_i and A_t are related as follows:

$$\frac{A_t}{A_i} = \frac{2\sin(\phi_m)\cos(\phi_s)}{\sin(\phi_s + \phi_m)}. \tag{26}$$

If A_r is the amplitude of the light reflected at the boundary and ϕ_r is the angle of reflection with respect to the normal, then

$$\frac{A_r}{A_i} = -\frac{\sin(\phi_s - \phi_m)}{\sin(\phi_s + \phi_m)}. \tag{27}$$

Using Snell's law, we recast these into a form that uses the input angle (angle in the substrate) only:

$$\frac{A_t}{A_i} = \frac{2\cos(\phi_s)}{\cos(\phi_s) + \frac{n_m}{n_s}\sqrt{1 - \left[\frac{n_s}{n_m}\sin(\phi_s)\right]^2}}, \tag{28}$$

$$\frac{A_r}{A_i} = \frac{\cos(\phi_s) - \frac{n_m}{n_s}\sqrt{1 - \left[\frac{n_s}{n_m}\sin(\phi_s)\right]^2}}{\cos(\phi_s) + \frac{n_m}{n_s}\sqrt{1 - \left[\frac{n_s}{n_m}\sin(\phi_s)\right]^2}}. \tag{29}$$

The transmitted intensity is given by

$$T = \frac{n_m\cos(\phi_m)}{n_s\cos(\phi_s)}\left[\frac{A_t}{A_i}\right]^2 \tag{30}$$

or

$$T = \frac{4n_m n_s \cos(\phi_s)\sqrt{1 - \left[\frac{n_s}{n_m}\sin(\phi_s)\right]^2}}{\left[n_s\cos(\phi_s) + n_m\sqrt{1 - \left[\frac{n_s}{n_m}\sin(\phi_s)\right]^2}\right]}. \tag{31}$$

The reflected intensity R is given by

$$R = \left[\frac{A_r}{A_i}\right]^2. \tag{32}$$

As an example of the use of the Fresnel equations, in a typical edge-lit hologram application, we may want the object beam to enter the film perpendicular to the film/substrate interface. To achieve good light usage efficiency, the goal for the reference beam light is to select the index of refraction of the film and its substrate, and the angle of incidence of the reference beam to achieve significant transmission from the substrate into the film. If we use off-the-shelf holographic film, we have no control over the index of refraction of the film. Therefore, it makes sense to select the substrate and the angle of the reference beam to match the film and the transmission we desire. In Fig. 3.4, the Fresnel equations are used to plot the intensity of an s-polarized reference beam that gets transmitted from the substrate into the film (ignoring absorption), for a film index of 1.5, and reference beam angles in the substrate of 85°, 87°, and 89° with respect to the normal to the interface between the film and the substrate.

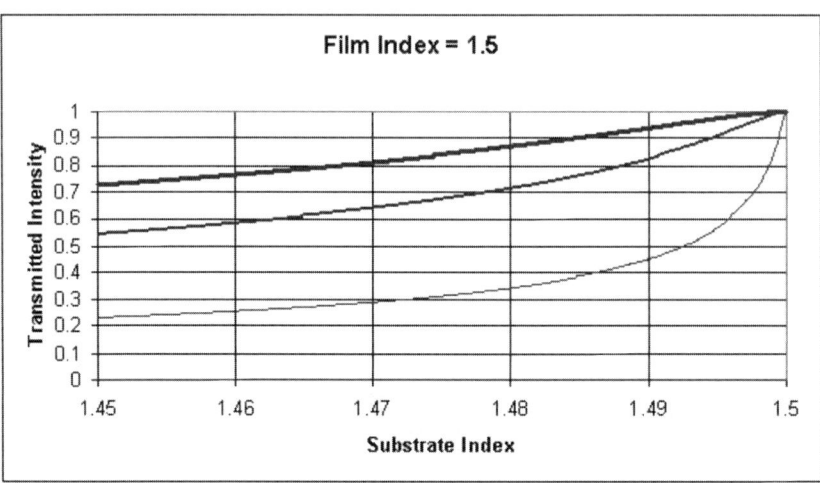

FIGURE 3.4. Substrate index of refraction versus transmitted intensity for a film index of refraction of 1.5, and reference beam input angles of 85°, 87°, and 89° shown as thickest to thinnest lines, respectively.

If, for example, your goal is to get at least 90% of the reference beam transmitted into the film, we see that for an 85° reference beam, the index of refraction of the substrate must be chosen to be between 1.49 and 1.50. The index match between substrate and film gets much tighter as the reference beam angle approaches 90°. If an index matching glue or other substance is used between the substrate and film, its index of refraction should be greater than or equal to the index of refraction of the substrate, and less than or equal to the index of refraction of the film.

3.5   The Future

Edge-lit holography is still in its infancy. Much research and development is still needed to move the field out of the laboratory and into the marketplace. A few suggestions for further work are given below. It is an area with great commercial potential, which is still as of this writing untapped. The difficulty with holography is that the entry fee is very high. As opposed to, say, an Internet software developer, where a few thousand dollars buys a computer and software to get the work done, equipment for making commercial grade holograms is expensive. Between lasers, optics, and optical bench equipment, perhaps a clean room, employee salaries, and so on, one needs to consider hundreds of thousands, or a several million dollars as the cost of entry into the field. This makes it prohibitive for enthusiastic startups without significant capital backing. That leaves the universities and already equipped industrial laboratories to take the initiative to move the field ahead. Following is an outline of some of the areas that need attention.

One of the most obvious projects is mass production. High-quality edge-lit holograms have been produced in small quantities, but to date, edge-lit holograms have not been mass-produced. There are two basic schemes to consider for mass production.

In the first, each hologram is a master. The film gets transported to an exposure station. Perhaps the film is temporarily laminated to a substrate, in which a reference beam travels through the substrate, and an object beam either travels through the substrate to the film (transmission hologram) or to the film directly (reflection hologram). The film may be on a roll, for example, and gets exposed, processed, delaminated, rolled to the next area to be exposed, and so on. Later the film may be cut and laminated to its permanent substrates. It is also possible to apply the unexposed film (in sheet or liquid form) to a permanent substrate, and then send it to an exposure station, where the object and reference beams are prealigned, and interact with the substrate and recording film in such a way as to expose the hologram. In both of these schemes, it would be easier and more practical to use a recording film that requires a dry postexposure process, such as heat and ultraviolet light, or no processing at all.

One of the caveats with any processing is that the film often changes its thickness. This change can be nonlinear. This then causes the fragile slanted fringes to be bent or warped, severely affecting the hologram output wavefront. Thus, another area of research is determining how the slanted fringes change with processing for different films and different processes.

In a second mass-production scheme, a master hologram is used, and copies are made from the master. In this case, contact copying could be considered. New holographic recording film is laid on the master hologram, usually using index matching fluid, which must closely match the index of the film, to establish proper contact between the master and the copy to be made. In one of a number of possible geometries, for a reflection edge-lit hologram, a replica of the object beam is then passed through the copy to the master, where the conjugate of the reference beam is produced by diffraction from the master hologram. That conjugate beam passes back through the copy at the appropriate angle to interfere with the object beam and reproduce a copy of the original set of slanted fringes, similar to the fringe set in the master. Great care must be taken to suppress additional fringe sets that may be formed from reflections off of other interfaces.

Additionally, for applications in which reasonably symmetric fringes are used, such as holographic optical elements for illuminators, nonholographic methods of production can be considered. In this case, slanted fringes are written with, for example, an electron beam entering the film at the appropriate angle (such as through a prism placed on the film). Another possibility is to use a lithographic-type mask and expose through the mask with light at the proper slant angle.[26]

Another area that needs further research is the effect of the absorption of the substrate and film on the fringe contrast, and how the reference beam light intensity diminishes from one end of the hologram to the other. A gradient intensity mask can be used in the reference or object beam to compensate for this intensity gradient during recording, as well as the effects of the Gaussian intensity profile of the laser beam. For edge-lit illuminators, uniformity, or a well-defined, tailored intensity distribution of the output light, is crucial to their success.

Recently, much progress has been made in full-color and computer-generated/computer-assisted holograms. In order for edge-lit holographic displays to become commercially viable on a large scale, the demands by the advertising and display industries for full, true colors and large sizes must be met. A goal, then, for the immediate future, would be to marry the color hologram and edge-lit technologies, and then work on increasing the overall sizes produced. One of the primary things holding such developments back is the lack of suitable holographic-recording materials. In edge-lit holography, because the reference beam travels a very long distance through the recording material, compared with the object beam, this puts considerable demands on the recording material, especially for full-color work. The recording material must have a small absorption coefficient. Another problem is that sensitizing dyes can fluoresce, causing exposure at unwanted wavelengths, and fogging the film, or using up its available bandwidth. The ideal recording material would also require dry processing, or no processing at all after proper exposure. It would need to be inexpensive, easily mass-produced in the unexposed state, and easily adaptable to mass-production exposure systems. This is a tall order and an expensive research proposition. However, the potential rewards are huge for any company that can succeed in making suitable recording materials.

Because mass-production of surface-relief holograms is a well-established industry, another goal is the development of slanted fringe surface-relief gratings that can accomplish similar outputs to volume slanted fringe gratings. This is more straightforward for simple illuminators in which the fringes are straight. The issues of noise and dispersion characteristics would need to be carefully addressed for the surface-relief solution.

Another area to look at is multiple gratings: stacking or laminating several together, or exposing multiple gratings within a single hologram, or several stacked holograms. For applications requiring multiple color outputs, such as "white light" illuminators, using the edge-lit technique, it would be much easier to make individual monochromatic holograms of different colors, and then stack them together, rather than make multiple color exposures within the same film. Additionally, by stacking several edge-lit holograms together with different orientations, one can consider illuminating them from multiple sides, yielding a plurality of images, which can be addressed one at a time or in combination. For illuminators, this would yield a brighter, more uniform output.

Many uses for edge-lit and waveguide holograms have been noted above, in the referenced documents, and in much literature not cited here. These uses fall into several categories, such as compact, broad area illuminators, 2D and 3D image displays and signs, and opto-electronic and communications systems. Within each category, one can think of multiple areas of application. For example, as noted above, thin, edge-lit broad area illuminators can be used in flat panel display front and backlighting systems, for compact sensors such as electronic fingerprinting devices, to illuminate other holograms, converting conventionally lit holograms to be edge-lit, and in industrial inspection systems, such as lighting in crevices and other tight spots. Many biomedical applications exist, such as for illumination in miniature biosensors, such as molecular fluorescence and affinity detectors. An adjunct of the fingerprint sensor could be, for example, a simple, inexpensive, handheld, miniature areal microscope that is placed on the skin to record the image of anomalies such as moles or cancers directly to a computer for image processing or comparing images taken at different time periods. In such a system, light is sent to the edge-lit hologram, which redirects the light to illuminate the skin. Light from the skin passes back through the hologram to a detection system, for example, which includes a charge-coupled display (CCD) or complementary metal-oxide-semiconductor (CMOS) detector array. The image is then passed to a computer for image storage and processing. We have only just begun to think of clever uses for edge-lit holograms.

As shown in this discussion, edge-lit holography is an exciting field, which can be applied to a vast number of industrial and commercial applications with enormous financial potential. However, there are still a number of hurdles to overcome. With so many applications that could benefit by the unusual properties of edge-lit holograms, one would hope that the field will advance rapidly in the early part of the 21st century. The author looks forward to the day when he can pick up this chapter, and check off all of the issues presented above as having been successfully accomplished.

References

[1] Y. Amitai, "Design of wavelength-division multiplexing/demultiplexing using substrate-mode holographic elements," *Opt. Comm.* **98**, 24–28 (1993).

[2] Y. Amitai, I. Shariv, M. Kroch, A. Friesem, and S. Reinhorn, "White-light holographic display based on planar optics," *Opt. Lett.* **18**(15), 1265–1267 (1993).

[3] H. Ando and K. Namba Shinji, "Light-guiding device having a hologram layer," U.S. Patent 5,268,985 (1993).

[4] R.D. Bahuguna and T. Corboline, "Prism fingerprint sensor that uses a holographic optical element," *Appl. Opt.* **35**(26), 5242–5245 (1996).

[5] S.A. Benton, S.M. Birner, and A. Shirakura, "Edge-lit rainbow holograms," *Proc. SPIE* **1212**, 149–157 (1990).

[6] S.A. Benton and S.M. Birner, "Self-contained compact multi-color edge-lit holographic display," U.S. Patent 5,121,229 (1992).

[7] S.M. Birner, "Steep reference angle holography: analysis and applications," M. Sc. Vis. Stud. thesis, Massachusetts Institute of Technology, (February, 1989).

[8] O. Bryngdahl, "Holography with evanescent waves," *JOSA* **59**(12), 145 (1969).

[9] O. Bryngdahl, "Evanescent waves in optical imaging," *Progress in Optics* **XI**, 167–221 (1973).

[10] H.J. Caulfield and Q. Huang, "Wide field of view transmission holography," *Opt. Comm.* **86**, 487–490 (1991).

[11] H.J. Caulfield, Q. Huang, A. Putilin, and V. Morozov, "Multimode waveguide holograms capable of using non-coherent light," U.S. Patent 5,295,208 (1994).

[12] H.J. Caulfield, Q. Huang, A. Putilin, and V. Morozov, "Side illuminated multimode waveguide," U.S. Patent 5,465,311 (1995).

[13] H.J. Caulfield, Q. Huang, A. Putilin, V. Morozov, and J. Shamir, "Waveguide hologram illuminators," U.S. Patent 5,515,184 (1996).

[14] H.J. Caulfield, Q. Huang, A. Putilin, V. Morozov, and J. Shamir, "Waveguide hologram illuminators," U.S. Patent 5,854,697 (1998).

[15] Z. Coleman, "Modern holographic recording and analysis techniques applied to edge-lit holograms and their applications," Doctoral Thesis, Loughborough University, 1997.

[16] Z. Coleman, M. Metz, and N. Phillips, "Holograms in the extreme edge illumination geometry," *Proc. SPIE* **2688**, 96–108 (1996).

[17] M. Drake, M. Lidd, and M. Fiddy, "Waveguide hologram fingerprint entry device," *Opt. Eng.*, **35**(9), 2499–2505 (1996).

[18] S. Eguchi, I. Igaki, H. Yahagi, F. Yamagishi, H. Ikeda, and T. Inagaki, "Uneven-surface data detection apparatus," U.S. Patent 4,728,186 (1988).

[19] W. Farmer, S. Benton, and M. Klug, "The application of the edge-lit format to holographic stereograms," *Proc. SPIE* **1461**, 215–226 (1991).

[20] Y. Fujimoto, M. Katagiri, N. Fukuda, and K. Sakamoto, "Fingerprint input apparatus," U.S. Patent 5,177,802 (1993).

[21] D.G. Hall, "Diffraction efficiency of waveguide gratings: Brewster's law," *Opt. Lett.* **5**(7), 315–317, (1980).

[22] M. Henrion, "Diffraction and exposure characteristics of the edgelit hologram," Master's thesis, MIT (1995).

[23] R. Hopkins, "*Military Standardization Handbook, Optical Design (MIL-HDBK-141)*," (1962).

[24] Q. Huang and H.J. Caulfield, "Edge-lit reflection holograms," *Proc. SPIE* **1600**, 182—186 (1991).

[25] Q. Huang and H.J. Caulfield, "Waveguide holography and its applications," *Proc. SPIE* **1461**, 303–312 (1991).

[26] Q. Huang, J.A. Gilbert, and H.J. Caulfield, "Substrate guided wave (SGW) holo-interferometry," *Proc. SPIE* **1667**, 172–181 (1992).

[27] Q. Huang and J.A. Gilbert, "Diffraction properties of substrate guided-wave holograms," *Opt. Eng.* **34**(10), 2891–2899 (1995).

[28] Q. Huang, "Substrate guided wave holography: analysis, experiments, and applications," Doctoal thesis, The University of Alabama in Huntsville (1994).

[29] S. Igaki, S. Eguchi, F. Yamagishi, H. Ikeda, and T. Inagaki,"Real-time fingerprint sensor using a hologram," *Appl. Opt.* **31**, 1794–1802 (1992).

[30] R. Kostuk, Y. Huang, D. Hetherington, and M. Kato, "Reducing alignment and chromatic sensitivity of holographic interconnects with substrate mode holograms," *Appl. Opt.* **28**(22), 4939–4944 (1989).

[31] T. Kubota, "Flat type illuminator and its applications," *Jpn. J. Opt.* **19**, 383–385 (1990).

[32] T. Kubota and M. Takeda, "Array illuminator using grating couplers," *Opt. Lett.* **14**, 651–652 (1989).

[33] T. Kubota, K. Fujioka, and M. Kitagawa, "Method for reconstructing a hologram using a compact device," *Appl. Opt.* **31**, 4734–4737 (1992).

[34] T. Kubota, "Creating a more attractive hologram," *Leonardo* **25**, 503–506 (1992).

[35] T. Kubota and H. Ueda, "Compact display system for hologram," *Proc. SPIE* **2866**, 207–214 (1996).

[36] L.H. Lin, "Edge Illuminated Holograms," *JOSA* **60**, 714A (1970).

[37] W. Lukosz and A. Wuthrich, "Holography with evanescent waves," *Optik* **41**(2), 194–211 (1974).

[38] W. Lukosz and A. Wuthrich, "Hologram recording and readout with the evanescent field of guided waves," *Opt. Comm.* **19**(2), 232–235, (1976).

[39] M. Metz, "Edge-lit holography strives for market acceptance," *Laser Focus World*, 159–163, (May 1994).

[40] M. Metz, C. Flatow, Z. Coleman, and N. Phillips, "The use of edge-lit holograms for compact fingerprint capture," *Proc. CardTech/SecurTech '95*, Washington, D.C., 222–228 (1995).

[41] M. Metz, Z. Coleman, N. Phillips, and C. Flatow, "Holographic optical element for compact fingerprint imaging system," *Proc. SPIE* **2659**, 141–151 (1996).

[42] M. Metz, C. Flatow, N. Phillips, and Z. Coleman, "Device for forming and detecting fingerprint images with valley and ridge structure," U.S. Patent 5,974,162 (1999).

[43] M. Miler, V.N. Morozov, and A.N. Putilin, "Diffraction components for integrated optics," *Sov. J. Quant. Electron.* **19**(3), 276–283 (1989).

[44] J.M. Miller, N. de Beaucoudrey, P. Chavel, J. Turunen, and E. Cambril, "Design and fabrication of binary slanted surface-relief gratings for a planar optical interconnection," *Appl. Opt.* **36**(23), 5717–5727 (1997).

[45] G. Moss, "Holographic indicator for determining vehicle perimeter," U.S. Patent 4,737,001 (1988).

[46] G. Moss, "Holographic display panel for a vehicle windshield," U.S. Patent 4,790,613 (1988).

[47] G. Moss, "Segmented 3–D hologram display," U.S. Patent 4,795,223 (1989).

[48] G. Moss, "Holographic thin panel display system," U.S. Patent 4,807,951 (1989).

[49] G. Moss, "Holographic rear window stoplight," U.S. Patent 4,892,369 (1990).

[50] H. Nassenstein, "Holographie und interferenzversuche mit inhomogenen oberflachen-wellen," *Phys. Lett.* A **28**, 249–251 (1968).

[51] H. Nassenstein, "Interference, diffraction and holography with surface waves ('Subwaves') I," *Optik* **29**, 597–607 (1969).

[52] H. Nassenstein, "Interference, diffraction and holography with surface waves ('Subwaves') II," *Optik* **30**, 44–55 (1969).

[53] H. Nassenstein, "Holography with surface waves," U.S. Patent 3,635,540 (1972).

[54] N. Nishihara and T. Kubota, "A compact display system for hologram using a holographic grating," *Proc. 3D Image Conference '95*, 19–23 (1995).

[55] H. Okamoto, H. Ueda, K. Taima, E. Shimizu, T. Nishihara, F. Iwata, and T. Kubota, "A compact display system for hologram," *Proc. SPIE* **2333**, 424–428 (1995).

[56] N.J. Phillips, C. Wang, and T.E. Yeo, "Edge-illuminated holograms, evanescent waves and related optics phenomena," *Proc. SPIE* **1600**, 18–25 (1991).

[57] N.J. Phillips and C. Wang, "The recording and replay of true edge-lit holograms," *IEE Conference Publication* **342**, 8–11 (1991).

[58] N.J. Phillips, C. Wang, and Z. Coleman, "Holograms in the edge-illuminated geometry-new materials developments," *Proc. SPIE* **1914**, 75–81 (1993).

[59] A. Putilin, V. Morozov, Q. Huang, and H. Caulfield, "Waveguide holograms with white light illumination," *Opt. Eng.* **30**(10), 1615–1619 (1991).

[60] A. Shimizu and K. Sakuda, "Simple measuring technique for the diffraction efficiency of slanted volume gratings at various wavelengths," *Appl Opt.* **36**(23), 5769–5774 (1997).

[61] L. Singher and J. Shamir, "Waveguide holographic elements recorded by guided modes," *Appl. Opt.* **33**(7), 1180–1186 (1994).

[62] K.A. Stetson, "Holography with total internally reflected light," *Appl. Phys. Lett.* **11**(7), 225–226 (1967).

[63] K.A. Stetson, "Improved resolution and signal-to-noise ratios in total internal reflection holograms," *Appl. Phys. Lett.* **12**(11), 362–364 (1968).

[64] K.A. Stetson, "An analysis of the properties of total internal reflection holograms," *Optik* **29**, 520–536 (1969).

[65] T. Suhara, H. Nishihara, and J. Koyama, "Waveguide holograms: a new approach to hologram integration," *Opt. Comm.* **19**(3), 353–358 (1976).

[66] J. Tedesco, "Edge-lit holographic diffusers for flat-panel displays," U.S. Patent 5,418,631 (1995).

[67] H. Ueda, K. Taima, and T. Kubota, "Edge-illuminated color holograms," *Proc. SPIE* **2043**, 278–286 (1993).

[68] H. Ueda, E. Shimizu, and T. Kubota, "Image blur of edge-illuminated holograms," *Opt. Eng.* **37**(1), 241–246 (1998).

[69] J. Upatnieks, "Compact holographic sight," *Proc. SPIE* **883**, 171–176 (1988).

[70] J. Upatnieks, "Method and apparatus for recording and displaying edge-illuminated holograms," U.S. Patent 4,643,515 (1987).

[71] J. Upatnieks, "Edge-illuminated holograms," *Appl. Opt.* **31**(8), 1048–1052 (1992).

[72] J. Upatnieks, "Compact hologram displays and method of making compact hologram," U.S. Patent 5,515,800 (1992).

[73] J. Wreede and J. Scott, "Display hologram," U.S. Patent 5,455,693 (1995).

[74] A. Wuthrich and W. Lukosz, "Holographie mit quergedämpften wellen II. Experimentelle Untersuchungen der Beugungswirkungsgrade," *Optik* **42**, 315 (1975).

[75] A. Wuthrich and W. Lukosz, "Holography with guided optical waves, I. experimental techniques and results.," *Appl. Phys.* **21**, 55–64 (1980).

[76] A. Wuthrich and W. Lukosz, " Holography with guided optical waves II, theory of the diffraction efficiencies," *Appl. Phys.* **22**, 161–170 (1980).

[77] J.-H. Yeh and R. Kostuk, "Free-space holographic optical interconnects for board-to-board and chip-to-chip interconnections," *Opt. Lett.* **21**(16), 1274–1276 (1996).

4

Subwavelength Diffractive Optical Elements

Joseph N. Mait, Dennis W. Prather,
and Neal C. Gallagher

4.1   Introduction

In contrast to the experience of Jonathan Swift's Gulliver, the photon has moved from a world with gargantuan surroundings to one whose surroundings are miniature. This journey has been made possible through the development of nanofabrication technologies capable of producing optical elements whose size is on the order of a wavelength and whose features are less than a wavelength. One's intuition and the common tools for optical analysis, e.g., ray tracing and scalar optics, fall short in this miniature world. Instead, one must use vector electromagnetic theory and interface boundary conditions to understand the behavior of so-called subwavelength elements.

The manipulation of electromagnetic fields with subwavelength-scale elements is common in radio and microwave applications. Quarter-wavelength antennas are used for radio applications, and even everyday microwave ovens use several subwavelength elements. Because the operating wavelengths are on the order of 10 cm, the fabrication of subwavelength structures is not mechanically demanding. In contrast, the fabrication of optical elements with features on the order of 0.1 μm or smaller can be very demanding.

Nonetheless, optical subwavelength structures have been mass-produced and sold commercially for many years in the form of audio compact discs (CDs). The subwavelength pits used to store information in a CD are shown in Fig. 4.1. To protect the information layer of a CD, its surface is normally coated with a thin layer of plastic, whose refractive index is approximately 2. Given that the wavelength of light used in a disc player is approximately 0.8 μm and that the pit depth is 0.11

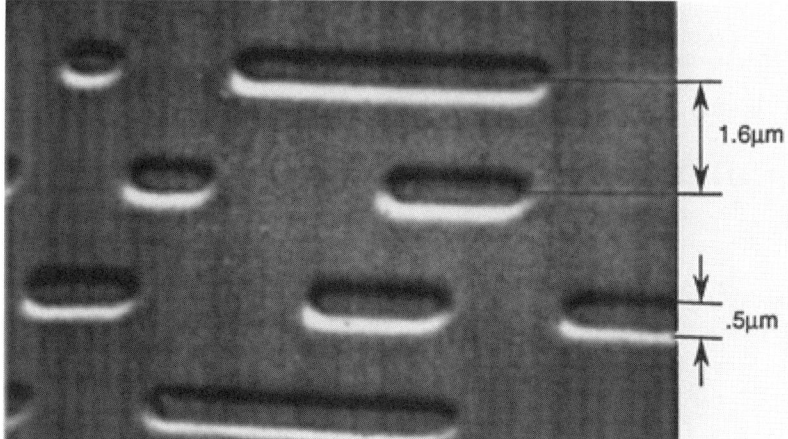

FIGURE 4.1. Detail of a CD at 10,600× magnification. The image is an electron micrograph of the disk's metalized layer without the clear plastic layer that protects the metalized surface of commercially available disks. Disk supplied by the Digital Audio Disc Corporation of Terre Haute, Indiana.

μm, the pit depth is one-quarter of the operating wavelength in the plastic. Each pit imposes a half-wavelength phase delay between the field incident on the CD and the one that is reflected.

If the lateral dimensions of the pit are also subwavelength, then unique diffractive effects occur and must be accounted for in analysis and design. In this chapter, we address these effects, which have broad application in beam deflection, splitting, and focusing.

The computer-aided design of elements that control a wavefield via diffraction began in the mid-1960s with the invention of computer-generated holography; so we begin in Section 4.2 with a review of the early developments of this field. A hologram is, in essence, a diffraction grating that is modulated in amplitude and phase. The development of the computer and associated peripherals made it possible to produce holographic patterns without the need to generate interference fringes optically. However, the fabrication technology of the time was capable of producing elements with only low spatial frequency. Through advances in fabrication, it is now possible to produce elements that generate high spatial frequencies and to control even the evanescent waves of an incident field. In Sections 4.3 and 4.4, we discuss tools developed to analyze such structures, and in Section 4.5, we discuss the state of the art in their design.

4.2   Computer-Generated Holography and Fabrication

Since the first computer-generated hologram (CGH) was demonstrated at IBM in 1966 by Lohmann and Paris,[1] many methods have been proposed for their design.[2]

Lohmann's detour-phase hologram consisted of an array of transparent rectangular apertures on an opaque background (or vice versa), whose size and position were proportional to the amplitude and phase of the complex-valued field one wished to encode. The array of rectangles were drawn first by hand, but soon thereafter by computer. An alternative technique proposed by Lee used a modulated fringe pattern as opposed to rectangles to encode the desired information.[3] Because they modify only the amplitude of an incident wavefield, these first CGHs are now referred to as binary-amplitude CGHs.

One major drawback with binary-amplitude CGHs is that a maximum of only about 10% of the available light forms the desired image in the first diffracted order. To increase diffraction efficiency, it is possible to use a binary-phase CGH, which modifies an incident field by adding either π or 0 to its phase. In contrast to a binary-amplitude CGH, which multiplies an input wavefield by either 0 or 1, a binary-phase CGH multiplies the wavefield by -1 and 1. This roughly doubles the amount of light that is diffracted and quadruples the diffraction efficiency into the first-order image from 10% to 40%.

One of the first binary-phase CGHs was a detour-phase hologram etched into a silicon substrate through integrated circuit fabrication techniques.[4, 5] The relationship between the etch depth d and the desired phase delay θ is

$$d = \left(\frac{\lambda}{n_s - 1} \right) \left(\frac{\theta}{2\pi} \right),$$ (1)

where λ is the illumination wavelength and n_s is the refractive index of the substrate. Because the substrate was overcoated with aluminum to make it reflective, the rectangular pits were etched only one-quarter wavelength deep to realize a π-phase shift on reflection.

The transition from amplitude to phase-only elements significantly improved the performance of diffractive elements. However, the most significant trend has been toward ever-larger diffractive elements with ever-decreasing feature sizes. This trend has been made possible through advances in fabrication, as well as advances in computer power. In the late 1960s and early 1970s, most masks for CGHs were made by photoreduction of a pattern onto a piece of film or a glass plate. Feature sizes were on the order of 10 μm. Advances in conventional lithography and direct-writing techniques pushed the minimum feature down to approximately 1 μm.

However, to realize features less than 1 μm required the application of new fabrication technologies. The first diffractive element written using electron beam (e-beam) lithography was fabricated in 1976 with 1.0-μm features.[6] Since then, considerable advances have been made in the e-beam fabrication of diffractive elements. For example, the correction of proximity effects from the e-beam[7, 8] makes it possible to produce subwavelength diffractive structures in the visible spectrum.[9] Further, the use of high-energy, electron-beam–sensitive glass has made it possible to fabricate multilevel phase elements using e-beam writing, not just binary-phase.[10, 11] E-beam lithography is now the preferred method for fabricating elements with features on the order of 0.1 μm.

With the advances in fabrication technology came a greater appreciation for the limitations of scalar diffraction theory for analysis and design.[12-14] These limits were confronted in 1983 when Sweeney and Gallagher demonstrated, albeit at microwave frequencies, the first diffractive element to alter the vector nature of an electromagnetic field, not just its amplitude and phase.[15] To heat a magnetically contained plasma, the element had to induce a space-variant polarization rotation of a 1-cm microwave source with near 100% efficiency. The proper design of such an element requires tools that can analyze and design arbitrary elements using rigorous vector-based diffraction theory. However, such tools were unavailable at the time and Sweeney and Gallagher designed their element using a scalar-based phase synthesis algorithm.[16-18] Nonetheless, their element increased system performance over that provided by the previous nondiffractive implementation.

Since the 1980s, considerable progress has been made in the rigorous analysis and design of diffractive optical elements (DOEs). One of the most significant results has been the high diffraction efficiencies produced by binary diffractive elements that have subwavelength features. Whereas scalar theory predicts a diffraction efficiency of only 40% for binary elements, binary subwavelength deflectors with efficiencies over 80%, predicted in the early 1990s,[19] were recently demonstrated.[20] In fact, the presence of only a few subwavelength features is sufficient to overcome the scalar-based predictions on diffraction efficiency.[21] In the remainder of this chapter, we discuss some recent developments in the rigorous analysis of diffractive elements and their application to the design of elements with subwavelength features.

4.3   Rigorous Diffraction Models and Analysis

To determine the fields diffracted by a structure, one must solve Maxwell's equations subject to the constraint that the tangential and normal field components are continuous along the boundary of the structure. However, only when the boundary conditions are applied in a global fashion can one determine a closed-form analytic solution. Thus, analytic solutions exist for only a few simple geometries, in particular, those that can be represented in, or conformally mapped into, separable coordinate systems. [22-24]

Closed-form solutions for many problems can be determined if one assumes that the medium within which the wavefield propagates is linear, isotropic, homogeneous, nondispersive, and nonmagnetic. Although these assumptions reduce the solution of Maxwell's equations to the solution of a single scalar equation, which is more tractable, the equation is only an approximation. In particular, the scalar assumptions ignore any coupling between the electric and magnetic fields, especially along the boundary of a structure.

As indicated in Section 4.2, scalar diffraction theory is an adequate and efficient tool, so long as diffractive features are large. However, as feature sizes decrease, the approximations on boundary conditions become less valid, and one must use more rigorous techniques to solve Maxwell's equations. Consequently, the rigorous electromagnetic analysis of diffractive elements with subwavelength features requires computational electromagnetic methods.

The first tools developed for the analysis of subwavelength structures were limited to infinitely periodic structures, i.e., gratings.[25–32] The grating assumption allows the electromagnetic fields to be expanded in terms of known eigenfunctions, which provides a convenient means for decomposing the diffracted fields.

The two most popular techniques for analyzing gratings are rigorous coupled wave analysis (RCWA) and the modal method. However, technically, these two methods are different formulations of the same approach.[25] Both represent the fields in a diffractive structure as a sinusoidal expansion, both ensure continuity between the fields interior and exterior to the structure by applying boundary conditions, and both determine the coefficients in the expansion by inverting a linear system. However, the RCWA is unique in its use of state-space methods that transform a second-order differential equation into two coupled first-order differential equations, which simplifies the solution of the eigenvalue equation.

However, the use of a sinusoidal expansion, which is infinitely periodic, limits the application of such methods to periodic structures. Many diffracting structures of interest—lenses, for example—are not periodic. Although, in theory, eigenfunction expansion methods can be applied to nonperiodic structures, by necessity, the number of terms in the expansion is large. Instead, researchers have developed alternative analysis techniques that, for example, apply Maxwell's equations to spatially sampled volumes,[10, 33–35] spatially sampled surfaces,[36–41] and spatially tiled volumes.[42] Combinations of these techniques[43] and combinations with temporally sampled equations have also been reported.[34, 35] In the following, we review finite-difference methods, the finite element method, and boundary integral methods.

In the finite-difference (FD) and finite-difference time-domain (FDTD) methods, one constructs a lattice of nodes in the regions interior and exterior to the diffractive surface. (see Fig. 4.2a). The electromagnetic fields are determined at each node by approximating the differential form of Maxwell's equations as coupled difference equations. To account for the artificial truncation of the exterior region, which produces numerically generated reflections, one can apply absorbing boundary conditions (ABCs) at the periphery of the exterior region.[44–47] The accuracy of the results generated is a function of the spatial sampling rate of the nodal lattice, as well as the temporal sampling rate. The finer the sampling, the more accurate the results become. The advantages of FD and FDTD methods over others is their ability to model inhomogeneous and anisotropic media. In addition, the FDTD is capable of modeling the transient behavior of the fields. The computational cost of such a calculation is proportional to the number of nodes in the computational space. The properties of the FD method, as well as the properties of the techniques discussed below, are summarized in Table 4.1.

In contrast to the FD and FDTD methods, which determine the electromagnetic fields at a point, the finite element method (FEM) determines the fields over a region (see Fig. 4.2b). In the FEM, one reduces the wave equation to a set of

TABLE 4.1. Applications comparison between methods for rigorous analysis of diffractive optical elements. Sampling rate was determined assuming less than 5% error in the magnitude response. All methods are capable of analyzing lossy materials.

Numerical technique	Inhomogeneous materials	Anisotropic materials	Arbitrary contours	Absorbing boundaries	System of equations	Sampling rate	Memory requirements
Finite Difference	yes	yes	no	required	sparse	$\lambda/15$	low
Finite Element	yes	yes	yes	required	sparse	$\lambda/20$	low
Boundary Element	no	no	yes	not required	full	$\lambda/10$	high
Hybrid FEBEM	yes	yes	yes	not required	partial	$\lambda/20$	moderate

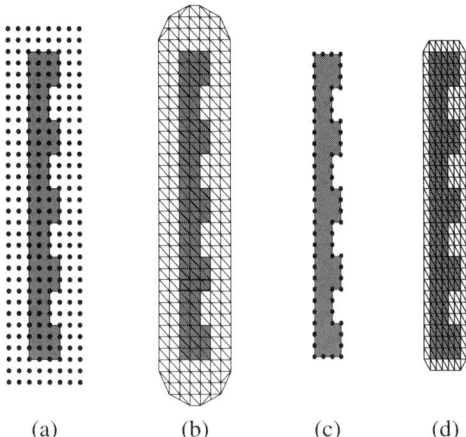

FIGURE 4.2. Representation of the interior, exterior, and boundary of a diffractive optical element for different rigorous analysis methods. (a) Sampled space for the FD and FDTD methods. (b) Tiled space with absorbing boundary conditions for FEM. (c) Sampled contour for boundary integral methods. (d) Tiled interior and sampled contour with boundary-matching conditions for FEBEM.

linear algebraic equations that describes the fields over an array of finite-sized elements that tile the solution space. Fundamental to the FEM is the use of shape interpolation functions to approximate the field values within each element. For modeling complex geometries, one can tile the solution space nonuniformly.

The accuracy of the results generated by the FEM is a function of the tiling, i.e., the number and shape of the elements, and the manner in which one applies ABCs. As in the FD methods, one must apply ABCs at the artificial boundary that defines the solution space.[48–52] However, there are only a few well-defined problems for which the use of ABCs in the FEM yield acceptable results.[10] In general, ABCs can lead to unpredictable errors in the determination of the observed fields.[43, 53, 54] Although the effect can be minimized by extending the range of the solution space, this increases the computational effort and memory requirements needed for analysis,[43] and it makes the FEM less attractive than other techniques for the synthesis applications discussed in Section 4.5.

Boundary integral methods solve the electric- or magnetic-field integral equations for the surface distribution induced on the boundary of a diffractive structure by an incident field. Because radiation conditions are implicit within the integral formulation, ABCs are not needed (see Fig. 4.2c). Once the boundary integral equations are constructed, they must be solved using numerical techniques. Two common techniques are the method of moments[55, 56] (MOM) and the boundary element method[57, 58] (BEM). In the MOM, the unknown field values in the integral equations are expanded using a set of scaled basis functions, e.g., a rectangular or triangular pulse. In contrast, the field values in the BEM are expanded using shape interpolation basis functions, i.e., polynomials of fixed order. For the same order of analysis, that is, the same number of nodes per element, the BEM provides better accuracy than does the MOM.

A hybrid method well suited for the analysis of DOEs is the finite element-boundary element method (FEBEM).[43] This technique uses an FEM analysis in the interior of the DOE and a BEM analysis in its exterior (see Fig. 4.2d). By applying an exact electromagnetic boundary condition at the periphery of the solution space, the FEBEM alleviates the problem of the artificial backreflections typical of the FEM.

It is important to point out that the methods described above require high spatial sampling of the surface or volume to ensure that calculated results are accurate to within a few percentage points of error. For example, to ensure accurate results using the BEM, the surface contour must be sampled at $\lambda/10$ intervals. A 1D diffractive element with a 1000-wavelength aperture, therefore, requires at least 10,000 samples. Dependent on the material and structural properties of the element, the matrices necessary to solve the electromagnetic equations require between 0.8 and 3.2 Gb of storage; roughly on the order of the square of the number of samples. However, the number of floating point operations is proportional to the cube of the number of samples, or 10^{12}. For a 2D profile, the memory requirements increase 16-fold, and the number of floating point operations increases 64-fold. Thus, the accurate and rigorous analysis of a structure whose dimensions are wavelengths long will tax the resources of most desktop-computing platforms and many high-end platforms. The rigorous analysis of realistic diffractive structures, therefore, requires innovative methods to reduce computational complexity.

4.4   Simplified Rigorous Analysis

As mentioned in Section 4.3, the number of nodes or voxels used to represent aperiodic, finite extent structures has a direct impact on computational resources. The methods we describe in this section reduce the computational costs of rigorous analysis by exploiting symmetry and employing realistic assumptions on the nature of the electromagnetic fields. The methods include the modeling of axisymmetric structures, e.g., lenses, field stitching for modeling electrically large structures, and a hybrid scalar-vector technique for modeling structures that contain a mixture of superwavelength and subwavelength features.

4.4.1   Axi-symmetric Modeling

As shown in Fig. 4.3, the exploitation of axial symmetry in an FDTD analysis can reduce the number of nodes, and hence the computational costs, associated with a 3D analysis. This simplification is especially critical if one wishes to design small, low f/number lenses for integration with vertical cavity surface-emitting lasers.

The dimensionality reduction is achieved by exploiting the 2π-azimuthal periodicity of axially symmetric structures. Using a Fourier series expansion, one

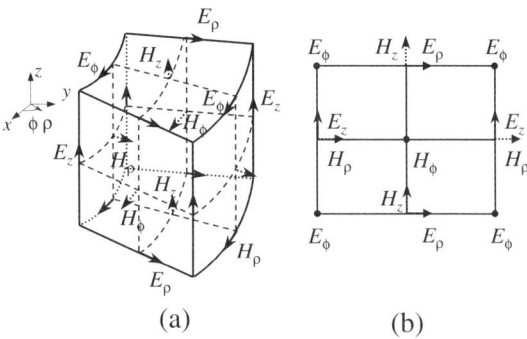

FIGURE 4.3. Exploitation of axial symmetry to reduce a 3D problem to two dimensions. (a) Three-dimensional FDTD grid for three-dimensional analysis. (b) Two-dimensional FDTD grid for 3D analysis of an axially symmetric structure.

can transform the representation of the electric and magnetic fields in three spatial variables to electric and magnetic azimuthal modes that are functions of only two spatial variables. Although in principle the expansion requires an infinite number of nodes, the extent of the series depends on the nature of the fields. In practice, the maximum number of modes required is dependent on the incident angle of illumination and the size of the diffractive structure. In the simplest case, a normally incident plane wave can be represented by its +1 and −1 orders. Waves incident at large off-axis angles require many more.

Although the 3D mesh is reduced to two dimensions, the computational costs are nonetheless slightly greater than those of a 2D problem. Whereas a 2D problem requires the solution of only three components, the reduction of a 3D problem to two dimensions still requires the determination of six electromagnetic field components. For this reason, the axisymmetric approach is often referred to as a "2.5-dimensional" method.

4.4.2 Field Stitching

The distinction between a rigorous analysis and a scalar-based one is the degree to which coupling between electromagnetic fields is considered. A rigorous analysis considers the intercoupling between all points on the surface of a diffractive structure, whereas scalar, at the other extreme, considers no coupling. The computational costs of any analysis are therefore related in some degree to modeling this coupling. The basis of field stitching is the approximation of global coupling effects by local ones.[59]

Consider, for example, that the computational costs of a BEM analysis are proportional to the number of coupling coefficients, which is the square of the number of elements. A linear increase in the size of the structure, therefore, leads to a quadratic increase in the number of coupling coefficients. However, as an alternative to analyzing one large structure, one can analyze many small patches of the original structure (see Fig. 4.4). So long as the number of patches is less

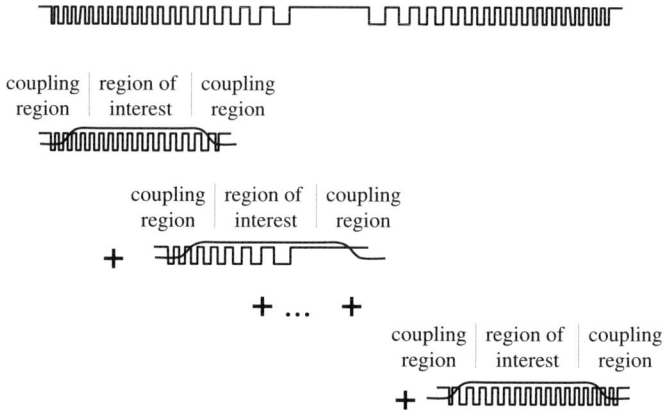

FIGURE 4.4. Representation of field stitching.

than the number of elements in a single patch, this approach yields a savings in memory.

However, the savings in memory must be balanced by the cost of summing the fields from the patches. We refer to this summation as field stitching. An examination of the coupling between two points on a diffractive surface reveals that the coupling between points spatially separated by tens of wavelengths is negligible in comparison to the coupling between points separated by only a few wavelengths. Thus, around any arbitrarily sized patch from a diffractive structure there exists a shoulder region where the coupling between points in the shoulder and points on the edge of the patch diminishes gradually. Prather et al.[59] have shown that a shoulder width of 20 wavelengths is sufficient to account for the coupling between the patch and the rest of the structure.

To determine the fields using field stitching, one first divides the diffractive surface into elemental regions. Along the perimeter of each region, one adds a shoulder region 20 wavelengths in width. Although the fields are calculated over the entire region (i.e., elemental region plus shoulder), only the fields contained with the original region are kept. The overlapping coupling regions between adjacent analysis windows are the threads that stitch the elemental fields into a single output field. One then propagates this single field to the observation plane.

4.4.3   Hybrid Scalar-Vector

So long as the field stitching is performed with care, the methods one uses to determine the fields within one elemental region can be independent of the methods used in any other region. This allows for further reductions in computational cost, because only those regions that contain subwavelength features need to be analyzed rigorously. Application of scalar methods are valid in those regions that contain features much larger than a wavelength. Thus, as shown in Fig. 4.5, one can use a hybrid scalar-vector method to concatenate scalar- and vector-based solutions.[60]

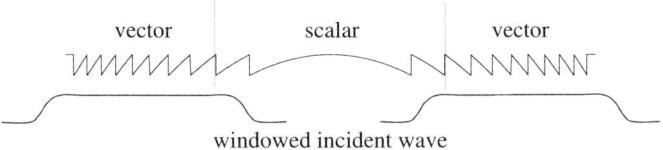

FIGURE 4.5. Hybrid scalar-vector modeling.

Improved analysis capabilities leads to greater understanding of the physics of subwavelength diffractive optics, and greater understanding leads to a desire for controlling fields. In the next section, we discuss the development of tools for the design of subwavelength diffractive optics.

4.5   Design

The present state of design of diffractive elements with subwavelength features is similar to that of the early years of computer-generated holography. Researchers are just beginning to understand how to encode electromagnetic field parameters using subwavelength features, and the first tentative steps have been taken toward the development of design algorithms.

We note that subwavelength diffractive structures have been used either as materials with unique properties, e.g., artificially birefringent, or as an element for wavefront transformation, e.g., beam deflection, beam splitting, and beam focusing. In the spirit of this volume, we consider only the latter application. Design of gratings for the former can be found in Kuittinen et al.[61] We begin first with some general comments on design that we apply to the design of subwavelength elements.

The essence of design is to map some desired function onto a prescribed form. For example, one may wish to generate an array of equal intensity spots using an element whose surface-relief profile is binary. The profile that one designs depends on the electromagnetic models used to describe the interaction between fields and matter, the models used for field propagation, and the performance metric used to describe the desired behavior.[62] The profile also depends on how one incorporates fabrication constraints into the design algorithm, and we discuss direct and indirect approaches to address this problem.

4.5.1   Direct Design

In a direct approach to imposing fabrication constraints, the constraints are incorporated into a parametric representation of the diffractive element, e.g., using multilevel quantized phases, discrete locations for phase transitions, or both.[63] The response of the element, characterized by the performance metric, is optimized by modifying the element parameters.

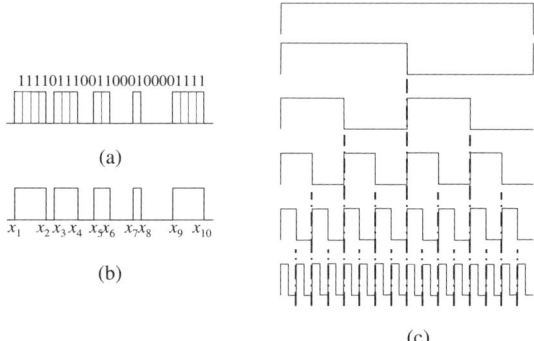

11110111001100010000111
(a)

x_1 $x_2 x_3 x_4$ $x_5 x_6$ $x_7 x_8$ x_9 x_{10}
(b)

(c)

FIGURE 4.6. Representation of binary diffractive elements. (a) Binary word. (b) Vector of phase-transitions. (c) Wavelet expansion. The entire set of wavelet functions necessary to represent the structure in (a) and (b) is shown.

For example, as shown in Fig. 4.6a, a binary diffractive element can be represented as a binary word. The bit size is typically assumed to be the smallest feature that can be fabricated. The value of the bit indicates whether a substrate is etched to a prescribed depth at that point. Design consists of navigating through the large, but finite, number of binary words in an efficient manner to find one that optimizes the performance metric in a global manner.

Two algorithms that have become standards in scalar design for navigating the solution space are simulated annealing[64–66] and the iterative Fourier transform algorithm[16–18, 67, 68] (IFTA). The distinction between these two algorithms is that fields are forward- and inverse-propagated in the IFTA, which allows modifications to be made to both the diffractive structure and its response; whereas in simulated annealing, fields are only forward-propagated and modifications are made only to the diffractive structure. For this reason, we refer to the IFTA as a bidirectional algorithm and simulated annealing as a unidirectional algorithm.

Bidirectional algorithms are typically not as time-consuming as unidirectional algorithms, because two domains can be exploited to converge to a solution. Projections-onto-convex-sets and generalized projections are bidirectional algorithms.[68] Other unidirectional algorithms include gradient search, simplex, and genetic methods.[70]

The computational efficiency of scalar-based design algorithms is due to the fact that field propagation can be modeled as a Fourier (or Fresnel) transform and that the etch depth d is proportional to phase θ, as given by (1). If one attempts to extend these algorithms to the design of diffractive elements with subwavelength features, one finds that the link between a binary structure and its response is no longer a simple one. Intuition and simple transforms no longer suffice. Instead, one must use one of the analysis routines discussed in Section 4.3 to determine a structure's response.

Further, because the vector equations for diffraction cannot be inverted in a simple manner, designers cannot avail themselves of the advantages of bidirectional

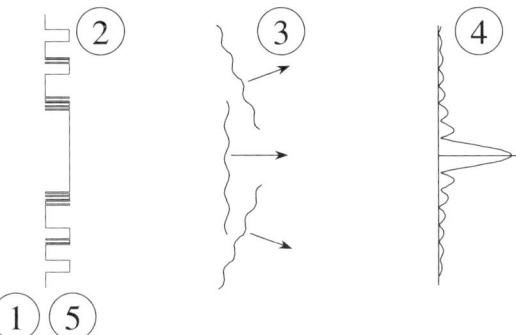

FIGURE 4.7. Unidirectional algorithm for subwavelength diffractive element design. The sequence of steps is (1) initial element, (2) field determination, (3) field propagation, (4) field evaluation, and, finally, (5) modified element.

algorithms and must use unidirectional design algorithms instead. The steps in a unidirectional algorithm are indicated in Fig. 4.7. Fabrication constraints are imposed directly, and field determination and propagation are both performed rigorously.

Given the complexity of a single analysis, the parametric representation of a diffractive element is critical to computational efficiency. The fewer parameters one uses to describe the diffractive element, the fewer number of analyses one has to perform as part of an optimization. If fact, although simulated annealing has been used to design a binary subwavelength diffractive lens,[71] when rigorous analysis is required, a binary bitmap representation is perhaps the least efficient.

Thus, alternative representations have been used to reduce the number of parameters and, thereby, reduce the computational burden. As shown in Fig. 4.6b, one representation is as a vector of phase transition locations. In this representation, the number of regions must be fixed beforehand. The phase-transition vector has been used in simulated annealing and gradient search routines to design deflectors, lenses, and array generators.[72–74]

More recently, a wavelet representation has been used to describe the diffractive element.[75] The emphasis that wavelets place on scale allows one to capture both the macroscale and microscale characteristics of an element (see Fig. 4.6c). For example, the large-scale wavelets of a diffractive lens embody information about the zonal boundaries, whereas the small-scale wavelets embody information about the structure within the zones. The former produces the desired diffractive behavior; the latter improves the fidelity of the response.

Each of these direct methods, which are dependent on representation, benefit from the use of field stitching. Because the parametric changes affect only the local structure, it is unnecessary to perform a rigorous analysis on the entire structure after each change. Instead, field stitching can be used to perform a local analysis in the region where the structure has changed. Further, because the low-order wavelets of a wavelet-based representation contain large features, the hybrid scalar-vector analysis is especially well suited for design using wavelets.

4.5.2   Indirect Design

The methods described in Section 4.5.1 apply fabrication constraints directly to the structure, and structure representation is critical. However, as an alternative, one can reduce the computational burden of a design by imposing fabrication constraints in an indirect manner.[62] This indirect, or two-step, approach requires an intermediate function that is related to both the desired response and the diffractive structure. In scalar-based designs, the complex-wave amplitude function serves as the intermediate function. The link between the complex-wave amplitude and the diffractive structure is given by (1), and the link between complex-wave amplitude and response is a Fourier transform. In the first design step, one determines the complex-wave amplitude that optimizes the performance metric. In the second step, fabrication constraints are imposed on the complex-wave amplitude. The two steps can be described, respectively, as transformation and encoding, or mapping.

The complex-wave amplitude function can also be used as an intermediate function in subwavelength designs. The complex-wave amplitude can be linked to the response by scalar propagation, but the link between subwavelength structure and complex-wave amplitude must be determined using rigorous analysis. Characterizing this relationship has been a goal of several researchers[61, 76–78] yet, it is still not fully understood. If the relationship can be determined, it can be inverted and used to encode the complex-wave amplitude function using subwavelength features.

As represented in Fig. 4.8, the indirect approach allows the complex-wave amplitude function to be determined independent of the fabrication constraints. Thus, any optimization algorithm can be used in this step. Fürer et al.[79] and Mait et al.[80] used simulated annealing and gradient search, respectively, and Mellin and Nordin[81] used an iterative algorithm similar to the IFTA. In all of these methods, propagation is facilitated with the angular spectrum. In fact, the key to Mellin and Nordin's iterative angular spectrum algorithm is the replacement of the angular spectrum's evanescent fields, which are lost through forward and inverse propa-

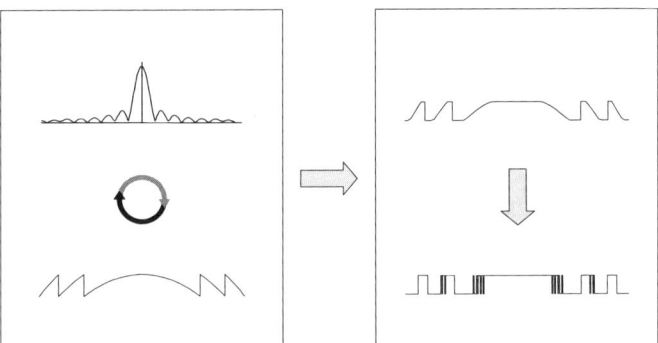

FIGURE 4.8. Indirect algorithm for subwavelength diffractive element design. The complex-wave amplitude that generates the desired response in some optimal fashion is determined in the first step. In the second step, the complex-wave amplitude is encoded as a binary subwavelength diffractive element.

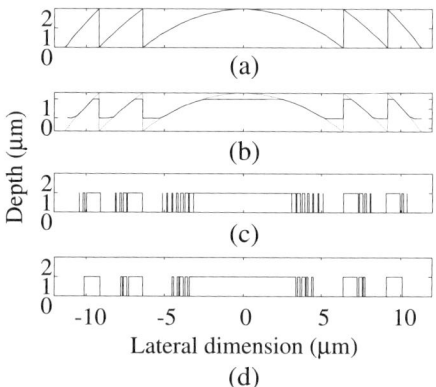

FIGURE 4.9. Procedure for encoding binary subwavelength diffractive element with area encoding. (a) Desired phase function. (b) Clipped phase function due to etch depth limitations. (c) Area-encoded binary subwavelength element. (d) Spatially quantized structure due to constraints on minimum feature.

gation, with the evanescent fields generated by the initial structure. Although the angular spectrum is generally applied only to scalar analysis,[82] it can be used to analyze diffractive structures that have high spatial frequencies.[83]

The complex-wave amplitudes generated by these algorithms (see, for example, Fig. 4.9a) are then encoded using a lookup table. It is at this point that fabrication constraints are considered. Through (1), the etch depth still limits the maximum phase the diffractive element can obtain. If the desired phase exceeds the maximum phase anywhere, it must be clipped as represented in Fig. 4.9b, which reduces diffraction efficiency.[84] Binary-encoding produces the structure represented in Fig. 4.9c. However, the minimum feature width of the fabrication process imposes an implicit spatial quantization on the binary structure as represented in Fig. 4.9d. Recent results indicate that the effect of this is small.[84]

The encoding approach described above has been used successfully to design far-field arrays generators and lenses. Recently, Mait et al.[9] demonstrated the lens shown in Fig. 4.10 using area-modulated annuli with subwavelength widths. For operation at 633 nm, the lens has a measured 65-μm focal length and 36-μm aperture ($f/1.75$). Ring pitch is approximately 160 nm, and the minimum ring width is less than 60 nm. The lens was fabricated in SiO_2 with an etch depth of 400 nm and, in terms of minimum feature and element size, represents the present state-of-the-art in subwavelength fabrication.

Sampling is implicit in encoding, and to ensure high diffraction efficiency, the sampling period should be on the order of the wavelength in the substrate. Lalanne et al. has shown that the sampling distance that yields the highest diffraction efficiency occurs when only a single mode propagates in the surface relief.[20]

All encoding techniques presented to date are based on grating analyses. For example, to encode an arbitrary phase function by pulsewidth modulation, even one that is not periodic, researchers use a lookup table that relates the phase trans-

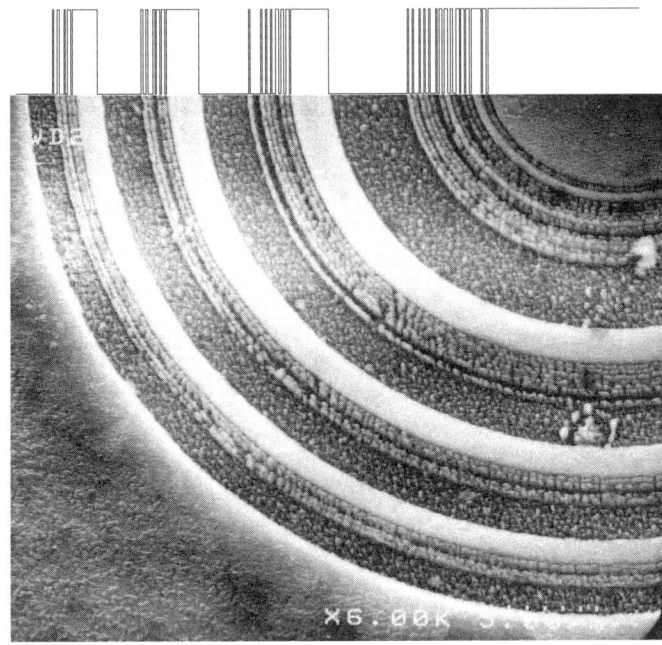

FIGURE 4.10. Binary subwavelength diffractive lens encoded with annular rings. The lens has a measured focal length of 65 μm and a measured diameter of 36 μm. The smallest annular width is less than 60 nm, and the etch depth is 400 nm. A radial representation of the binary profile used to fabricate the lens is shown for comparison. Note that all features in the profile are evident in the structure.

mission of a binary subwavelength grating to its fill factor. Although the distance over which the local grating assumption is valid has not yet been determined, the errors introduced using this method do not appear to be large. Other modulation methods include pulse-position and pulse-frequency, as well as a combination of these.[21, 72, 73, 85–89]

The encoding can also be performed in two spatial dimensions, e.g., pulse-area modulation as opposed to simply pulsewidth. It is interesting to note that an element encoded using subwavelength area modulation resembles a Lohmann detour-phase hologram. However, in contrast to the detour-phase hologram, area is now proportional to phase and not amplitude. Nonetheless, it appears that techniques first developed over 30 years ago to control the diffraction of a wavefield have served and continue to serve the community well.

4.6   Final Remarks

Advances in computer and fabrication technology have made it possible to produce diffractive elements that have features on the order of or less than an optical wavelength. In this chapter, we have presented some of the concomitant advances in analysis and design that are necessary for this technology to become as widespread as large-scale diffractive elements have recently become. It is encouraging that despite the reliance on rigorous techniques to account properly for boundary conditions and field coupling, design algorithms developed with scalar-based diffraction are still applicable. However, limits on the application of these methods have not yet been determined. For example, the binary encoding technique described performs well so long as the desired phase transmission varies slowly in space. How rapidly the phase can change before the encoding technique fails to reproduce it faithfully is unknown. Nonetheless, the advances we have presented make it possible to produce passive optical elements whose physical scale matches that of active optoelectronic elements and to enable, thereby, even greater optoelectronic integration and miniaturization than heretofore possible.

Acknowledgments

The authors wish to acknowledge the many fruitful interactions with their colleagues and coworkers, especially Mark Mirotznik, Catholic University; Shouyuan Shi and Xiang Gao, University of Delaware; Axel Scherer, California Institute of Technology; Philippe Lalanne, Institut d'Optique; Greg Nordin, University of Alabama at Huntsville; David Mackie and Joseph van der Gracht, U.S. Army Research Laboratory; and Jonathan Simon, Montgomery Blair High School.

References

[1] A.W. Lohmann and D.P. Paris, "Binary Fraunhofer holograms generated by computer," *Appl. Opt.* **6**, 1739–1748 (1967).

[2] S.-H. Lee, ed., *Selected Papers on Computer-Generated Holograms and Diffractive Optics*, SPIE Milestone Series **MS-33** (SPIE, Bellingham, WA, 1992).

[3] W.-H. Lee, "Binary synthetic holograms," *Appl. Opt.* **13**, 1677–1682 (1974).

[4] N.C. Gallagher, J.C. Angus, F.E. Coffield, R.V. Edwards, and J.A. Mann, "Binary phase digital reflection holograms: Fabrication and potential applications," *Appl. Opt.* **16**, 413–417 (1977).

[5] J.C. Angus, F.E. Coffield, R.V. Edwards, J.A. Mann, R.W. Rugh, and N.C. Gallagher, "Infrared image construction with computer-generated reflection holograms," *Appl. Opt.* **16**, 2798–2799 (1977).

[6] O.K. Ersoy, "Construction of point images with the scanning electron microscope: a simple algorithm," *Optik* **46**, 61–66 (1976).

[7] D.E. Zaleta, W. Däschner, M. Larsson, B.C. Kress, J. Fan, K.S. Urquhart, and S.-H. Lee, "Diffractive optics fabricated by electron-beam direct write methods," S.-H. Lee, ed., in *Diffractive and Miniaturized Optics*, SPIE Critical Reviews of Optical Science and Technology **CR 49**, 117–137 (1993).

[8] M. Ekberg, F. Nikolajeff, M. Larsson, and S. Hård, "Proximity-compensated blazed transmission grating manufacture with direct-writing e-beam lithography," *Appl. Opt.* **33**, 103–107 (1994).

[9] J.N. Mait, A. Scherer, O. Dial, D.W. Prather, and X. Gao. "Diffractive lens fabricated with binary features less than 60 nm," *Opt. Lett.* **25**, 381–383 (2000).

[10] P.D. Maker, D.W. Wilson, and R.E. Muller, "Fabrication and performance of optical interconnect analog phase holograms made by electron beam lithography," in R.T. Chen and P.S. Guilfoyle, eds., *Optoelectronic Interconnects and Packaging*, SPIE Critical Reviews of Optical Science and Technology **CR 62**, 415–430 (1996).

[11] W. Daschner, P. Long, R. Stein, C. Wu, and S.-H. Lee, "General aspheric refractive micro-optics fabricated by optical lithography using a high energy beam sensitive glass gray-level mask," *J. Vacuum Sci. Technol. B* **14**, 3730–3733 (1996).

[12] R.C. Enger and S.K. Case, "Optical elements with ultrahigh spatial-frequency surface corrugations," *Appl. Opt.* **22**, 3220–3228 (1983).

[13] D.A. Gremaux and N.C. Gallagher, "Limits of scalar diffraction theory for conducting gratings," *Appl. Opt.* **32**, 1948–1953 (1993).

[14] D.A. Pommet, M.G. Moharam, and E.B. Grann, "Limits of scalar diffraction theory for diffractive phase elements," *J. Opt. Soc. Am. A* **11**, 1827–1834 (1994).

[15] N.C. Gallagher and D.W. Sweeney, "Computer generated microwave kinoforms," *Opt. Eng.* **28**, 599–604 (1989).

[16] P.M. Hirsch, J.A. Jordan, and L.B. Lesem, U.S. Patent 3,619,022 (9 November 1971).

[17] R.W. Gerchberg and W.O. Saxton, "A practical algorithm for the determination of phase from image and diffraction plane pictures," *Optik* **35**, 237–246 (1972).

[18] N.C. Gallagher and B. Liu, "Method for computing kinoforms that reduces image reconstruction error," *Appl. Opt.* **12**, 2328–2335 (1973).

[19] W. Farn, "Binary gratings with increased efficiency," *Appl. Opt.* **31**, 4453–4458 (1992).

[20] Ph. Lalanne, S. Astilean, P. Chavel, E. Cambril, and H. Launois, "Blazed-binary subwavelength gratings with efficiencies larger than those of conventional échelette gratings," *Opt. Lett.* **23**, 1081–1083 (1998).

[21] E. Noponen, J. Turunen, and F. Wyrowski, "Synthesis of paraxial-domain diffractive elements by rigorous electromagnetic theory," *J. Opt. Soc. Am. A* **12**, 1128–1133 (1995).

[22] H.C. van de Hulst, *Light Scattering by Small Particles* (Dover Publications, New York, 1981).

[23] C.A. Balanis, *Advanced Engineering Electromagnetics* (Wiley, New York, 1989).

[24] A. Ishimaru, *Electromagnetic Wave Propagation, Radiation, and Scattering* (Prentice Hall, Englewood Cliffs, NJ, 1991).

[25] T.K. Gaylord and M.G. Moharam, "Analysis and applications of optical diffraction by gratings," *Proc. IEEE* **73**, 894–937 (1985).

[26] K. Knop, "Rigorous diffraction theory for transmission phase gratings with deep rectangular grooves," *J. Opt. Soc. Am. A* **68**, 1206–1210 (1978).

[27] R. Petit, ed., *Electromagnetic Theory of Gratings*, (Springer-Verlag, Berlin, 1980).

[28] D. Maystre, "Rigorous vector theories of diffraction gratings," in E. Wolf, ed., *Progress in Optics XXI*, (North Holland, New York, 1984), 1–67 .

[29] L. Li, "Multilayer modal method for diffraction gratings of arbitrary profile, depth, and permittivity," *J. Opt. Soc. Am. A* **10**, 2581–2591 (1993).

[30] Ph. Lalanne and G.M. Morris, "Highly improved convergence of the coupled-wave method for TM polarization," *J. Opt. Soc. Am. A* **13**, 779–784 (1996).

[31] J. Turunen, "Form-birefringence limits of Fourier-expansion methods in grating theory," *J. Opt. Soc. Am. A* **13**, 1013–1018 (1996).

[32] L. Li, "Use of Fourier series in the analysis of discontinuous periodic structures," *J. Opt. Soc. Am. A* **13**, 1870–1876 (1996).

[33] G.R. Hadley, "Numerical simulation of reflecting structures by solution of the two-dimensional Helmholtz equation," *Opt. Lett.* **19** 84–86 (1994).

[34] D.B. Davidson and R.W. Ziolkowski, "Body-of-revolution finite-difference time-domain modeling of space-time focusing by a three-dimensional lens," *J. Opt. Soc. Am. A* **11**, 1471–1490 (1994).

[35] D.W. Prather and S. Shi, "Formulation and application of the finite-difference time-domain method for the analysis of axially-symmetric DOEs," *J. Opt. Soc. Am. A* **16**, 1131–1142 (1999).

[36] D.W. Prather, M.S. Mirotznik, and J.N. Mait, "Boundary element method for vector modeling diffractive optical elements," I. Cindrich and S.-H. Lee, eds., in *Diffractive and Holographic Optics Technology II, Proc. Soc. Photo-Opt. Instrum. Eng.* **2404**, 28–39 (1995).

[37] K. Hirayama, E.N. Glytsis, T.K. Gaylord, and D.W. Wilson, "Rigorous electromagnetic analysis of diffractive cylindrical lenses," *J. Opt. Soc. Am. A* **13**, 2219–2231 (1996).

[38] D.W. Prather, M.S. Mirotznik, and J.N. Mait, "Boundary integral methods applied to the analysis of diffractive optical elements," *J. Opt. Soc. Am. A* **14**, 34–43 (1997).

[39] G. Koppelmann and M. Totzeck, "Diffraction near fields of small phase objects: Comparison of 3-cm wave measurements with moment-method calculations," *J. Opt. Soc. Am. A* **8**, 554–558 (1991).

[40] A. Wang and A. Prata, "Lenslet analysis by rigorous vector diffraction theory," *J. Opt. Soc. Am. A* **12**, 1161-1169 (1995).

[41] D.W. Prather, S. Shi, M.S. Mirotznik, and J.N. Mait, "Vector-based analysis of axially symmetric and conducting diffractive lenses," in *Diffractive Optics and Micro-Optics* **10**, 1998 OSA Technical Digest Series (OSA, Washington, DC, 1998), 94–96.

[42] B. Lichtenberg and N.C. Gallagher, "Numerical modeling of diffractive devices using the finite element method," *Opt. Eng.* **33**, 3518–3526 (1994).

[43] M.S. Mirotznik, D.W. Prather, and J.N. Mait, "A hybrid finite-boundary element method for the analysis of diffractive elements," *J. Mod. Opt.* **43**, 1309–1322 (1996).

[44] T.G. Moore, J.G. Blaschak, A. Taflove, and G.A. Kriegsmann, "Theory and application of radiation boundary operators," *IEEE Trans. Antennas Propagat.* **AP-36**, 1797–1812 (1988).

[45] J.P. Berenger, "A perfectly matched layer for the absorption of electromagnetic waves," *J. Comp. Physics* **114**, 185–200 (1994).

[46] A. Boag and R. Mittra, "A numerical absorbing boundary condition for finite difference and finite element analysis of open periodic structures," *IEEE Trans. Microwave Theory* Tech. **MTT-43**, 150–154 (1995).

[47] D.M. Sullivan, "A simplified PML for use with the FDTD method," *IEEE Microwave Guided Wave Lett.* **6**, 97–99 (1996).

[48] B. Engquist and A. Majda, "Absorbing boundary conditions for the numerical simulation of waves," *Math. Computat.* **31**, 629–651 (1977).

[49] P.P. Silvester, D.A. Lowther, C.J. Carpenter, and E.A. Wyatt, "Exterior finite elements for 2-dimensional field problems with open boundaries," *IEE Proc.* **124**, 1267–1279 (1977).

[50] A. Bayliss and E. Turkel, "Radiation boundary conditions for wave-like equations," *Commun. Pure App. Math.* **33**, 707–725 (1980).

[51] J.M. Jin, J.L. Volakis, and V.V. Liepa, "Fictitious absorber for truncating finite element meshes in scattering," *IEE Proc. H* **139**, 472–476 (1992).

[52] U. Pekel and R. Mittra, "An application of the perfectly matched layer PML concept to the finite element method frequency domain analysis of scattering problems," *IEEE Microwave Guided Wave Lett.* **5**, 258–260 (1995).

[53] B. Stupfel and R. Mittra, "A theoretical study of numerical absorbing boundary conditions," *IEEE Trans. Antennas Propagat.* **AP-43**, 478–486 (1995).

[54] J.L. Yao-Bi, L. Nicolas, and A. Nicolas, "2D electromagnetic scattering by simple shapes: a quantification of the error due to open boundary," *IEEE Trans. Mag.* **M-29**, 1830–1834 (1993).

[55] R.F. Harrington, *Field Computation by Moment Methods* (Krieger, Malabar, FL, 1968).

[56] J.J. Wang, *Generalized Moment Methods in Electromagnetics* (Wiley, New York, 1991).

[57] S. Kagami and I. Fukai, "Application of boundary-element method to electromagnetic field problems," *IEEE Trans. Microwave Theory Tech.* **MTT-32**, 455–461 (1984).

[58] K. Yashiro and S. Ohkawa, "Boundary element method for electromagnetic scattering from cylinders," *IEEE Trans. Antennas Propagat.* **AP-33**, 383–389 (1985).

[59] D.W. Prather, S. Shi, and J.S. Bergey, "A field stitching algorithm for the analysis of electrically large finite aperiodic diffractive optical elements," *Opt. Lett.* **24**, 273–275 (1999).

[60] D.W. Prather and S. Shi, "Combined scalar-vector method for the analysis of diffractive optical elements," *Opt. Engr.* **39**, 1850–1857 (2000).

[61] M. Kuittinen, J. Turunen, and P. Vahimaa, "Subwavelength-structured elements," in J. Turunen and F. Wyrowski, eds., *Diffractive Optics for Industrial and Commercial Applications*, (Akademie Verlag, Berlin, 1997), 303–323.

[62] J.N. Mait, "Understanding diffractive optical design in the scalar domain," *J. Opt. Soc. Am. A* **12**, 2145–2158 (1995).

[63] E. Noponen, J. Turunen, and A. Vasara, "Parametric optimization of multilevel diffractive optical elements by electromagnetic theory," *Appl. Opt.* **31**, 5910–5912 (1992).

[64] B.K. Jennison, J.P. Allebach, and D.W. Sweeney, "Iterative approaches to computer-generated holography," *Opt. Eng.* **28**, 629–637 (1989).

[65] J. Turunen, A. Vasara, and J. Westerholm, "Kinoform phase relief synthesis: a stochastic method," *Opt. Eng.* **28**, 1162–1167 (1989).

[66] M.R. Feldman and C.C. Guest, "High efficiency hologram encoding for generation of spot arrays, " *Opt. Lett.* **14**, 479–481 (1989).

[67] J.R. Fienup, "Iterative method applied to image reconstruction and to computer-generated holograms," *Opt. Eng.* **19**, 297–306 (1980).

[68] F. Wyrowski, "Digital holography as part of diffractive optics," *Rep. Prog. Phys.*, 1481–1571 (1991).

[69] A. Levi and H. Stark, "Restoration from phase and magnitude by generalized projections," in H. Stark, ed., *Image Recovery: Theory and Application*, (Academic Press, New York, 1987), 277–320.

[70] D.E.G. Johnson, A.D. Kathman, D.H. Hochmuth, A.L. Cook, D.R. Brown, and B. Delaney, "Advantages of genetic algorithm optimization methods in diffractive optic design," in S.-H. Lee, ed., *Diffractive and Miniaturized Optics*, SPIE Critical Reviews of Optical Science and Technology **CR 49**, 54–74 (1993).

[71] D.W. Prather, J.N. Mait, M.S. Mirotznik, and J.P. Collins, "Vector-based synthesis of finite, aperiodic subwavelength diffractive optical elements," *J. Opt. Soc. Am. A* **15**, 1599–1607 (1998).

[72] E. Noponen, A. Vasara, J. Turunen, J.M. Miller, and M.R. Taghizadeh, "Synthetic diffractive optics in the resonance domain," *J. Opt. Soc. Am. A* **9**, 1206–1213 (1992).

[73] Z. Zhou and T.J. Drabik, "Optimized binary, phase-only, diffractive optical element with subwavelength features for 1.55 μm," *J. Opt. Soc. Am. A* **12**, 1104–1112 (1995).

[74] E. Noponen, "Synthesis of diffractive-lens arrays with wavelength-scale features: global optimization of lens profile," in *Diffractive Optics*, European Optical Society Topical Meetings Digest Series **22** (EOS, Jena, Germany, 1999), 178–179.

[75] J. Simon, D.M. Mackie, D.W. Prather, and S. Shi, "Vector-based optimization of two-dimensional axially-symmetric DOEs using a wavelet basis," 1999 Annual Meeting of the Optical Society of America, Santa Clara, CA (September 1999).

[76] H. Haidner, J.T. Sheridan, and N. Streibl, "Dielectric binary blazed gratings," *Appl. Opt.* **32**, 4276–4278 (1993).

[77] D.H. Raguin and G.M. Morris, "Antireflection structured surfaces for the infrared spectral region," *Appl. Opt.* **32**, 1154–1167 (1993).

[78] M.E. Warren, R.E. Smith, G.A. Vawter, and J.R. Wendt, "High-efficiency subwavelength diffractive optical element in GaAs for 975 nm," *Opt. Lett.* **20**, 1441–1443 (1995).

[79] F. Fürer, M. Schmitz, and O. Bryngdahl, "Diffraction efficiency improvement of diffractive lenses by Gaussian-beam illumination," *Opt. Express.* **1**, 234–239 (1997).

[80] J.N. Mait, M.S. Mirotznik, and D.W. Prather, "Parametric design of subwavelength diffractive elements," 1999 Annual Meeting of the Optical Society of America, Santa Clara, CA (September 1999).

[81] S.D. Mellin and G.P. Nordin, "Rigorous electromagnetic analysis and the limits of scalar design of finite aperture diffractive phase optical elements," in *Diffractive Optics*, European Optical Society Topical Meetings Digest Series **22** (EOS, Jena, Germany, 1999), 34.

[82] J.W. Goodman, *Introduction to Fourier Optics* (McGraw-Hill, New York, 1968).

[83] J.E. Harvey, C.L. Vernold, A. Krywonos, and P.L. Thompson, "Diffracted radiance: a fundamental quantity in nonparaxial scalar diffraction theory," *Appl. Opt.* **38**, 6469–6481 (1999).

[84] J.N. Mait, D.W. Prather, and M.S. Mirotznik, "Design of binary subwavelength diffractive lenses by use of zeroth-order effective-medium theory," *J. Opt. Soc. Am. A* **16**, 1157–1167 (1999).

[85] E. Noponen and J. Turunen, "Binary high-frequency-carrier diffractive optical elements: electromagnetic theory," *J. Opt. Soc. Am. A* **11**, 1097–1109 (1994).

[86] E. Tervonen, J. Turunen, and J. Pekola, "Pulse-frequency-modulated high-frequency carrier diffractive elements for pattern projection," *Opt. Eng.* **33**, 2579–2587 (1994).

[87] M. Schmitz, R. Bräuer, and O. Bryngdahl, "Phase gratings with subwavelength structures," *J. Opt. Soc. Am. A* **12**, 2458–2462 (1994).

[88] E. Noponen and J. Turunen, "Complex-amplitude modulation by high-carrier-frequency diffractive elements," *J. Opt. Soc. Am. A* **13**, 1422–1428 (1996).

[89] V. Kettunen, P. Vahimaa, J. Turunen, and E. Noponen, "Zeroth-order coding of complex amplitude in two dimensions," *J. Opt. Soc. Am. A* **14**, 808–815 (1997).

5
Coherence-Gated Holograms

Christopher Lawson and Denise Brown-Anderson

5.1   Introduction

Imaging through highly scattering media using optical radiation has received much attention recently because of the potential applications in nondestructive evaluation of composite materials and in medical diagnostics. In materials science, imaging of the internal porous structure of certain types of composite materials may enable analysis of the structural characteristics of the composites. In medicine, noninvasive high-resolution optical imaging of the internal structure of human tissue would be very valuable for analysis of the structure and function of internal organs and for detecting small cancerous tumors.

Optical imaging techniques may potentially overcome some of the limitations of conventional techniques. Unlike x-rays, optical radiation is nonionizing and may be potentially useful for extracting tissue information only obtainable through spectroscopic techniques. It is also expected that with optical radiation, submicron resolutions may be possible. However, there is one major problem with optical radiation. Although nonultraviolet (UV) optical radiation is not strongly absorbed by biological tissue, it is very strongly scattered. The multiple scattering process in these inhomogeneous biomedical tissues can severely degrade image resolution by randomizing the optical image information in the transmitted or reflected optical signals. Hence, to achieve good optical image resolution (i.e., submicron or better resolution), it is necessary to differentiate between image-bearing photons and those that have been strongly scattered.

Many techniques have been investigated for imaging through turbid media. These techniques can be divided into five major categories: (1) optical coherence tomography (OCT), (2) time-gating techniques, (3) polarization techniques, (4) heterodyne techniques, and (5) holographic coherence-gating techniques. Sometimes several techniques are combined to optimize the speed, resolution, and imaging capabilities of the system.

This chapter focuses on holographic coherence gating. Holographic methods have been used for many years and have been useful for improving imaging through turbid media. One such method is wave-front correction by phase conjugation. For example, in 1962, Leith and Upatnieks,[1] the first to apply laser light to holography, put together an experiment to holographically record a probe wave that had passed through a turbid media and then allowed the phase-conjugate signal to pass back through the turbid media. This process corrected the wave-front distortion due to the original pass through the media.

Recently, researchers have used similar techniques with liquid crystal holograms for the reconstruction of simple plane waves having minimal spatial information content.[2] Liquid crystal holograms were also used to restore severely aberrated complex images.[3]

Others have used two-wave mixing with various types of nonlinear optical (NLO) materials. In two-wave mixing, two waves (a reference beam and an object or probe beam) interfere in the NLO material, and as a result, an interference pattern is created (just as in conventional holography). This interference pattern contains the information about the object to be imaged. Then, by blocking the object beam, the reference beam can be used to read out the hologram. Using this method, Hyde et al.[4] were able to demonstrate depth-resolved imaging of objects embedded in diffuse media with 30-μm transverse resolution. This method, although exhibiting excellent resolution, may not be optimum for applications that require instantaneous, single-shot image acquisition and is applicable mainly for use with materials with slow response times (i.e., photorefractive crystals). Similar time-gated holographic imaging has been achieved using multiple quantum wells (MQW),[5] but, in this case, a large voltage must be applied to the MQW, and an extra laser source is used, limiting the range of applicability of this technique.

In contrast to the two-wave mixing experiments mentioned above, single-shot, instantaneous imaging can be performed with degenerate four-wave mixing (DFWM) techniques. In DFWM, three waves from the same source interact inside an NLO medium, and as a result, a fourth wave is produced. This fourth wave, the phase conjugate signal, is a reconstructed image of the object. It is for this property that DFWM is sometimes referred to as real-time holography. It is also important to note that the image is only obtained when the least scattered light remains coherent with the reference beam, thus, the term "holographic coherence gate." DFWM has practical advantages over other wave mixing techniques, because there is no need to use more than one light source, as in nondegenerate four-wave mixing techniques, and no need to block one beam for readout purposes, as in two-wave mixing techniques. Another advantage is that there is no need for digital processing, as in electronic holography. As a real-time holographic technique, DFWM is faster than is two-wave mixing, making it useful for imaging biological tissue.

One disadvantage of using DFWM for imaging is that low contrast images are often obtained, because the severely scattered photons contribute to the ambient background in the phase conjugate signal. However, DFWM is a promising technique that has been used for imaging through turbid media,[6, 7] for holographic storage, and for 3D image reconstruction.[8]

The success of the DFWM technique hinges on the response of the NLO material. Candidate NLO materials include liquid crystals, photorefractive crystals, photorefractive polymers, semiconductors, and multiple quantum well devices. The nonlinear optical response of the materials used in the DFWM process affects the contrast of the images. Higher nonlinearities lead to higher contrast images. Additionally, for biological imaging applications, it is necessary to use materials with a response time on the millisecond time scale or faster. To this end, we have focused on photorefractive crystals and dye-doped liquid crystals. Both of these classes of materials exhibit strong optical nonlinearities on millisecond or faster time scales.

Photorefractive materials have been shown to be a promising media to generate phase conjugation. Photorefractive crystals (such as $BaTiO_3$, $LiNbO_3$, $Bi_{12}SiO_{20}$, and $Bi_{12}GeO_{20}$) have an NLO effect that is due to charge-migration, and thus, they tend to have very strong, but relatively slow, optical nonlinearities. These materials allow four-wave mixing with low powers[9] (on the order of milliwatts) and are useful for applications such as phase conjugate laser resonators, optical computing systems, and phase conjugate interferometry.

Liquid crystals have nonlinearities due to a combination of orientational and thermal effects. Although they tend to respond faster than photorefractive crystals, liquid crystal optical nonlinearities are weaker. Dye-doped liquid crystals have been used for a number of applications, including holographic storage,[10] real-time holography,[11] aberration correction,[3] optical switching,[12] and optical limiting and sensor protection.[13] We have also been able to use dye-doped liquid crystals for imaging through turbid media.[14]

In the following sections, imaging results obtained with different NLO materials under various experimental conditions will be discussed. Section 5.2 will describe multipulse imaging with a $45°$-cut Rh-doped $BaTiO_3$ photorefractive crystal. The achievable resolution for this coherence-based imaging technique will be discussed. Section III will describe single-pulse imaging results obtained for two types of liquid crystals. In one case, depth-resolved imaging is demonstrated. In another case, for the first time, single-pulse imaging through a turbid media with dye-doped liquid crystals is demonstrated. The achievable resolution with each material will also be described. Finally, Section 5.4 will provide a summary of the imaging results and a discussion of the future of holographic coherence gating.

5.2   Imaging with Photorefractive Crystals

This section details experimental results of a DFWM coherence-based imaging technique that can be used to provide 3D subsurface imaging of the internal structure of objects embedded in a scattering media. A Ti:Sapphire self-mode-locked laser (Lexel Laser Inc., Fremont, CA, Model 480) operating at 780 nm with a

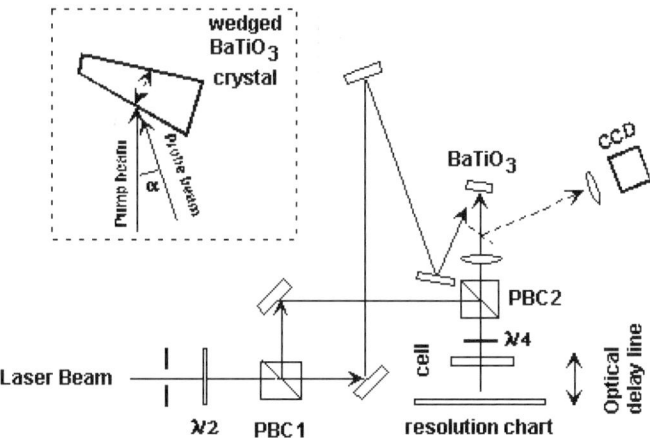

FIGURE 5.1. Experimental DRWM imaging setup. Self-pumped phase conjugation in the BaTiO$_3$ crystal is obtained via Fresnel reflection of the pump beam from the internal face of the crystal.

temporal pulse width of ∼120 fs was used as a low-coherence light source. The average output power of the 100 MHz mode-locked pulse train was about 40 mW. The laser beam was expanded with a telescope to a spot size of ∼4 mm in diameter. As a NLO media, a 1 mm thick, 45°-cut Rh-doped BaTiO$_3$ crystal[15, 16] (Deltronic Crystal Industries, Inc., Dover, NJ) was used.

5.2.1   Experimental Setup

The experimental setup is shown in Fig. 5.1. The laser beam is split into a forward pump and a probe beam by the polarizing beamsplitter cube PBC1. A half-wave plate adjusts the splitting ratio of the intensities of the pump and probe beams. The polarizing beamsplitter PBC2 and the quarter-wave plate form the backscattered signal from the test object. The translation stage behind the quarter-wave plate holds the test object and serves as an optical delay line. Backscattered light from different depths of the test object forms the probe beam in the DFWM scheme. The probe beam at the output of PBC2 has the same polarization as the pump beam (extraordinary polarization with respect to the BaTiO$_3$ crystal orientation, parallel to the plane of the Fig. 5.5).

As one can see from the insert of Fig. 5.1, the photorefractive crystal was used as a self-pumped phase conjugator. It utilizes a DFWM configuration, in which the backward pump beam is formed via Fresnel reflection of the forward pump beam from the back face of the photorefractive crystal (the reflection coefficient is about 17%). The two polished surfaces of the photorefractive crystal were slightly tilted with respect to each other (the angle of the wedge was about 1°) to avoid cross-talk from multiple reflected beams, which can occur when the continuous wave (cw) regime of the laser source is used.[17] It is worth noting that in wave coupling experiments with photorefractive crystals, it is usually difficult to avoid multiple

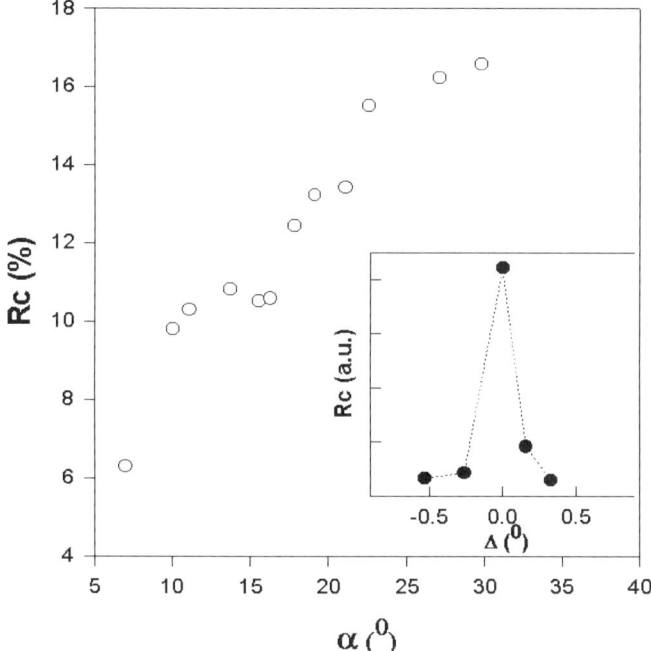

FIGURE 5.2. Conjugate reflectivity as a function of angular separation of the probe and pump beams is measured with Ti:Sapphire laser, $w_{pump}/w_{probe} \sim 1.2$. The insert illustrates the accuracy requirements for proper alignment of the photorefractive crystal in self-pumped phase conjugation geometry at 780 nm.

internal reflections of the beam inside the crystal, unless the crystal is submerged into a refractive index matched oil. In the case of low-coherence holography with a two-wave mixing geometry,[4] these multiple reflected beams contribute only to the incoherent background noise (helping to wash out the hologram formed via the two-wave mixing process). In contrast, here, utilization of the Fresnel reflection from the back face of the self-pumped phase conjugator enables instantaneous reconstruction of the image by means of real-time holography.

5.2.2   NLO Properties of the BaTiO₃ Crystal

In the first experiments, the intensity of the conjugate signal was measured by replacing the charge-coupled display (CCD) camera shown in Fig. 5.1 with a silicon photo detector. A probe beam in that case was formed with a flat mirror (which served as a test object) and without the focusing lens in the probe beam arm.

Figure 5.2 illustrates the NLO properties of the photorefractive crystal and the performance of the self-pumped phase conjugator described above. Measurements of the intensity of the conjugate signal as a function of the angular separation of the pump and probe beams were performed with a Ti:Sapphire laser operating in a long coherence length, cw regime. As can be seen from Fig. 5.2, the phase conjugate reflectivity for the 45°-cut BaTiO$_3$ crystal decreases from 17% to about 6% when the angle α is decreased from 30° to 7°. Additional measurements, which we performed with a He:Ne laser (at 633 nm) as a pumping source, show that the maximum gain of the phase conjugate signal was observed at an angular separation of about 40°. Using a formal analogy between real-time four-wave mixing and holography,[18] it can be seen that the optimal wave mixing geometry for the self-pumped phase conjugator corresponds to a grating spacing $\Lambda \sim 1.2$ μm, which is formed via the interference of the forward pump and the probe beams in the photorefractive crystal. This value for the optimal grating spacing for our self-pumped phase conjugator is in good agreement with experimental data obtained for two-wave mixing holographic imaging experiments reported recently.[19] For the best signal gain in the BaTiO$_3$-based NLO coherent filter in the near infrared (IR) spectral region, the optimal angular separation should be large ($\alpha \sim 52°$ at 780 nm). As it will be explained below, the use of such large angles in low-coherence imaging experiments results in the increase of undesirable "beam walk-off" effects in the NLO coherent filter.

The conjugate signal is sensitive to the accuracy of the alignment of the crystal relative to the input pump beam (because the backward pump is generated from the Fresnel reflection from the crystal). The insert of Fig. 5.2 shows that even a small angular misalignment, Δ, of the back face of the crystal from the optimum orientation leads to significant decreases in the phase conjugate signal (a decrease of a factor of 2–3 for $\Delta \sim$ 5–6 in). This type of self-conjugator is very sensitive to the vibrations of the photorefractive crystal and is different from the case of the "vibration resistant" passive phase conjugators analyzed elsewhere.[20]

5.2.3   Coherence-Based Imaging with the BaTiO$_3$ Crystal

The real-time holographic images shown in Figs. 5.3 and 5.4 were obtained using a Ti:Sapphire laser as a pumping source but operating with different temporal coherence lengths in each case. An Air Force resolution chart was used as a test object, and the resolution chart patterns were imaged with a 5-cm focal length lens onto the photorefractive crystal with a magnification of approximately unity. A COHU 4800 CCD camera with a lens in front was used to monitor the phase conjugate image.

In accordance with the optical scheme in Fig. 5.1, the conjugate image of the test object is only formed whenever the difference in the path lengths of the "object" beam (for given depth) and the (forward) pump beam is less than the coherence length l_c of the pumping source. In the first case (Fig. 5.3), the cw regime of the Ti:Sapphire laser was used. Using this long coherence length regime of the light source (the position of the delay line to within ~2 cm), non-depth-resolved image acquisition was demonstrated. In contrast, the image in Fig. 5.4 was obtained with the self mode-locked regime of the laser (120-fs transform limited pulses), which is characterized by a very small coherence length ($l_c \sim 40$ μm). Consequently, in this case, a depth-resolved profile with a depth resolution of about 20 μm is acquired.

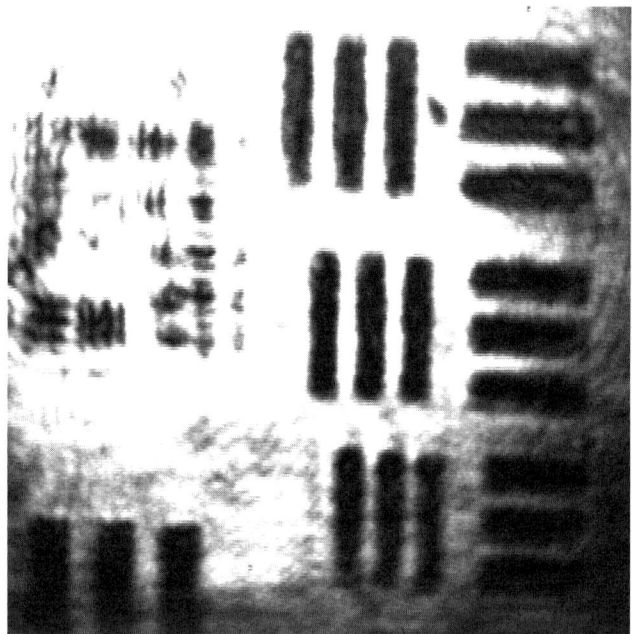

FIGURE 5.3. Real-time holography via DFWM on $BaTiO_3$. Conjugate images of the resolution chart are obtained with the cw regime of Ti:Sapphire laser. Angular separation between pump and probe beams is $\sim 9°$.

As can be seen from Fig. 5.4, for the case of depth-resolved imaging, the imaging field-of-view (FOV) is significantly narrowed compared with that obtained with the long coherence length light source. This FOV narrowing is explained by beam-walk-off effects,[21] which occur in low-coherence holographic DFWM imaging systems. The interacting beams in this system are coherent only in a narrow area of their intersection, which reduces the "working" volume of NLO material as the angle between coupling beams is increased. Hence, the effective area of NLO coherent filter in that case is determined by the angular separation of the coupling beams, rather than by the beam cross-section or by the thickness of NLO material. In particular, for "thick" photorefractive crystals, the exposed volume of the crystal can only increase undesirable beam fanning effects. To increase the FOV in low coherence-based imaging systems, it is necessary to decrease the angular separation between the beams.

5.2.4 Imaging Through a Scattering Media with the BaTiO₃ Crystal

Figure 5.5 shows experimental results that demonstrate low-coherence reflectometry through turbid media obtained with a self-pumped phase conjugator on $BaTiO_3$. The resolution chart in this case was placed behind a 1-mm-thick scattering cell

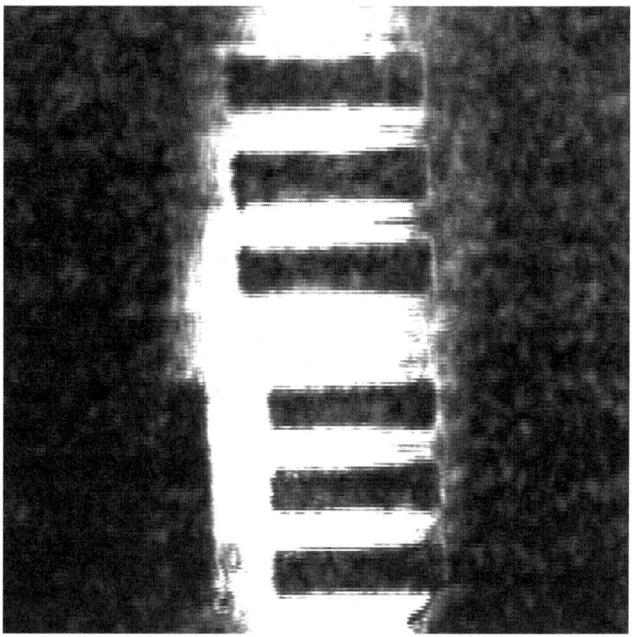

FIGURE 5.4. Real-time holography via DFWM on BaTiO₃. Depth-resolved image obtained with self-mode-locked regime of Ti:Sapphire laser. Angular separation between pump and probe beams is ∼ 9°.

containing a suspension of polystyrene microspheres in water (0.460 μm at 0.15% concentration). Using a Mie-scattering program,[22] we are able to calculate the size parameter and relative refractive index for a given sphere refractive index, medium refractive index, radius, and free-space wavelength; these parameters are then used to calculate extinction coefficients used for determining the number of mean free paths corresponding to a given concentration of polystyrene microspheres in water. Hence, the results in Fig. 5.9 correspond to 4 mfp for the double pass through the 1-mm-thick cell. Image subtraction techniques were used in this case to decrease the contribution of the background. This was achieved by recording the background noise after blocking the probe beam. The recording time of the image was about 30 s.

There are two important notes to make about imaging with photorefractive crystals. First, although a DFWM process was used for these coherence-based holograms, the holograms were not created in real time. As noted above, the recording time of the images was approximately 30 seconds. Second, the best images were acquired for large angles between the writing beams (i.e., $\alpha \approx 50°$). These large angles between the writing beams lead to image FOV narrowing (see Fig. 5.4). As will be shown in the next section, both of these problems can be remedied by using dye-doped liquid crystal films as the NLO media.

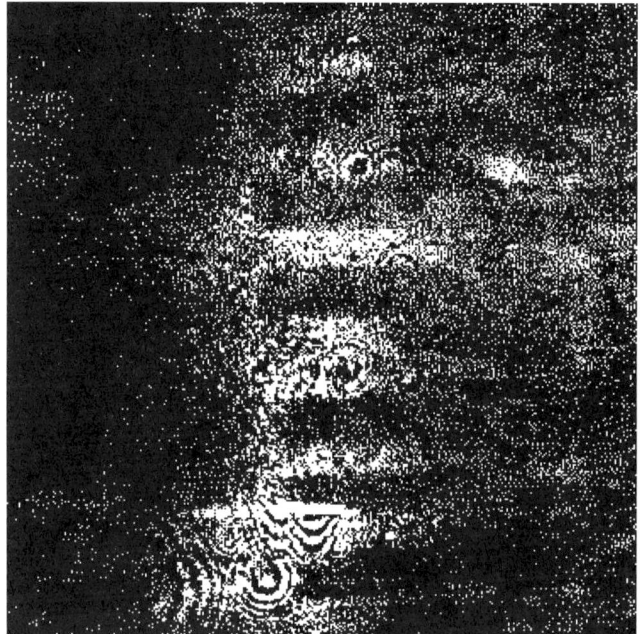

FIGURE 5.5. Self-pumped phase conjugation on $BaTiO_3$ crystal is obtained via Fresnel reflection of the pump beam from the internal face of the crystal. Bars correspond to 3.56 lp/mm.

5.3   Imaging with Dye-Doped Liquid Crystals

Nematic liquid crystals have recently been shown to exhibit strong nonlinearities[23] and high DFWM phase conjugate reflectivities.[10, 24, 25] The nonlinearity of these materials can also be enhanced by adding dye or applying external electric fields. Furthermore, they are largely birefringent and positively anisotropic (K15 has $\Delta n = 0.21$ and $\Delta \varepsilon = 12$). The parameters Δn and $\Delta \varepsilon$ are the refractive index change and the change in dielectric permitivity, respectively. In the isotropic phase (above the nematic-isotropic transition temperature, T_c), liquid crystals can become optically clear, and thus, increasing the interaction length can further enhance their NLO properties. Also, in this phase, the molecules are randomly oriented and are highly susceptible to external fields; thus, they tend to align themselves in the direction of the optical field.[24] Below, single-pulse imaging results with dye-doped nematic liquid crystals are described. First, depth-resolved imaging is demonstrated for the case when there is no scattering media, the DFWM scheme. And, second, imaging through a scattering media is demonstrated. Furthermore, the achievable resolution and the potential of these materials as low-coherence imagers are discussed.

FIGURE 5.6. Experimental setup for 2D depth-resolved imaging with two spatially separated objects.

5.3.1   Depth-Resolved Imaging with Dye-Doped Liquid Crystals

Figure 5.6 shows the experimental setup for DFWM based on 2D imaging experiments. The first depth-resolved 2D imaging experiments were performed with an alexandrite laser (Light Age, Inc., Somerset, NJ) at 750 nm. This laser provides a relatively short coherence length of about 5 mm when it operates in Q-switch regime (60-ns pulsewidth). By working in the long pulse mode, the coherence length of this laser can be increased to a few centimeters. The angle between the probe beam and the forward pump beam for 2D imaging experiments was approximately 6°. The probe beam arm includes a polarizing cube and quarter-wave plate. The translation stage located behind the quarter-wave plate serves as an optical delay line. Mounted on the translation stage were two spatially separated objects: the letter "A" (approximate size of 3 mm) drawn on a microscope slide and a U.S. Air Force resolution chart. A COHU 4800 CCD camera was used to register the conjugate signal of the image of the reflected (backscattered) light from the two spatially separated objects. The distance between two objects was ∼5 mm, and they were moved synchronously with the translation stage.

A thin layer of nematic liquid crystal E7 (provided by EM Industries, Inc., Hawthorne, NY) was used as a real-time holographic-recording medium. In this experiment, we also utilized thermally induced changes in the index of refraction of the liquid crystal. The sensitivity of the thermally induced optical refractive index changes was enhanced by the addition of an IR dye (BDN, provided by Exciton Inc., Dayton, OH). The liquid crystal layer, which had a thickness of 100 μm, was arranged between two microscope slides. To obtain planar alignment of

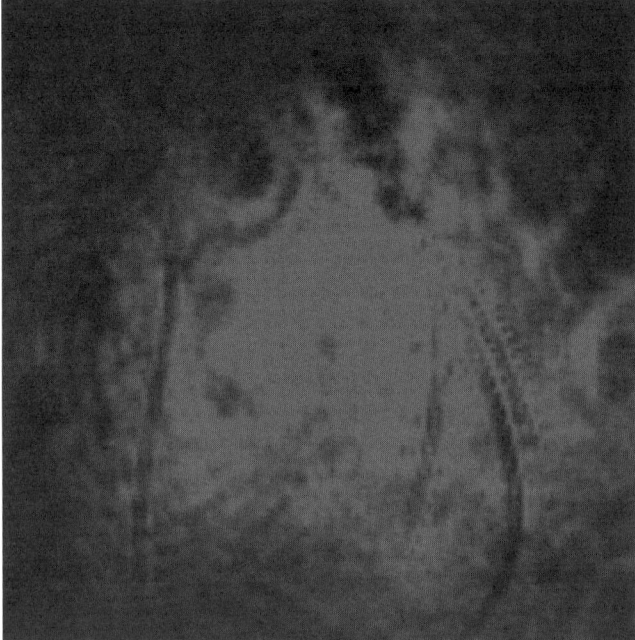

FIGURE 5.7. Image of background light, which is produced when the OPLs of both objects do not match the OPL of the forward pump beam.

the nematic liquid crystals, the interior surfaces of the glass slides were coated with polyvinyl alcohol and unidirectionally buffed using a cloth brushed attached to a motorized translation stage.

Figures 5.7, 5.8, and 5.9 are examples of images acquired with a single shot of the laser pulse and corresponding to different positions of the optical delay line. The average power of the laser was about 500 mW, at a 25-Hz repetition rate. The spot size on the liquid crystal was about 0.5 cm^2. The probe beam energy density was about 5 mJ/cm^2. Figure 5.7 corresponds to the condition when the optical path length (OPL) for both reflected (backscattered) signals from both objects does not match the OPL of the forward probe beam arm. In this case, no conjugate signal is generated, and the CCD camera registers only the background signal. Figure 5.8 corresponds to the matching condition for OPL of the signal reflected from the microscope slide and the laser pulses in the forward probe beam arm. In other words, if this OPL mismatch is within the coherence length of the laser source, then DFWM coherence filtering provides reconstruction of the image of the letter "A" only, and the signal reflected (backscattered) from the second object does not contribute to the conjugate signal. The image in Fig. 5.9 corresponds to the same condition of optical delay line as in Fig. 5.8. The difference is that the depth resolution is degraded by increasing the coherence length of the laser source (i.e., switching the alexandrite laser from Q-switch regime to long pulse mode). This increase in the coherence length of the laser pulses allows creation of a real-time

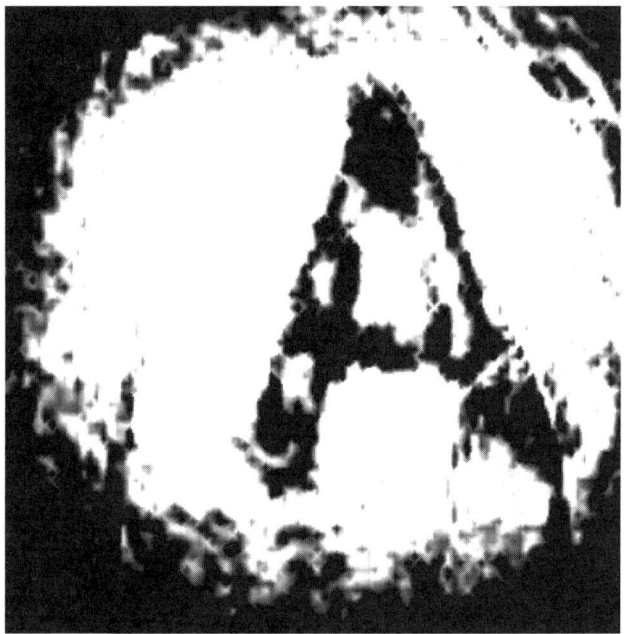

FIGURE 5.8. Image of the letter "A" produced when the OPL of object "A" matches the OPL of the forward pump beam.

hologram of both objects in the thin NLO layer and simultaneous reconstruction of these images via DFWM processes. The letter "A" can be seen on the right side, and the lines of the U.S. Air Force resolution chart can be seen on the left side of the image. The quality of the images obtained in these first experiments was limited by two factors: first, by the fact that the alexandrite laser was not operating in the single transverse mode, and, second, by the relatively low resolution of the digitized video images that were stored as arrays of only 120×120 pixels.

5.3.2   Imaging Through a Scattering Media

A DFWM experimental setup similar to Fig. 5.6 was used to carry out the following dye-doped liquid crystal-based imaging experiments.[14] Again, a Q-switched alexandrite laser (Light Age, Inc.) served as our light source. The laser power was 250 mW, the repetition rate was 20 Hz, and the spot size was 3.2 mm, which yields an energy density of about 0.16 J/cm^2. The probe beam is formed as a result of reflection off the object (a U.S. Air Force resolution chart), which is attached to a mirror. A translation stage holds the mirror and a quarter-wave plate. The translation stage also serves as an optical delay line. A 125-cm focal length lens was placed in the probe beam arm and used to focus the object onto the liquid crystal film with 1.3 times demagnification. The beamsplitter S1 splits the laser beam into a forward and backward pump beam. The energy ratio of the probe to the forward

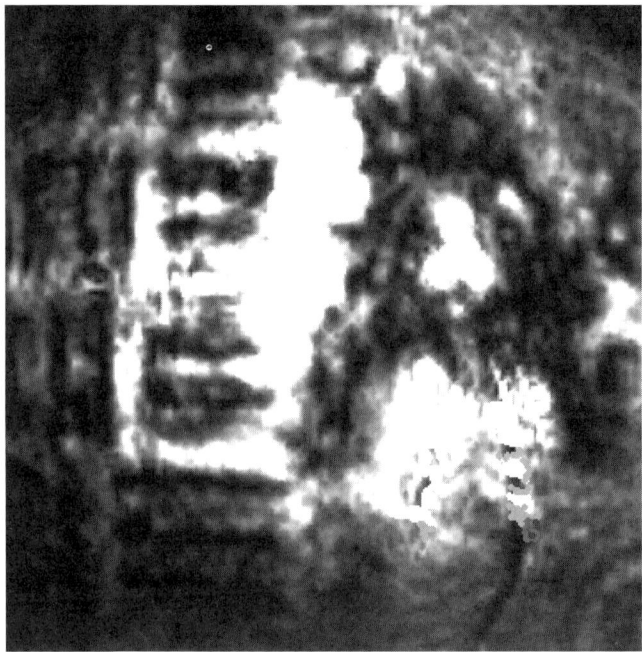

FIGURE 5.9. Demonstration of instantaneous depth-resolved imaging. Depth resolution is degraded so that images of both objects appear.

and backward pump beams was 0.7:1:1. The angle between the forward pump and the probe beam was about 6°. Beamsplitter S2 was used to direct the phase conjugate image to the imaging system, which consisted of an aperture, a focusing lens, and a Sony (Tokyo, Japan) CCD camera.

For these measurements, a planar-aligned K15 nematic liquid crystal (EM Industries) doped with an IR dye (HITC from Exciton) was used. Planar alignment was again achieved by coating the interior surfaces of two microscope slides with polyvinyl alcohol, after which they were unidirectionally buffed to help create the parallel alignment of the liquid crystal molecules with respect to the microscope slides. Using a mylar spacer, the liquid crystal was then arranged between the two slides, giving a film thickness of about 100 μm. Using a Shimadzu model UV-3101PC spectrophotometer, the film had a measured absorbance of 1 at the working wavelength (797 nm) of the alexandrite laser. This absorbance corresponds to a linear absorption coefficient of 230 cm^{-1}.

As was the case for the photorefractive crystal-based NLO imaging technique, the FOV of the liquid crystal film-based NLO technique depends on the angular dependence of the phase conjugate signal, as shown in Fig. 5.10. It can be seen that the Rc signal increases sharply as α decreases from 26° to 6°. The maximum phase conjugate reflectivity was about 0.5%. An angle of 6° between the forward pump and the probe beam generated a diffraction grating with a period ($\Lambda = \lambda/2 \sin \alpha$) of 3.8 μm. This sharp increase in R_c at small angles is in good qualitative

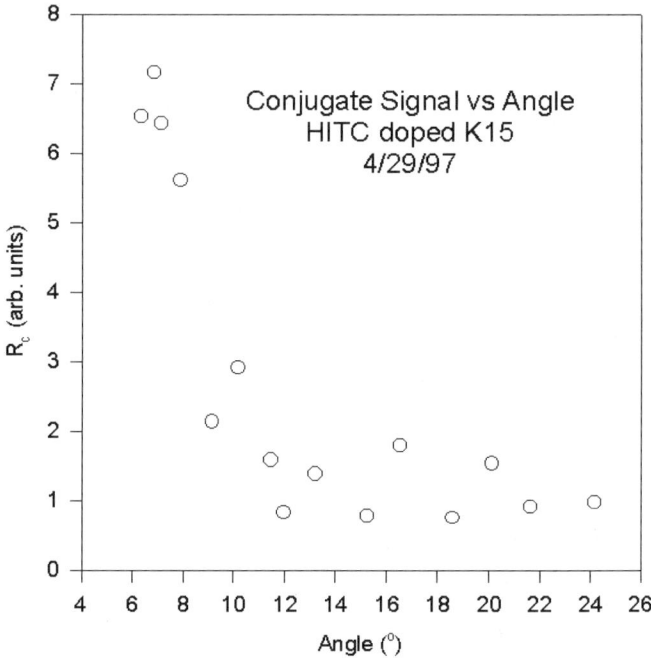

FIGURE 5.10. Intensity of the phase conjugate signal vs. angle between probe and pump beams. DFWM was performed on dye-doped liquid crystal film. Energy density of the laser beam was 0.16 j/cm^2.

agreement with the dependence of the first-order probe-diffraction efficiency of liquid crystal films observed elsewhere.[26] In accordance with the two-wave mixing measurements performed on the same type of liquid crystal (but doped with C_{60} molecules) as used in our experiments, the optimal diffraction efficiency of the NLO film corresponds to a grating constant $\Lambda_{opt} \sim 50$–$70\,\mu m$. This grating spacing is in marked contrast to the $\Lambda_{opt} \sim 1\,\mu m$ grating spacing for the BaTiO$_3$ crystal. By comparing the data for the liquid crystal films (Fig. 5.10) and the photorefractive crystal (Fig. 5.2), it can be seen that the maximum phase conjugate reflectivity occurs at much smaller beam crossing angles for liquid crystals. And because larger FOVs are obtainable for smaller beam crossing angles, NLO imagers based on dye-doped liquid crystal films may potentially exhibit wider imaging FOVs than do those based on photorefractive crystals and can thus be useful for low-coherence imaging applications.

The results in Figs. 5.11 and 5.12 show images of the resolution chart patterns that were taken with a single laser pulse. There was an angle of 6° between the probe and the forward pump beam. In Fig. 5.11, the finest bars correspond to 14.3 lp/mm and demonstrate an achievable transverse resolution of about 70 µm. The depth resolution, however, was relatively low (about 5 mm), because the long pulse mode of the alexandrite laser was used.

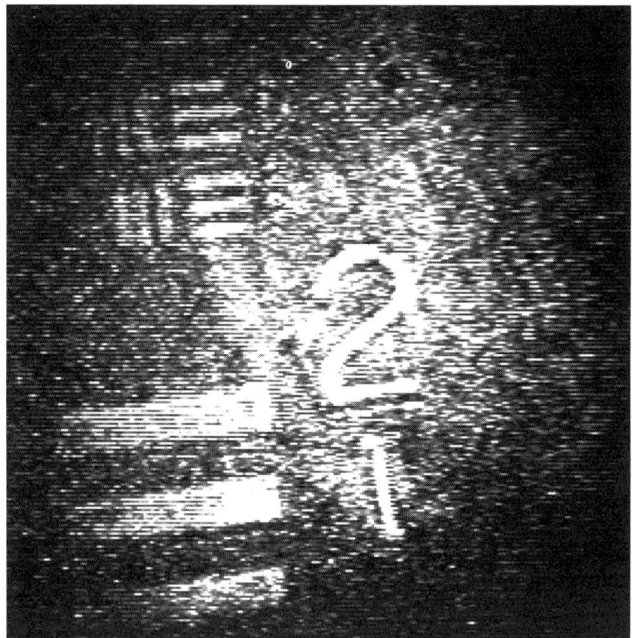

FIGURE 5.11. Two-dimensional image of the U.S. Air Force resolution chart. Finest bars correspond to 14.3 lp/mm. The energy density of the probe beam was 0.16 j/cm^2.

Investigations of the 2D imaging resolution through a turbid media were performed by inserting a 1-mm-thick cell with a suspension of polystyrene microspheres in water between the object (a U.S. Air Force resolution chart) and the quarter-wave plate in the experimental configuration mentioned above. Figure 5.12 shows an image of the U.S. Air Force resolution chart acquired with a single shot of the laser beam and without the scattering media present. The bars correspond to 1.41 lp/mm (about 700 µm) on the resolution chart. The NLO media used for this experiment was the same HITC-doped K15 liquid crystal (EM Industries) mentioned earlier. The ratio of the intensities of the probe beam to the forward and backward pump beams was 0.7:1:1, and the laser beam spot size on the liquid crystal surface was about 3.5 mm. Figure 5.13 corresponds to a scattering media with 0.09% concentration (2 mfp) of polystyrene microspheres in water. As can be see from Fig. 5.13, instantaneous imaging on a liquid crystal layer through a weakly scattering turbid media using the DFWM approach was demonstrated.

5.4 Conclusions

Noninvasive imaging through turbid media is important not only for diagnostic imaging of medical or biological tissue, but also for characterizing the internal porous structure of certain types of composite materials. Because these materials

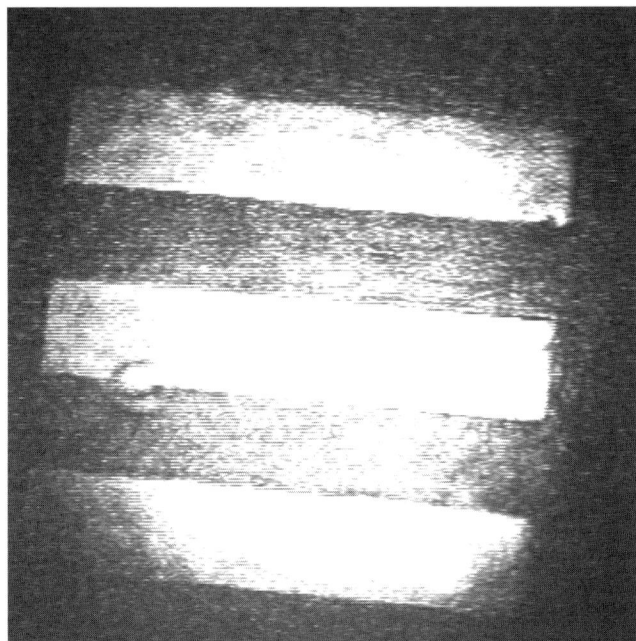

FIGURE 5.12. Image of a section of the U.S. Air Force resolution chart that corresponds to 1.41 lp/mm; there is no scattering media in the probe beam arm.

scatter incident radiation very strongly, it is often necessary to differentiate between the diffusely scattered photons and the image bearing photons. For this purpose, many techniques have been investigated, including OCT, time-gating techniques, polarization techniques, heterodyne techniques, and coherence-gating techniques. DFWM, for example, has shown to be a promising coherence-gating technique for imaging through turbid media. With a suitable NLO material, the DFWM technique has shown to be an even better method for coherence imaging.

In this chapter on coherence-gated holograms, a novel imaging technique has been described. With this technique, 2D imaging is accomplished via DFWM in the near-IR spectral region. With three different NLO materials, imaging resolutions on the micron scale were achieved. First, a $45°$-cut $BaTiO_3$ photorefractive crystal was used as a self-pumped phase conjugator (pumped with femtosecond Ti:Sapphire laser pulses) to obtain 2D depth-resolved images through a 4-mfp scattering media. The depth resolution in this case was about 20 μm. Next, a BDN dye-doped E7 nematic liquid crystal (100-μ m thickness, planar aligned) was used in conjunction with an alexandrite laser (750 nm, Q-switched, 25-Hz repetition rate) to obtain single-pulse cross-sectional images. In this case, the coherence length of the laser source limited the depth resolution, which was about 5 mm. Finally, a HITC dye-doped K15 nematic liquid crystal was used to provide, for the first time, single-shot image acquisition through a 2-mfp scattering media. The achievable transverse resolution was shown to be about 70 μm. Analysis of the materials as NLO filters

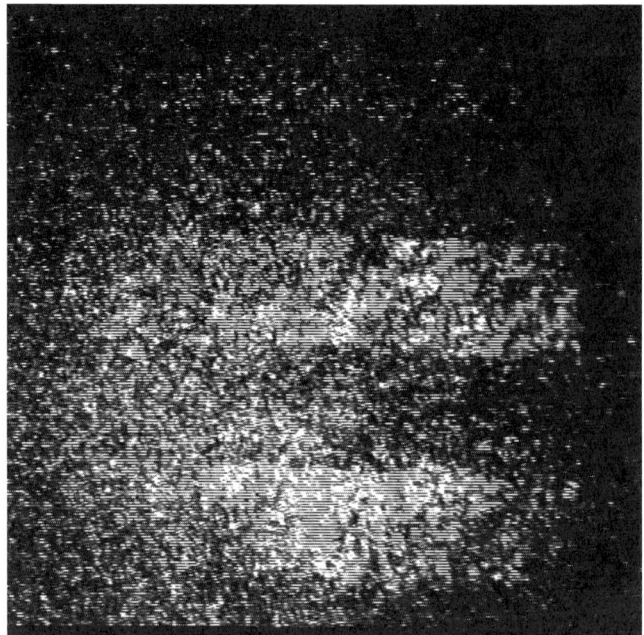

FIGURE 5.13. Image of a section of the U.S. Air Force resolution chart that corresponds to 1.41 lp/mm; scattering media consists of a 0.09% concentration of polystyrene microspheres in water.

showed that liquid crystal-based nonlinear optical filters could be used for wave-mixing geometries that require small angular separations of the interaction beams. These results also show that liquid crystal films exhibit high diffraction efficiencies at small angles and thus provide large imaging FOVs. Hence, they may serve as potential candidates for NLO filters in various applications that involve 2D noninvasive imaging.

To ultimately make the DFWM technique practical for use in clinical studies, it will be necessary to compete with other techniques by improving the depth and spatial resolution, which depends on the phase conjugate reflectivity of the DFWM process in the nonlinear optical materials. Typical resolutions have been on the submicron scale with other optical techniques, such as OCT. One way to improve the resolution is to enhance the NLO properties of the NLO material used in the DFWM technique. In particular, with the liquid crystals mentioned earlier, the addition of dye was used to enhance their NLO properties. The electric field-assisted NLO properties of nematic liquid crystals also make them a candidate for applications involving imaging through turbid media. It has been shown that DC voltages applied parallel to the director axis of the homeotropically aligned dye-doped liquid crystal film increased the sensitivity of the DFWM technique (i.e., increased the phase conjugate reflectivity) by an order of magnitude.[27] Other NLO materials should also be investigated for their potential to provide high phase conjugate reflectivities in large FOV imaging applications. The long-term applicability of nematic liquid crystals, as well as other NLO materials, for practical NLO wave-mixing-based imagers will continue to be a fruitful research area.

Acknowledgments

The authors would like to acknowledge support from the U.S. Air Force Office of Scientific Research under Grant F49620-99-1-0036.

References

[1] E.N. Leith and J. Upatnieks, "Reconstructed wavefronts and communication theory," *J. Opt. Soc. Am.* **52**, 1123–1130 (1962).

[2] I.C. Khoo, "Nonlinear optical properties of liquid crystals for optical imaging processes," *Opt. Eng.* **25**, 198–201 (1986).

[3] C. Karaguleff and G.L. Clark, Sr., "Optical aberration correction by real-time holography in liquid crystals," *Opt. Lett.* **15**, 620–622 (1990).

[4] S. Hyde, N. Barry, R. Jones, J. Dainty, and P. French, "High resolution depth resolved imaging through scattering media using time-resolved holography," *Opt. Commun.* **122**, 111–116 (1996).

[5] R. Jones, N.P. Barry, S.C. Hyde, J.C. Dainty, and P.M.W. French, "Time-gated holographic imaging using photorefractive multiple quantum well devices," *Proc. SPIE* **2981**, 192–199 (1997).

[6] V. Fleurov, D. Brown, and C. Lawson, "Low-coherence reflectometry through scattering media with degenerate four-wave mixing in a thin nonlinear optical layer," *J. Nonlinear Opt. Phys. Mater.* **6**, 95–102 (1997).

[7] V. Fleurov, C. Lawson, D. Brown, A. Dergachev, S. Mirov, and R. Lindquist, "Low-coherence reflectometry based on DFWM in a thin liquid layer," *Proc. SPIE* **2853**, 126–134 (1996).

[8] B.P. Ketchel, G.L. Wood, R.J. Anderson, and G.J. Salamo, "2-D image reconstruction using strontium barium titanate," *Appl. Phys. Lett.* **71**, 7–9 (1997).

[9] A. Bledowski and W. Krolikowski, "Anisotropic four wave mixing in cubic photorefractive crystals," *IEEE J. Quantum Electron.* **24**, 652–659 (1988).

[10] F. Simoni, O. Francescangeli, Y. Reznikov, and S. Slussarenko, "Dye-doped liquid crystals as high-resolution recording media," *Opt. Lett.* **22**, 549–551 (1997).

[11] A.G. Chen and D.J. Brady, "Real-time holography in azo-dye-doped liquid crystals," *Opt. Lett.* **17**, 441–443 (1992).

[12] S.Z. Janicki and G.B. Schuster, "A liquid crystal opto-optical switch: nondestructive information retrieval based on a photochromic fulgide as trigger," *J. Am. Chem. Soc.* **117**, 8524–8527 (1995).

[13] I.C. Khoo, M.V. Wood, M.Y. Shih, and P.H. Chen, "Extremely nonlinear photosensitive liquid crystals for image sensing and sensor protection," *Opt. Express* **4**, 432–442 1999.

[14] D. Brown, V. Fleurov, P. Carroll, and C. Lawson, "Coherence-based imaging through turbid media by use of degenerate four-wave mixing in thin liquid crystal films and photorefractives," *Appl. Opt.* **37**, 5306–5312 (1998).

[15] J.E. Ford, Y.F. Fainman, and S.H. Lee, "Enhanced performance from 45°-cut BaTiO$_3$," *Appl. Opt.* **22**, 4808–4815 (1989).

[16] B.A. Wechsler, M.B. Klein, C.C. Nelson, and R.N. Schwartz, "Spectroscopic and photorefractive properties of infrared-sensitive rhodium-doped barium titanate," *Opt. Lett.* **19**, 536–538 (1994).

[17] J. Zhang, X. Lu, L. Zhang, X. Mu, Q. Jiang, Z. Shao, H. Chen, and M. Jiang, "Conjugate fidelity and multiple reflection waves in self-pumped phase conjugation with doped (K$_{0.5}$Na$_{0.5}$)$_{0.2}$(Sr$_{0.75}$Ba$_{0.25}$)$_{0.9}$ Nb$_2$O$_6$ crystals," *Opt. Commun.* **132**, 574–582 (1996).

[18] A. Yariv, "Phase conjugate opticals and real-time holography," *IEEE J. Quantum Electron.* **14**, 650–660 (1978).

[19] N. Barry, R. Jones, S.C.W. Hyde, J.C. Dainty, P.M.W. French, S.B. Triverdi, and E. Diegues, "Evaluation of photorefractive holographic imaging through turbid media," *Proc. CLEO* **11**, 42–45 (1997).

[20] M. Cronin-Golomb, J. Palaski, and A. Yariv, "Vibration resistance, short coherence length operation, and mode-locked pumping in passive phase conjugate mirrors," *Appl. Phys. Lett.* **47**, 1131–1133 (1985).

[21] N. Abramson, *Light in Flight* (SPIE Optical Engineering Press, Bellingham, WA, 1996), Chap. 4.

[22] C.F. Bohren and D.R. Huffman, *Absorption and Scattering of Light by Small Particles* (Wiley, NY, 1983), Chap. 8.

[23] L. Li, H.J. Yuan, G. Hu, and P. Palffy-Muhoray, "Third order non-linearities of nematic liquid crystals," *Liquid Crystals* **16**, 703–712 (1994).

[24] I.C. Khoo, *Liquid Crystals: Physical Properties and Nonlinear Optical Phenomena* (Wiley-Interscience, NY, 1995), Chap. 10.

[25] I.C. Khoo, "Orientational photorefractive effects in nematic liquid crystal films," *IEEE J. Quantum Electron.* **32**, 525–534 (1996).

[26] I.C. Khoo, "Holographic grating formation in dye-doped and fullerene C$_{60}$–doped nematic liquid crystal films," *Opt. Lett.* **20**, 2137–2139 (1995).

[27] D.B. Anderson, PhD. dissertation, University of Alabama at Birmingham (1999).

6

Sculpturing of Three-Dimensional Light Fields by Iterative Optimization

Joseph Shamir and Rafael Piestun

6.1   Introduction

Conventional optics is mainly involved in the generation of light fields over a single plane. This is the case, for example, in imaging applications and diffractive optics. However, recent years witnessed the possibility of generating synthetic light fields satisfying given requirements within 3D domains.

The ability to control the wave-field (WF) distribution in a region of 3D space is directly applicable in different areas of science and technology. Applications include beam shaping and scanning, optical energy delivery, wave-front correction, apodization, depth of focus enhancement, holographic displays, point spread function design, imaging, diffuser and aspheric design, and the design of diffractive optics in general. In addition, the concept can be applied in diverse areas, such as high-accuracy measurements, computer vision, optical storage, atom optics, micro-optical systems, optical information processing, and photonic interconnection networks.

The objective of this chapter is to review this exciting field of research by providing the physical basis and procedures, describing special applications, and speculating on future prospects.

6.2   Holography and Three-Dimensional Imaging

The issue of 3D WF distributions is closely related to holographic techniques, and therefore, it is appropriate to start the discussion with some notes on holography and 3D imaging. Considering the popular interest in the subject of 3D imaging, it is also important to distinguish between myth and reality.

Wavefront reconstruction was invented by Dennis Gabor[1-3] in 1948 and later reinvented, independently, by Yuri Denisyuk,[4] who considered the subject from a different point of view. Gabor coined the name hologram to his invention, which implies that this kind of recording contains all the information about the object recorded. Holography, however, went almost unnoticed until the invention of the laser, the coherent radiation that made it possible to record and reconstruct high-quality holograms.[5, 6]

Exposing the public to the concept of holography toward the end of the 1960s ignited the imagination and generated a wasted amount of stories and movies in which the principal actors were supposed to represent holographic images. Is this a realistic possibility? Obviously, considering the way these holographic characters are presented, the answer is negative. The rest of this section is devoted to a discussion of this question, and it is relation to a second question: In that case, what is possible?

To somewhat narrow the scope of the discussion, we restrict ourselves to the treatment of real holographic images like those represented by the above-mentioned characters. This is in contrast with the more common virtual images that can only be viewed through the hologram like observing the character through a window. Figure 6.1 represents the reconstruction of such a real image. To observe this image, the hologram must be in the line of sight too. Otherwise, no light diffracted by the hologram can reach the eye of the viewer. A way around this difficulty is to fill the region of the image with a light-scattering medium (some kind of smoke or a block of diffusing glass). Either of these solutions is not suitable to put a holographic image among living people. Even in this case, it is impossible to view the image with our back to the hologram, when we block the light that is supposed to reconstruct the image. Again, we could fill all the walls of the room with holograms and then we would be able to observe the image from almost any position in the room, but this is complicated. We shall briefly discuss the issue of multiple holograms toward the end of the chapter.

To present the discussion of the above paragraph in a more scientific way, we start by stating that a holographic reconstruction is not a real 3D image but an illusion of one. The reason is as follows: Let us consider all of the information that

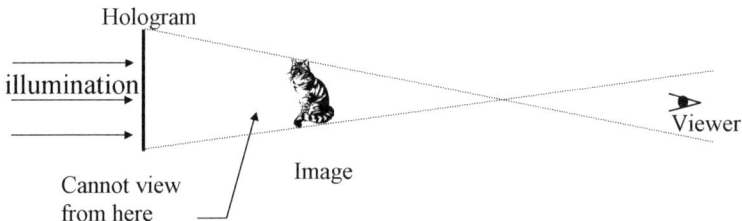

FIGURE 6.1. The real image reconstructed from a hologram can be only viewed from a region where the hologram is seen behind the object.

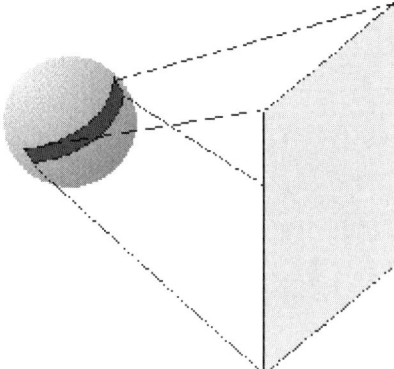

FIGURE 6.2. Projection between a plane and a section of spherical surface.

one can ideally store on an infinite plane and map it, mathematically, into a sphere in some arbitrary phase space (for example, the 3D Fourier transform domain). It turns out that all of this information can be mapped on the shell of this sphere[7] (See Fig. 6.2 and Section 6.3). Inverting and generalizing the argument, we may say that it is impossible to store all of the information from a 3D region of space onto a 2D surface. This fact should be obvious also from dimensional considerations. Nevertheless, we can create the illusion of 3D image storage on a 2D hologram if we relax the above statement and do not require the storage of all of the information from a 3D region of space.

To relax the storage requirements, we should consider the question: What is the necessary information needed to reconstruct a view of a 3D object? To answer this question, we revert to another terminology that was used by Gabor to describe holography,[1–3] that is, *reconstruction of wavefronts*. The idea behind this terminology is that we view an object by detecting the complex amplitude distribution over an electromagnetic wavefront that carries all of the available information about the object. This wavefront is recorded by a hologram and then can be reconstructed. What remains to be determined is the information that is available in such a wavefront, and this takes us back to the previous discussion of storing 3D information on a surface. By its definition, a wavefront is a surface, and therefore, it can be stored, in principle, on another surface, a hologram, because there is no dimensionality change involved. In short, a light wavefront also stores information of 3D objects on a 2D wavefront, and therefore, it cannot carry all of the information about the 3D object, and to facilitate an apparently good reconstruction, we do not have to store all of the 3D information because it does not exist on a wavefront.

The objective of this chapter is to analyze this process and employ it to sculpture 3D distributions of light following predefined specifications.

An effective way to generate specific WF distributions that did not exist previously in the real world is by the use of diffractive optical elements (DOEs).[8, 9] Conventional DOEs, however, are designed to generate a predetermined function or light distributions over a single transverse plane in free space. In addition, typical DOEs require a large number of microstructured resolution elements, and when amplitude modulation is used, classic DOEs generate the reconstructed function off the optical axis.

This chapter describes how proper design procedures can lead to DOEs that possess relatively low information content but are still capable of generating WF distributions within 3D domains, either in off-axis or on-axial modes. The available degrees of freedom can be optimized according to given physical constraints. To attain this goal, the desired characteristics of the WF must be expressed in a suitable mathematical form over the region of interest and its relation to the boundary represented by the DOE.[7, 10]

6.3   Fundamental Limitations for Generating Three-Dimensional Wave-Fields

In this section, we consider the basic mathematical properties of 3D distributions of WFs. What can be said about any 3D WF is important to understand the limitations encountered in the synthesis problem. This section puts on a mathematical basis some of the considerations we have previously discussed.

6.3.1   Limitations in the Spectral Domain

As noted in the previous section, it is useful to visualize the basic limitations for attaining 3D WF distribution in the spectral domain, i.e., what the properties of the 3D Fourier transform of the WF are.

Let us consider an optical system and the half free-space behind its output plane, as represented in Fig. 6.3. Regardless of the composition of the system, the WF distribution in the free-space portion will be described by the proper wave equation. At this stage, we do not consider technological limitations and assume that the boundary conditions, imposed at the output of the system, may lead to an ar-

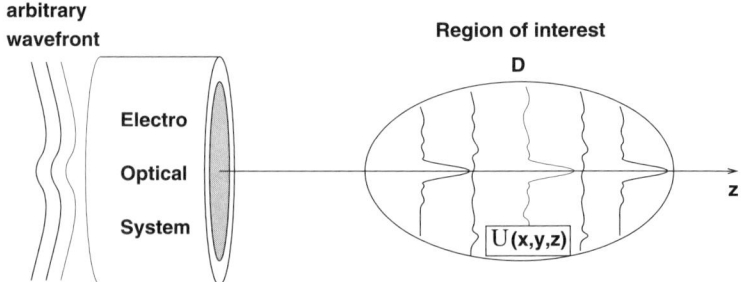

FIGURE 6.3. The problem under consideration: Specify the system and the incident field to generate a given 3D distribution within a finite domain.

bitrary WF distribution in the immediate vicinity following this plane. This field distribution is limited by a finite effective bandwidth and support. The finite effective support is a direct consequence of the finite aperture of the system and is ultimately limited by the finite energy carried by any realizable wave. The bandwidth is limited by the wavelength of the radiation, and it is generally assumed that the spatial frequency is further limited by the different apertures in the system. The energy propagated beyond this cutoff frequency may usually be neglected.

A wide class of optical problems in free space can be treated by considering the light field as a scalar quantity.[11–13] Thus, we consider scalar monochromatic (coherent) WFs of the form

$$\mathcal{U}(\mathbf{r}, t) = U(\mathbf{r}) \exp(-iwt) \tag{1}$$

in the half free space $z > 0$. $\mathbf{r} = (x, y, z)$ denotes the 3D position vector, t denotes the time, and w denotes the angular temporal frequency.

We refer to the situation in which a monochromatic scalar field, emerging from an optical system, propagates through the space $z \geq 0$ free of sources. In this region, $U(\mathbf{r})$ satisfies the homogeneous Helmholtz equation

$$\nabla^2 U + k^2 U = 0, \tag{2}$$

where $k = 2\pi/\lambda = w/c$ is the wavenumber.

Let us consider now the 3D field distribution $U(x, y, z)$. Its 3D Fourier transform(FT) representation is

$$U(x, y, z) = \int \int \int_{-\infty}^{\infty} u(f_x, f_y, f_z) \exp[i2\pi(f_x x + f_y y + f_z z)] \, df_x \, df_y \, df_z. \tag{3}$$

Substitution into (2) leads to

$$\int \int \int_{-\infty}^{\infty} [k^2 - (2\pi f_x) - (2\pi f_y) - (2\pi f_x)] u(f_x, f_y, f_z) \tag{4}$$

$$\exp[i2\pi(f_x x + f_y y + f_z z)] \, df_x \, df_y \, df_z = 0.$$

The family of functions $\exp[i2\pi(f_x x + f_y y + f_z z)]$ is orthogonal. Thus, assuming a nontrivial solution $[u(f_x, f_y, f_z) \neq 0]$, the necessary condition is that the spatial frequencies satisfy

$$f_x^2 + f_y^2 + f_z^2 = (1/\lambda)^2, \tag{5}$$

i.e., the spatial frequencies lie on the surface of a sphere of radius $1/\lambda$ that is related to Ewald's sphere [see Fig. 6.4].[7, 14, 15]

We have determined the spectral support of any 3D WF distribution satisfying the wave equation. To calculate the explicit form of the 3D spectrum, we use the general solution of (2) given by the operator equation[12, 13]

$$U(x, y, z) = \mathcal{F}^{-1} \exp[iz2\pi \sqrt{(1/\lambda)^2 - (f_x^2 + f_y^2)}] \mathcal{F} f(x, y), \tag{6}$$

where $\mathcal{F}$ represents the FT operator and $f(x, y) = U(x, y, 0)$ is the field at $z = 0$.

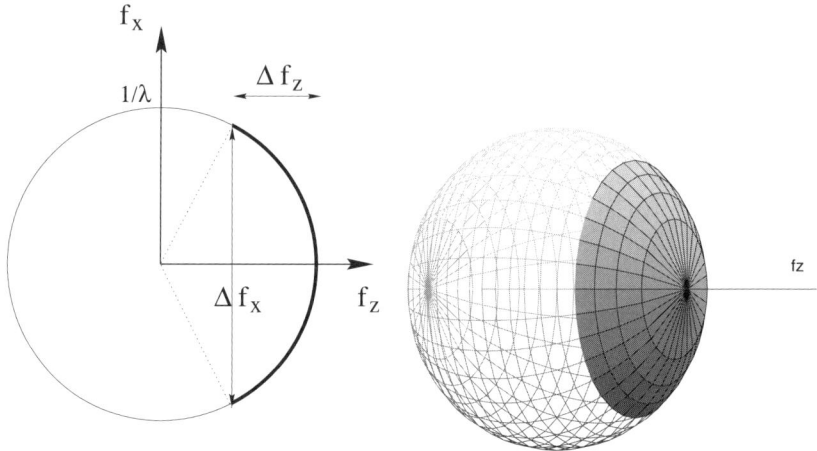

FIGURE 6.4. Spectral domain of 3D WFs. (a) Definition of parameters. (b) 3D representation. The domain is limited to the surface of a sphere.

Explicitly, this is expressed as

$$U(x, y, z) = \int\int_{-\infty}^{\infty} F(f_x, f_y) \exp[i 2\pi (f_x x + f_y y + f_z z)]\, df_x\, df_y, \quad (7)$$

where

$$f_z = \sqrt{(1/\lambda)^2 - (f_x^2 + f_y^2)} \quad (8)$$

and

$$F(f_x, f_y) = \mathcal{F} f(x, y) = \int\int_{-\infty}^{\infty} f(x, y) \exp[-i 2\pi (f_x x + f_y y)]\, dx\, dy. \quad (9)$$

In this form, the representation of a WF is also known as the angular spectrum representation.[12] Equation (7) can be written as

$$U(x, y, z) = \int\int\int_{-\infty}^{\infty} F(f_x, f_y)\delta\left(f_z - \sqrt{\lambda^{-2} - f_x^2 - f_y^2}\right)$$
$$\times \exp[i 2\pi (f_x x + f_y y + f_z z)]\, df_x df_y df_z. \quad (10)$$

Comparison between (10) and (3) indicates that attainable distributions must satisfy

$$u(f_x, f_y.f_z) = F(f_x, f_y)\ \delta\left(f_z - \sqrt{\lambda^{-2} - f_x^2 - f_y^2}\right). \quad (11)$$

This relation means that $u(f_x, f_y.f_z)$ must be zero everywhere in the spatial frequency space except from a spherical surface of radius $1/\lambda$. Moreover, it gives the relation between the 3D FT of the field distribution and the 2D FT of the 2D distribution in one plane.

The considerations above assumed an infinite aperture at $z = 0$. However, finite apertures within the optical system limit the angular spectrum to an effective transversal cutoff frequency $f_{tc} = \Delta f_x/2 < 1/\lambda$. This can be appreciated in Fig.

6.4. Due to the relation between transverse and longitudinal spatial frequencies, the limitation on the longitudinal frequencies is given by a bandpass of width $\Delta f_z = 1 - \sqrt{(1/\lambda)^2 - f_{tc}^2}$. Thus, practical systems usually utilize only part of the whole spherical domain.

A better use of this domain is achieved by designing a system collecting light from multiple directions. Such a kind of system can be implemented, for example, using nonplanar or multiple holograms. A wider set of distributions can be also attained if frequency multiplexing is implemented using different wavelengths. In this way, there is access to spherical domains with different radii. However, the implementation of these more complex systems is usually cumbersome, whereas the basic limitations are easy to extend from the one-sided illumination case. Therefore, these systems are not further discussed in the present work.

It should be noted that a similar approach can be applied to find the limitations in the spectral domain for vector WFs.

6.3.2   Spectral Domain of Intensity Distributions

In many practical situations, it is the intensity that is of main interest, and not the complex field. This choice leaves the phase as a free parameter, increasing the freedom in the synthesis process. Let us analyze, then, the spectral domain of intensity distributions.[14]. The intensity is simply

$$I(x, y, z) = |U(x, y, z)|^2. \tag{12}$$

Then, the FT satisfies

$$i(f_x, f_y, f_z) = \mathcal{F}I = u(f_x, f_y, f_z) \star u(f_x, f_y, f_z), \tag{13}$$

where $\star$ represents correlation. The domain of $I(f_x, f_y, f_z)$ is a volume of revolution as represented in Fig. 6.5. An additional constraint exists: Because $I(x, y, z)$ is real, its FT is conjugate symmetric: $i(f_x, f_y, f_z) = i^*(f_x, f_y, f_z)$. The transverse cutoff frequency for the intensity is $f_{tc}^I = \Delta f_x$, and the longitudinal cutoff is $f_{lc}^I = \Delta f_z$.

6.3.3   The Degrees of Freedom in a 3D Domain

The degrees of freedom of a function is the number of independent complex numbers needed to construct the function on a given interval. According to the sampling theorem, the field $U(x, y, z_0)$ is specified if it is sampled at the Nyquist frequency. Once $U(x, y, z_0)$ is known, $U(x, y, z)$ is determined for every $z > z_0$. Therefore, the spatial number of degrees of freedom [16] in an area W_0 at $z = 0$ is

$$F_0 \sim \frac{W_0 \Delta \Omega}{\lambda^2}, \tag{14}$$

where $\Delta \Omega = \lambda^2 \Delta f_x \Delta f_y$ is an element of the solid angle of all possible directions of propagation. The number F_0 remains invariant for any $z \geq 0$. This result is also known as the conservation of the space-bandwidth product.

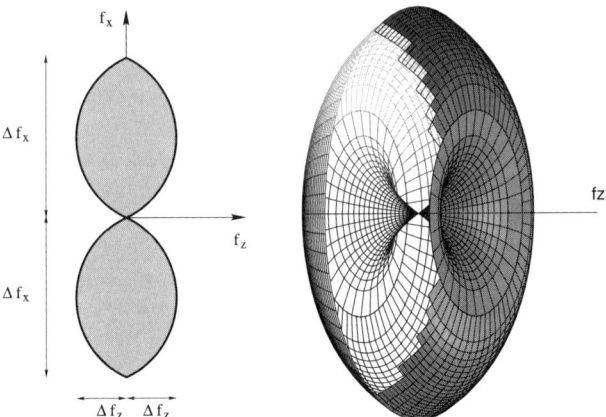

FIGURE 6.5. Spectral domain of 3D intensity distributions. (a) Definition of parameters. (b) 3D representation. The domain is limited to the volume inside the depicted surface of revolution.

As a consequence, the complex amplitude can be completely prescribed within only one transverse plane. However, because usually only the intensity is specified, there is some freedom to fix the intensity distribution over one more plane, although the existence of such solutions is not guaranteed.[17]

Another interesting and more rigorous approach to the problem of degrees of freedom is based on the prolate spheroidal functions. In this case, the number of degrees of freedom is related to the number of eigenvalues of the finite Fourier transform that are above a certain threshold determined by the noise levels.[18, 19]

Let us count now the number of degrees of freedom of a 3D domain. According to the transverse and axial bandwidths introduced in Section 6.3.1, the number of samples needed in a domain of crossection W and depth Δz is

$$F_{3D} = W\Delta f_x \Delta f_y \Delta z \Delta f_z. \tag{15}$$

The number of degrees of freedom of such a volume turns out to be impressive as the dimensions grow. However, this number is not achievable; we already know that only F_0 degrees of freedom are available. As a compromise, we can trade transversal extent with depth in the 3D domain to achieve a number $F_{3D} \approx F$. Obviously, this is not enough to assure the feasibility of a given 3D WF, because the ultimate condition stems from the wave equation. Moreover, the F_{3D} degrees of freedom needed to specify a WF in the 3D domain cannot be independently controlled even if $F_{3D} \approx F$. The situation is similar to the case of the synthesis of the intensity distribution in two planes;[17] the existence of an exact solution cannot be guaranteed in every case. Nevertheless, the present considerations give an indication of the direction in which the synthesis problem can be treated.

A rigorous approach to the problem of degrees of freedom for the generation of 3D fields can be formulated by considering the corresponding orthogonal modes within the DOE and the 3D domain.[20, 21] According to this approach, a set of orthogonal functions exists within the DOE domain that generates a corresponding set of orthogonal 3D fields within the 3D domain. The energy efficiency for the generation of each field mode decreases to zero as the order of the modes tends to infinity. As a consequence, the effective number of modes that can be utilized in the presence of noise of any kind is finite. These modes define the degrees of freedom available to the generation of a 3D WF.

It should be noted that in practical situations, the theoretical number of degrees of freedom, F_0, cannot usually be exploited due to the limited resolution of the recording media and the limited modulation range and resolution of amplitude and phase. The practical number of degrees of freedom can be derived by dividing the available aperture space by the area of a resolution element, $\delta x \delta y$, to obtain

$$F_p = \frac{W}{\delta x \delta y}. \tag{16}$$

The problem of attaining maximal advantage of the available degrees of freedom for given technological limitations is related to the coding process of information.

6.4 Specification of the Wave-Field Distribution in Three-Dimensional Domains

From the Fourier relation of (10), it is obvious that, in principle, any field distribution on a single transverse plane can be implemented as far as it satisfies the resolution and energy constraints. The technological limitations will determine for each case the practical feasibility or complexity of the task. A similar statement is also true for the specification of the field distribution along any axis in space. However, the set of 3D WFs that can be attained is only a subset of the set of all 3D distributions that can be associated with an appropriate 2D wavefront distribution. This problem can be mitigated in various ways:

(1) *Considering intensity distributions.* The phase freedom leads to a spectral domain that lies in a volume. However, this domain is limited and the function should be conjugate symmetric. From the point of view of the degrees of freedom, which are duplicated, we could be tempted to state that the intensity can be specified in two planes. But this is not always true. This possibility is strongly dependent on the distance between the two planes and the distributions under consideration. For example, it is not possible to produce, with one-sided illumination, a completely dark transverse plane after an illuminated plane.

(2) *Restricting the space domain.* Additional degrees of freedom can be achieved by considering only a portion of the 3D space. The restrictions are similar to those considered for the phase freedom; i.e., in general, the existence of a specific field distribution cannot be guaranteed.

(3) *Relaxing the specification with tolerances.* Specifying the WF within some tolerances leads to additional freedoms that may lead to the feasibility of special WFs that otherwise would be impossible to produce.

(4) *Establishing a hierarchy among different requirements.* With this approach, the most important aspects of the desired WF distribution may be attained, whereas less important features are only approximated.

These approaches are further investigated in Section 6.6.

6.5   DOE Design as an Optimization Problem

Our objective is to design a DOE that, when properly illuminated will generate a predetermined WF distribution within a given region of space. Viewing this as an electromagnetic problem, we wish to design the boundary conditions that result in the desired solution. This is effectively an inverse problem, and there is no *a priori* guarantee that a physical solution exists. As a consquence, we must seek a solution as close as possible to the desired one, which turns out to be an optimization problem.

Optimization problems are usually approached by iterative methods due to the difficulty in obtaining an analytic representation of the process. There are also several additional incentives to motivate the application of iterative processes. One of these incentives is that optimization problems can be also solved within an actual physical system. If run in such a system, an appropriate iterative algorithm can, in principle, accommodate to various environmental situations and physical parameters that are not well defined in a mathematical sense. This behavior is inspired by nature. For example, animal eyes appear to work, at least partly, by learning in an iterative fashion. The optical equipment (cornea, lens, retina, etc.) is more-or-less unchangeable or "given." Seeing involves not just the optical equipment, but also the neural equipment (optic nerve, brain, etc.) that operates in conjunction with the optical one. This neural network contains numerous readily adjustable variables (connections) that evolve as the individual learns to see. This arrangement is very desirable because it can readily accommodate gross defects in the optical system. Good performance does not necessarily require good optics, a situation that may be advantageous for many applications.

To implement an optimization algorithm, some performance criteria must be defined. These performance criteria can be assembled into a *cost function* that quantifies the overall performance of the system. The definition of this cost function is the heart of any optimization algorithm, and therefore, its proper definition is most critical for achieving the desired results. For our application, the cost function must provide a measure of how far we are from the desired WF distribution, and we shall refer to it as a *distance function.*

In the following, we provide a brief review of some optimization procedures that were found useful for DOE design.

6.5.1   Simulated Annealing

We seek a global energy minimum in a piece of metal by annealing. The metal is heated to a high temperature, where patches within it are mobile. It is then slowly

FIGURE 1.1

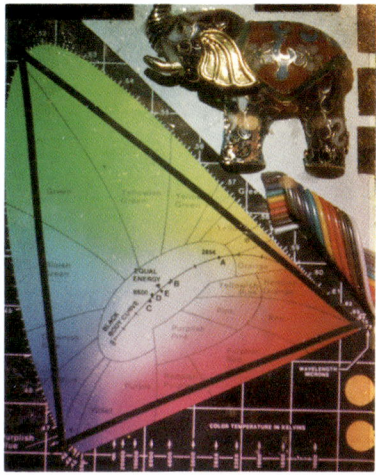

FIGURE 1.5

FIGURE 1.7

FIGURE 1.8

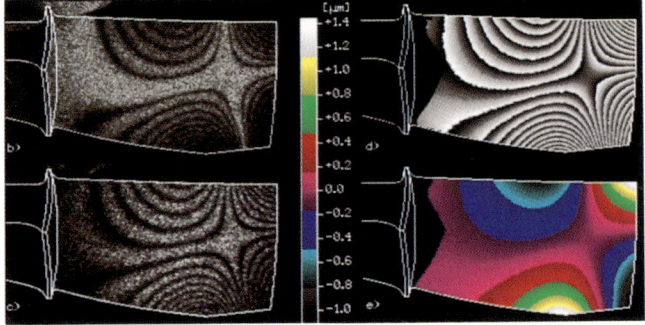

FIGURE 9.10(g)

cooled to allow it to settle into its lowest energy state. Simulated annealing (SA) [22] uses the same approach. In analogy to the natural process, the distance function here is a non-negative energy, the *energy function*, which is to be minimized. The energy depends on the set of free variables that are to be manipulated. We start with any random set of parameters and then stochastically perturb the variable set by a large amount (energy measure) at high "temperature" and by lower amounts at lower temperatures. If the perturbed variables lower the energy, we accept the perturbation. If the perturbed variables increase the energy, we may or may not accept the perturbation. That choice is determined stochastically from a probability distribution function governed by the temperature. At high temperature, the probability of accepting an energy-increasing perturbation is high, and it decreases with the lowering of the temperature. The temperature is slowly decreased until a steady-state minimum is achieved. This is a simplified overview of what is really not a method but a family of methods.

Adapted to our application, the main aspects of SA can be presented mathematically as follows: Assuming that the DOE is to be illuminated by a plane wave, we denote its desired optical transfer function by h. The procedure starts with an arbitrary initial solution $h^{(0)}$ (either random or some rough estimation of the desired function), and then the solution is iterated to reduce the energy function that must be appropriately defined. At the tth iteration, we induce a random change in the elements of $h^{(t)}$ to obtain the $(t+1)$th iteration of the function h. This changes the energy function, $d^{(t)}$ by an amount $\Delta d^{(t+1)}$ given by

$$\Delta d^{(t+1)} = d\left(h^{(t+1)}(i, j)\right) - d\left(h^{(t)}(i, j)\right). \tag{17}$$

The new function, h, is accepted if $\Delta d^{(t+1)} < 0$, and it may be also accepted if $\Delta d^{(t+1)} \geq 0$ based on the acceptance probability

$$Pr_{\text{accept}} = \exp\left(\frac{-\Delta d^{(t+1)}}{T}\right), \tag{18}$$

where T is the temperature parameter.

The procedure is now repeated starting from the new function $h^{(t+1)}$, decreasing the temperature slowly as the process continues. The cooling rate and the steps of the random perturbation are important parameters that depend on the specific process implemented. Because the achievement of the global minimum is only guaranteed after an infinite number of iterations, the process is terminated when an adequately low energy is obtained.

6.5.2   Genetic Algorithms

In biological evolution, it is not the individual but a population (species) that evolves. The success of the individual (*phenotype*) gives it an improved chance of breeding. In breeding, the genetic structure (*genotype*) of the offspring is made of genotypic contributions from both parents. In addition, errors (*mutations*) occasionally occur. The offspring then competes for the right to reproduce in the next generation. Thus, a gene pool evolves that not only governs future generations, but also bears within its memory information about its origin.

Optimization procedures based on the above ideas are called genetic algorithms (GA) [23]. In GA, a genome, or vector, is specified as a way to describe the system. It contains the information needed to describe the system. A figure of merit is then evaluated for each member of a pool of genomes. Winners are selected for genetic exchange (usually called "crossover") and mutation. Losers are usually dropped from the pool to keep the pool size constant.

GA usually has the following features:

- A chromosomal representation of solutions to the problem, which is usually binary.
- An evaluation function that gives the fitness of the population, which, in our case, will be a distance function.
- Combination rules (Genetic operators) to produce new structures from old ones—*reproduction, crossover, and mutation.*

There are several variants of GA. One form of these algorithms was used for generating spatial filters for pattern recognition that are also a kind of DOEs. The algorithm to generate such filters in a hybrid electro-optical system[24, 25] can be summarized as follows:

(1) Start:
Select at random a population of m members (binary functions) $\{h_1, h_2, \ldots, h_m\}$. In our case, these functions represent possible transfer functions for the DOE to be designed.

(2) Use each of these functions to generate a corresponding WF, and evaluate its distance functions from the desired WF, d_i $\{i = 1, 2, \ldots, m\}$.

(3) Compute the average value of the cost function

$$\theta = \frac{1}{m} \sum_{i=1}^{m} d_i. \tag{19}$$

(4) Set a discrete time parameter, t, to zero, and define a probability P for a mutation to occur and set it to some $P_{\max}$.

(5) Crossover:
Select the function h_l that corresponds to the minimal distance function, d_l. Pick from the population a function h_j at random. The two functions, h_l and h_j are the parents to be used for generating an offspring function. Select a random integer k between 0 and n, where n is the dimension of the vectors h. Create the offspring function, h_c, by taking the first k elements from one of the parents, randomly, and the remaining $n - k$ elements from the other parent.

(6) Mutate:
Induce a mutation (inverting the sign of the elements "1" to "0" or "0" to "1") with probability P on each element of the offspring vector h_c.

(7) Reproduce:
Pick at random a function h_d from the population subject to the constraint: $d_d \geq \theta$. Replace h_d in the population with the new offspring h_c, and update

the average value of the distance function,

$$\theta \to \theta + \frac{1}{m}(d_c - d_d). \tag{20}$$

(8) Setting parameters:
    Set the new parameters, $t \to t + 1$ and $P \to P_{\max}\left(\frac{1}{t}\right)^r$.
(9) Terminate the procedure when the distance function becomes adequately small
    (define small). (Alternatively, the process can be terminated after a predeter-
    mined number of iterations. The latter criterion is useful if there is no *a priori*
    knowledge about the expected behavior of the distance function.) Else, if
    $P > P_{\min}$, go to **5**; otherwise, go to **1**.

Selection of the parameters r, $P_{\min}$, $P_{\max}$ depends on the particular problem at
hand.

6.5.3   Projection Methods

Projection-based algorithms are useful for the solution of feasibility problems [26]
and constitute an extremely powerful class of optimization algorithms.

In this formulation, each requirement or information regarding the problem is
expressed as a set of the solution space. Each component of the intersection of this
family of constraints represents an acceptable or feasible solution. If the sets do
not intersect, the objective is to find the closest point to them according to a given
criterion.

The constraint sets represent conditions that must be fulfilled by the solution. In
our case, the constraints are determined by the desired WF distribution, the physical
characteristics of the experimental system, and possibly some other demands.

In the *Projections Onto Convex Sets* (POCS) algorithm,[27] the distance function
is measured from the function h^t obtained at the tth iteration to sets of various
constraints defined on the problem. For the POCS algorithm to work, some functional relationships are necessary and the physical constraints must be precisely
known and defined.

The POCS algorithm is an iterative process that is applied by transfering a
function from one domain to another (for example, the FT domain and the image
domain), and in every domain, it is projected onto one or several constraint sets.
The procedure is repeated in a cyclic manner until the solution converges to a
function h that satisfies simultaneously all the constraints. If all the constraint sets
are closed and convex, and they have at least one common domain, then the process
converges weakly.

In our context, it is difficult, and may sometimes be impossible, to fulfill these
conditions; so the convergence is not always guaranteed. Therefore, a more advanced procedure is needed for these applications. More powerful procedures are
based on parallel projection methods[28, 29] that are able to handle situations in which
not all constraints are convex and in which some of the constraints are inconsistent
with each other or do not correspond to a physical solution [30]. For applications

under these circumstances, generalized distance functions can be employed that will not increase from one iteration to the next. Although an exact solution may be inaccessible in a finite time, the algorithm can be terminated when the distance function attains a value that is defined to be adequately small. Thus, a good solution can be obtained even if the various constraints are inconsistent with each other. The solution obtained will have the shortest possible distance to all constraints, although no one of them may be completely satisfied. Because projection algorithms were found to be most efficient for our present purposes, they will be discussed further in relation to our specific application.

6.5.4   Discussion

Optimization procedures that are based on the accurate knowledge of all physical parameters involved can be implemented on high-precision digital computers. In their present form, these algorithms cannot be implemented if the processors have only a limited accuracy, if there are unknown parameters in the system, or if non-negligible noise exists in the processor. In an actual optical system, all of these factors are present. It has space-bandwidth limitations, it has optical aberrations, there are distortions, and there are dead zones in the SLMs and recording devices. In the presence of unknown parameters or parameters that are difficult to quantify, it is impossible to evaluate the exact distance from various constraints. Under these circumstances, one cannot have a unique, deterministic update rule for the function at each iteration. Therefore, if implementation on a hybrid electro-optical architecture is desired, one has to employ procedures in which unknown parameters may be also present. Stochastic algorithms, such as the SA and GA, are of this nature. During the design process, the DOE transfer function is updated in a random fashion and an update is accepted or rejected in view of its actual performance. For this reason, these stochastic algorithms are also immune to system aberrations and other distortions, such as those introduced by a nonideal SLM. Naturally, the design algorithm must be performed within the same system in which the results are to be used.

While implementing a DOE using a synthesis algorithm on a real physical system, we assume that the function h is written on a SLM whose pixel settings are the convenient variables. All other system parameters are treated as fixed, even if some of them are unknown. The incident optical beam quality, the behavior of the lenses, and the pixel-by-pixel detector performance are examples of these unknowns that we seek not to determine but to accommodate.

The projection-based algorithms appear to be the most powerful and have the fastest convergence among the algorithms reviewed in this section. These algorithms are best suited when all the constraint sets can be rigorously defined, such as in digital computer design of spatial filters, and DOEs to be implemented by using high-precision technologies. A relaxation of these requirements was achieved with the recently introduced projection onto fuzzy sets algorithm.[31]

The next section is devoted to a useful design procedure exploiting the capabilities of a projection algorithm.

6.6   Synthesis

Confronted with a synthesis problem, the first question that should be answered is whether there is a satisfactory solution of the wave equation that satisfies the given requirements. It is only afterward that we should enquire about the most suitable optical system to generate it. After a general architecture for the system is specified, the technological limitations can be identified and included as additional constraints on the 3D WF. For example, if there is an aperture in the system, it can be adequately represented as a limitation in the domain of the WF in that specific plane. Examples of other technological limitations are the limited space-bandwidth product, the restricted wavefront modulation capacity, the degree of coherence of the source, and so on.

As compared with the design of 2D WFs, one of the main difficulties concerning 3D synthesis problems is the uncertainty of the existence of a solution, even before the technological aspects are considered. Nevertheless, in case a solution satisfying all the requirements does not exist, it is still possible to look for a solution that approximates the requirements as much as possible. An appropriate selection of the admitted tolerances and of the relevant parameters is important in this respect. These aspects ultimately define the problem in a proper mathematical form. Moreover, even in those cases in which a solution does exist, it is not likely to be representable by an analytic expression. Thus, we are forced to search for a numerical solution in an optimization of the available resources under given requirements, tolerances, and priorities.

6.6.1   *General Statement of the Synthesis Problem*

A general optical system for WF generation is composed of classical and unconventional optical elements. Classical optical components are, for example, lenses, mirrors, prisms, polarizers, and apertures. Unconventional optical elements include wavefront modulating devices like fixed diffractive optical elements, specialized apodization filters, and dynamic spatial light modulators. In many situations, the preferred solution is a single diffractive element, in which advanced fabrication techniques can be employed to exploit the available degrees of freedom. Obviously, a reduction in the number of optical elements leads to lighter and more compact systems with higher reliability and lower cost.

The classical devices can be described by an operator representation that can be combined with the diffraction operator in different sections of the system. This approach is particularly interesting in the case of first-order optical systems[32–34], in which the combination of any number of classical elements leads to an operator that has exactly the same structure as the Fresnel diffraction operator.[13, 35]

The wavefront modulating devices are the core of our systems, because they can, in principle, generate any complex modulation of the transverse field. Therefore, the main task is to specify these wavefront modulation devices. We concentrate here on the specification of a single such device. The extension to multiple devices is almost straightforward.

Our objective can be stated as follows: *Given a known optical system composed of classsical optical devices and a wavefront modulation element, design this element, subject to given constraints, to obtain a desired field distribution within a given 3D domain.*

The situation is depicted in Fig. 6.3. The source can have, in principle, any arbitrary characteristics as far as they are known *a priori*. In this work, we concentrate mainly on the study of coherent sources.

The constraints on the modulation element originate from fabrication and performance considerations and are determined by answering questions, such as: Can it modulate amplitude or phase? With continuously varying or quantized values? Is it pixelated or continuous? What is the available SBP?

6.6.2   Procedure

Our aim is to find the solution that is closest to the given constraints, according to an appropriate measure. In general terms the procedure can be stated as follows:

(1) *Define a signal domain D behind the optical system and the desired WF distribution $U_0(x, y, z)$.* Note that in some situations, there is freedom to specify only the magnitude of the field (intensity distributions) or to define nonconnected domains. Moreover, the WF over all space outside of D is not predetermined. This provides significant flexibility to the whole procedure.

(2) *Define appropriate tolerances for the specified magnitudes: $\delta U(x, y, z)$.* These tolerances are used as additional freedoms within the procedure. Note that the tolerances are a function of the position and they do not have to be uniformly distributed.

(3) *Transform $U_0(x, y, z) \pm \delta U(x, y, z)$ within D into mathematical constraints: C_{U_0}. Use appropriate sampling lattices.* The sampling conditions are applied to avoid redundancy (see Section 6.3) and to achieve an adequate reconstruction of the WF, including distributions with spatially continuous constraints along the direction of propagation.

(4) *Transform the technological limitations and the general architecture of the optical system (including the wavefront modulation device) into additional constraints: C_T.*

(5) *Define a hierarchy among the different constraints and within features of each constraint.* For example, the existence of an aperture within the optical system imposes a finite-support constraint that cannot be ignored or approximated. A certain degree of efficiency cannot be neglected. On the other hand, certain aspects of the objective WF may be less relevant, depending on the specific application, as for example, stray light level.

The hierarchy can be mathematically expressed by a "weighted" algorithm and weighted norms, as will be shown in Secyion 6.6.3.

(6) *Specify the operator of the system.* This is the relation between the wavefront modulation function $f(x, y)$ and the 3D WF $U_f(x, y, z)$. In general, this is the

transfer operator of an optical system that reduces to the Fresnel diffraction in free space (provided this approximation is valid).

(7) *Define an appropriate distance function that represents the proximity of the 3D WF to the constraints defined in (3) and the hierarchy defined in (5)*:

$$d_w[U_f(x, y, z), U_0(x, y, z)].$$

(8) *Specify the wave-front modulation function $f(x, y)$ by solving the optimization problem*:

$$\text{Minimize } d_w[U_f(x, y, z), U_0(x, y, z)] \text{ subject to } U_f \in C_T.$$

The difficulty of the solution depends on the specific synthesis problem. In general, we deal with a large number of linear and nonlinear constraints, and the dimensionality of the problem is high (typically, above 10^7). In addition, the constraint sets may be nonconvex or nonconsistent. This is, in general, an ill-posed problem. A solution satisfying all constraints does not always exist. If it exists, it is not neccessarily unique and its dependance on the data may contain discontinuities.

It is pertinent to recognize the similarity of this problem to the inverse problems. However, although typically in inverse problems we deal with data collected from existing real fields, in the synthesis problem that we consider here, we have not *a priori* certainty regarding the existence of the objective WF. Moreover, in typical inverse problems, the data are affected by unavoidable given noise and a given measurement system, whereas in our synthesis problem, we have some freedom to specify the tolerances, the hierarchy of the data, and even the optical system.

6.6.3   Optimization: Set Theoretic Approach

According to the discussion in the previous sections, we are faced with the synthesis of a function employing some optimization algorithm. There are various optimization strategies that can be considered for the solution of the problem: steepest descendent methods, genetic algorithms, simulated annealing, and projections onto constraint sets, among others. However, usully we deal here with nonlinear, nonconvex, and nonconsistent constraints, and a large number of variables. Therefore, the solution is too computation intensive to be handled by most of them, which in some cases do not work at all. Then, the choice of an optimization algorithm and the way it is applied is a major issue. Best results are achieved when physical considerations are taken into account. According to our experience, the set theoretic approach and the projection-type algorithms proved to be efficient in this kind of problem. Thus, in the following, we concentrate on this method and its application to the 3D synthesis problem.

The set theoretic approach is useful in the design problem because it can deal with fuzzy propositions and inconsistent sets [7, 10, 26, 36]. As was previously observed, one of the main difficulties in the design of a required 3D distribution is that it is not clear *a priori* whether such a distribution is at all physically possible. The procedure presented in Section 6.6.2 overcomes this difficulty by the possibility of assigning high weights to important features and low weights to less important or

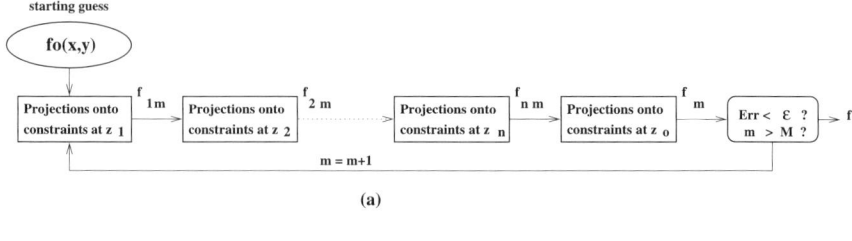

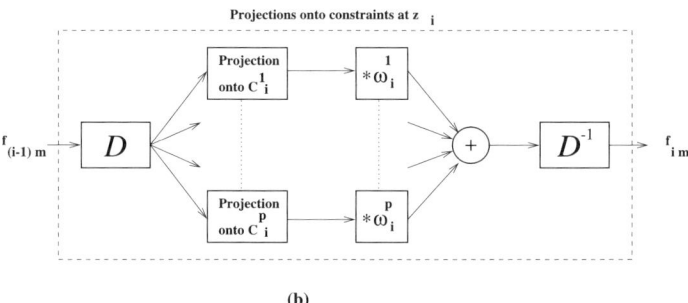

FIGURE 6.6. One realization of the block projections algorithm applied to 3D synthesis. (a) Global diagram shows iterative block projections onto constraints on planes at distances $z_0, z_1 \ldots z_n$ ($z_0 = 0$). The process starts with a random guess and stops when a good approximation to a solution is achieved or after a certain number of iterations M. (b) Detail of a block projection onto constraints at distance z_i. The p constraints are weighted according to $w_i^j \geq 0$, $j = 1 \ldots p$, where $\sum_{j=1}^{p} \lambda_i^j = 1$.

not precisely defined features. In this way, we use the flexibility of the algorithm that converges to a satisfactory solution, even if the prescribed distribution cannot be satisfied exactly by a physical field.

A typical implementation of the proposed algorithm is depicted in Fig. 6.6. This is a particular realization of the block projections algorithm described in Section 6.5.3, as it is applies to 3D WF synthesis.

We start from an arbitrary input function, which represents the field immediately behind the optical system. This field distribution is the result of the passage through the optical system, including the wavefront modulation device. We then project the input function onto the constraints at different transverse planes in succession.

We perform each block projection onto different constraints of a single transverse plane (Fig. 6.6). The mathematical projection is easier to perform in the propagated field domain than in the input domain. Therefore, we first perform a propagation $\mathcal{D}$ to a plane situated at distance z_i, followed by parallel projections onto different constraints on that plane. The results are weight-averaged before an inverse propagation $\mathcal{D}^{-1}$ is performed to complete the block projection.

The projection $\mathcal{P}_{z_i}$ of f onto C_{z_i} (a constraint on the field at distance z_i) is obtained by calculating the field at the distance z_i, projecting this function onto the constraint at that plane and finally performing the inverse propagation to obtain the corresponding field at the input:

$$\mathcal{P}_{z_i} f = \mathcal{D}^{-1} \bar{\mathcal{P}}_{z_i} \mathcal{D} f, \tag{21}$$

where $\bar{\mathcal{P}}_{z_i}$ represents the projection operator on the plane at the distance z_i onto the constraint C_{z_i}. This is possible because the diffraction propagator $\mathcal{D}$ satisfies Parseval's theorem.

The procedure is stopped after a predetermined number of iterations or when an acceptable approximation to the desired distribution is achieved.

6.6.4   Definitions

Let us consider n constraint sets C_i $(i = 1 \ldots n)$ in a Hilbert space; a signal f is a feasible solution if

$$f \in C = \bigcap_{i=1}^{n} C_i. \tag{22}$$

The problem is called *consistent* if C is nonempty. A *weighted norm* in $L^2(\mathbf{R}^m)$ is defined as

$$\|g\|_{\Omega_i} := \left[\int_{\infty}^{\infty} |g(\mathbf{x})|^2 \Omega_i(\mathbf{x}) d\mathbf{x}, \right]^{1/2} \tag{23}$$

where $\mathbf{x} \in \mathbf{R}^m$ and Ω_i is a real, non-negative, bounded weight function. The index i corresponds to each constraint set.

A vector $W = (w(1), \ldots, w(n))$ is defined as a *weight function* if $\sum_{i=1}^{n} w(i) = 1$.

A sequence of weight functions $\{w_k\}(k = 0, 1, 2, \ldots)$ is *fair* if for every i, infinitely many values of k exist for which $w_k(i) > 0$.

Let us consider the complex function $f(x, y)$. The *projection* $g = \mathcal{P}_i f$ of f onto C_i with respect to the respective weighted norm is defined by

$$\|f - g\|_{\Omega_i} = \mathbf{inf}_{v \in C_i} \|f - v\|_{\Omega_i}. \tag{24}$$

A *weighted projection* $g_w = \mathcal{P}_w f$ of f onto $\{C_i\}$ is defined as

$$\mathcal{P}_w f = \sum_{i=1}^{n} w(i) \mathcal{P}_i f. \tag{25}$$

A convenient (for our purpose) *cost function* is the *weighted squared error* defined as

$$\text{WSE} = \sum_{i=1}^{n} w(i) \|\mathcal{P}_i f - f\|_{\Omega_i}^2. \tag{26}$$

6.6.5   Algorithm: Block Projections

The iterative algorithm is implemented as [37]

$$f_{k+1} = f_k + \lambda_k [\mathcal{P}_{w_k}(f_k) - f_k], \tag{27}$$

where f_0 is an arbitrary initial function, $\{w_k\}$ is a fair sequence of weight functions, and $\{\lambda_k\}$ is a sequence of *relaxation parameters* with $\lambda_i \in (-1, 1)$.

The main iteration proceeds sequentially from block to block (k index), whereas within each block, the projections are performed in parallel. The condition that the weight function is fair guarantees that we do not stop projecting onto every constraint (although not necessarily on every iteration in k).

The special case in which $w_k(i) = 1$ for $k = qn + i$, ($q = 0, 1, \ldots$), leads to a serial algorithm equivalent to the projection onto convex sets (POCS).[27, 38] This algorithm is schematically represented in Fig. 6.7a. Note that the widely used Gerchberg-Papoulis and Gerchberg-Saxton[39–41] algorithm is a special case of the POCS algorithm.

Alternatively, if $w_k(i) \neq 0$ for all k and all i, we get a full parallel algorithm in which projections onto all sets are performed in every iteration (see Fig. 6.7b).

6.6.6   *Convergence*

We summarize the main convergence properties regarding this algorithm.

If the C_i are closed and convex, and their intersection is not empty, the algorithm weakly converges to a solution f such that[30, 37]

$$f \in C = \bigcap_{i=1}^{n} C_i. \tag{28}$$

If the constraints are convex but inconsistent, the parallel projections algorithm weakly converges to a global minimizer of the cost function WSE. This solution is the closest (according to the defined cost function) to satisfying the required constraints.[30]

For the parallel algorithm, also the following property holds:[29, 30] For any finite number of sets, not necessarily convex, provided $\lambda_k = \lambda \ \forall k$ and $\lambda \in [0, 1)$, the cost function is monotonically nonincreasing; i.e., $\mathrm{WSE}(f_{k+1}) \leq \mathrm{WSE}(f_k)$. This property holds even if the intersection of the constraint sets is empty.

6.7   Examples of Special Light Distributions

As indicated earlier, conventional DOEs are usually designed to generate a pre-determined function or light distribution over a single transverse plane,[8, 42] and they require a large number of microstructured resolution elements (of the order of 10^6 and above). In addition, if amplitude modulation is used, the information is encoded on a high-frequency spatial carrier, which leads to off-axis reconstruction.

The examples presented in this section, which were implemented with our new procedure, do not suffer from the above-mentioned drawbacks. These distributions are generated within a 3D domain (Fig. 6.3) in an on-axis mode (even in the case of amplitude modulation), and they require low information-content devices.

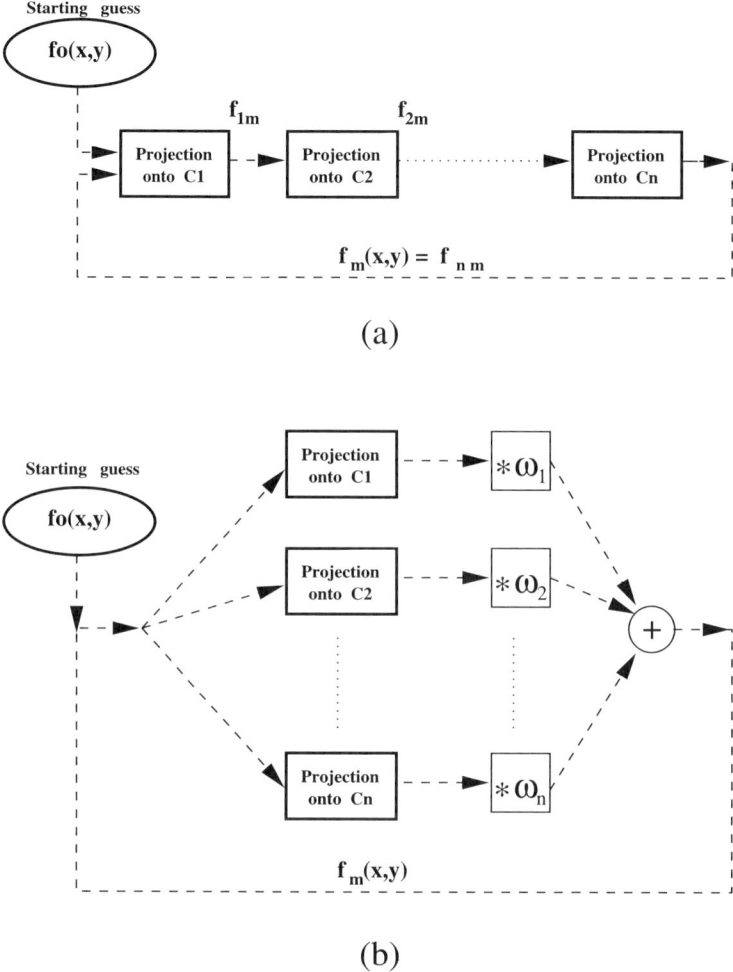

FIGURE 6.7. Two realizations of the block projections algorithm: (a) serial implementation (POCS). (b) full parallel implementation.

6.7.1   *Nondiffracting Beams*

An important issue in light propagation involves intensity patterns that maintain a well-defined shape and are not degraded by the effects of diffraction while propagating within a given domain. The best known examples are the so called "nondiffracting beams" (NDB).[43–51] They consist of spatial light distributions that present a prominent intensity peak along a certain propagation distance where it remains more or less constant. NDBs can be used for alignment and measurement systems as well as in optical processing systems with extended focal depth. Moreover, the concept can be extended to direct generation of patterns that remain constant along a predefined region of space.[52]

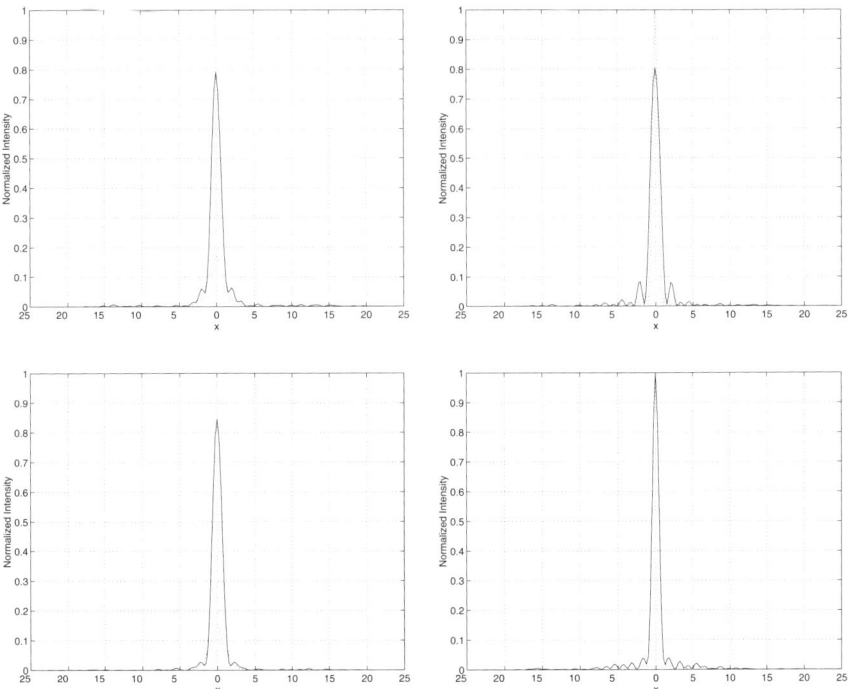

FIGURE 6.8. Low-sidelobe point-spread function with extended focal depth generated by a phase-only DOE. Intensity profiles at different locations within the focal region. The distance between transverse planes is λ/NA^2. The transverse coordinates (x) are in units of $\lambda/(NA)$.

The concept of NDB is useful to design the 3D point-spread function of a diffractive lens with extended focal depth. We seek to achieve a phase element with a constant point-spread function along an extended depth. In addition, we impose the condition of having low sidelobes to keep a good contrast. In Fig. 6.8, we present simulated wavefront characteristics that show the (squared) point-spread function of the obtained lens. The depth of focus is almost four times larger than it is in a regular lens with the same numerical aperture. The intensity peak is approximately constant over a distance of $9\lambda/(2NA^2)$ (Fig. 6.9), where NA is the numerical aperture of a comparable lens and λ is the wavelength. This special diffractive lens contains only 128×128 resolution elements (Fig. 6.10).

The second distribution introduced here (Fig. 6.11) presents two intense peaks inside a region of low background intensity. The desired distribution was synthesized to start at a distance of 1 m from the mask and to present a pair of NDBs along 20 cm. The distance between the peaks at $z = 1.08$ m was 600 μm, and their diameter at half peak intensity was 50 μm. The distance between the spots grew as $\lambda z/D$, where z is the distance from the mask and D is the aperture of the DOE. Completely parallel "beams" have also been designed.

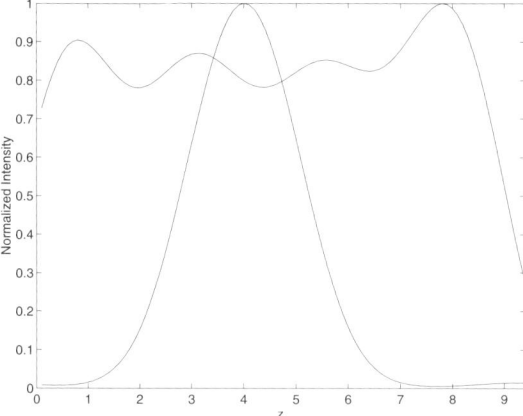

FIGURE 6.9. Amplitude variations of the central peak along the longitudinal axis (units of $\lambda/(2NA^2)$).

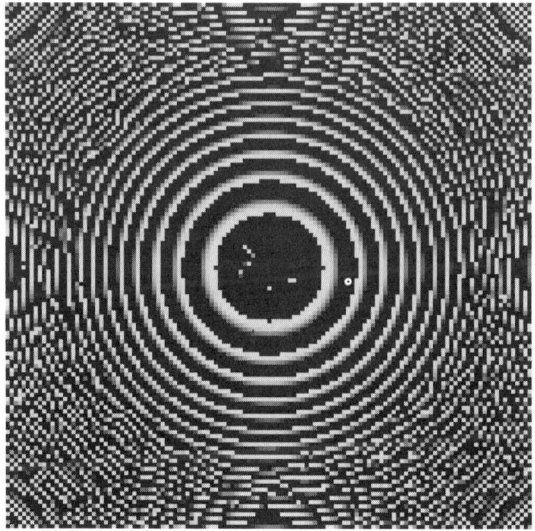

FIGURE 6.10. The diffractive element that corresponds to Figs. 8 and 9. The phase is represented by gray levels.

Although we call the distribution described above a "nondiffracting beam array" (NDBA), it differs significantly from a mere replication of Bessel beams in several ways: (1) The spot-size is determined by the whole aperture and, thus, requires half the space bandwidth of two independent Bessel beams with the same spot-size. (2) The "beams" are part of the same distribution; so their background lobe system is coupled. This is not the case if two Bessel beams are placed side by side. Moreover, it can lead to destructive interference if they are generated with the same coherent

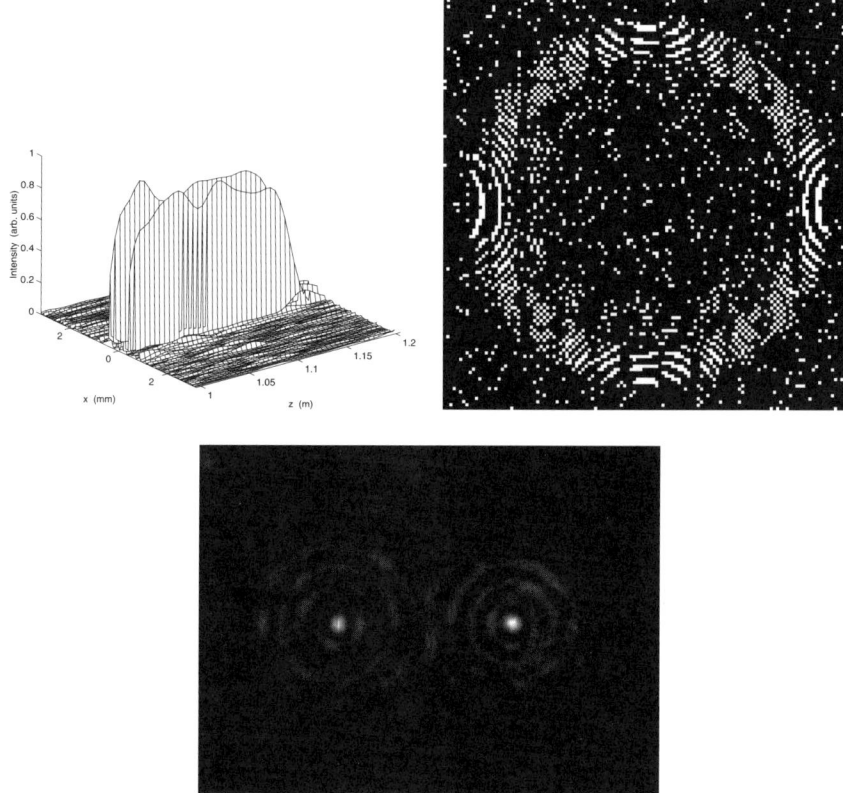

FIGURE 6.11. Nondiffracting beam array containing two intensity peaks: (a) simulated transversal profile along the propagation direction; (b) binary-amplitude DOE to generate it; (c) experimental result of the transverse distribution at distance $z = 108$ cm from the DOE (distance between peaks—600 µm, peak width at half intensity—50 µm).

source, which destroys the peak constancy. (3) The NDBA can be generated with a single diffractive element with low information content and finite aperture.

Fig. 6.11a shows the simulated transverse profile as a function of z. The corresponding DOE has 128×128 resolution elements and binary-amplitude modulation (Fig. 6.11b). The pixel size was 100×100 µm. As expected, it consists of a ring with some kind of modulation. The transverse intensity distribution photographed in the laboratory at the center of the axial interval is shown in Fig. 6.11c. The ring systems are analogous to those of Bessel beams, but because they are part of the same distribution, they are coupled.

6.7.2   Extinction and Regeneration

Previous studies[53] have shown that when the peak of a NDB is masked, it is regenerated at some distance behind the mask. This peculiar phenomenon is explained

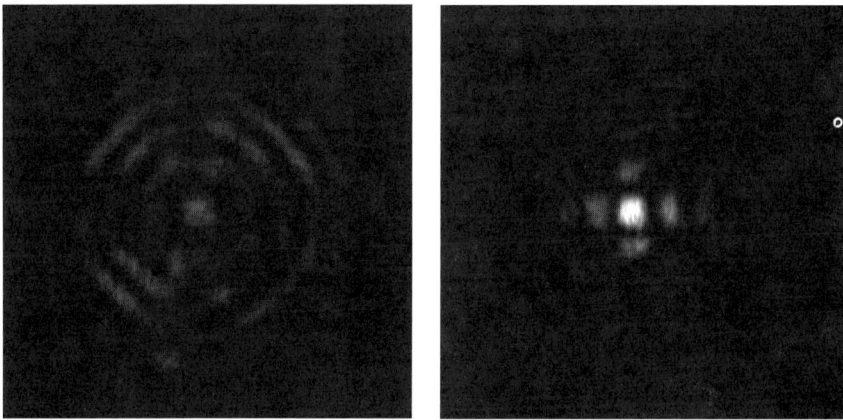

FIGURE 6.12. Dashed nondiffracting beam. (a) Distribution at $z = 135$ cm shows the local ring pattern generated along the valleys. (b) Distribution at $z = 142$ cm shows one of the four axial peaks (peak width at half intensity—54 μm).

by the fact that only a small part of the total energy of the wavefront resides in the peak, whereas the background distribution is responsible for supplying the energy lost by diffraction. However, when a NDB is implemented with a finite aperture, the peak will be extinguished at a finite distance, a natural consequence of energy conservation. It would be interesting to know whether it is possible to control these properties at will. Can we generate "beams" or intensity peaks that appear and disappear at different locations upon propagation in free space? The answer is positive with the limitations that will be discussed below.

In Fig. 6.12, we show experimental results that demonstrate the intensity distribution in a peak and in a valley of what we call a "dashed NDB."[7, 54, 55] This distribution possesses a peak that appears and disappears alternatively along a distance in the axial direction. In the experiment, four intensity peaks from $z = 1$ m to $z = 1.5$ m were obtained. The peak width was again under the classical diffraction limit. The extinction of the central peak is due to diffraction and interference with the lower background intensity. The regeneration is caused by the localized formation of a ring system in the background.

Figure 6.13 shows simulation results for the axial intensity profile in the correct plot, whereas the photos show the experimentally obtained intensity distributions over transverse planes at the marked positions. The intensity peaks are within the diffraction limit with a full-width at half-maximum under 54 μm. This on-axis DOE was implemented with only 128×128 binary amplitude elements with pixel separation 100×100 μm. The input wavefront was a plane wave with $\lambda = 633$ nm.

6.7.3 Helicoidal Beams

Intensity peaks do not necessarily have to move along straight lines. It is possible to conceive patterns that follow any trajectory along their propagation.[54−56] In

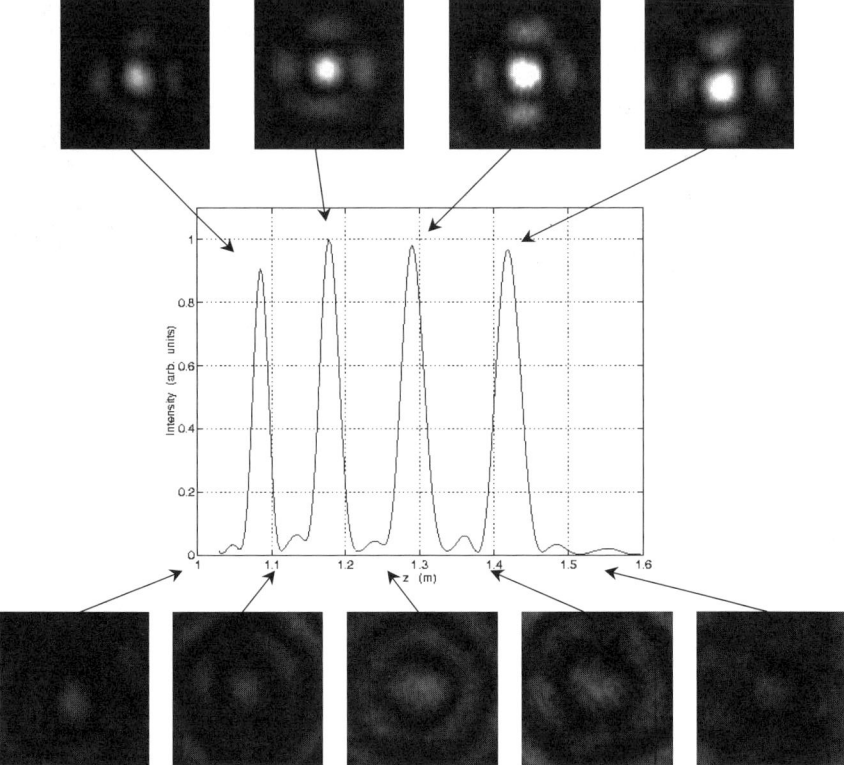

FIGURE 6.13. Axial behavoir of the dashed nondiffracting beam. The upper row shows the peaks of intensity, whereas the lower row shows the dark valleys along the axis.

order to produce such a general behavior, the phase of the wavefronts will play an important role as, for example, through the generation of phase singularities.

It is possible, for example, to generate a distribution with a diffraction-limited intensity peak rotating clockwise around the optical axis for one period in the axial direction. We call this kind of distribution a "helicoidal beam." The amplitude-modulated DOE is shown in Fig. 6.14. In this case, it was necessary to increase the information content and use a matrix of 256×256 resolution elements. It is interesting to note that if we use the definition for a geometric ray as the normal to the wavefront, in such a beam, it will trace a helicoidal trajectory.

6.7.4 Dark Beams

Dark beams are wave fields carrying information in dark regions. Specially tailored dark beams can be synthesized in domains of 3D space to suit required applications. The concept is demonstrated here with the design of dark beams that maintain a constant notch while propagating a predetermined distance. The energy is concentrated around the dark region with negligible sidelobes. These beams are

FIGURE 6.14. Amplitude-only DOE that generates a helicoidal beam.

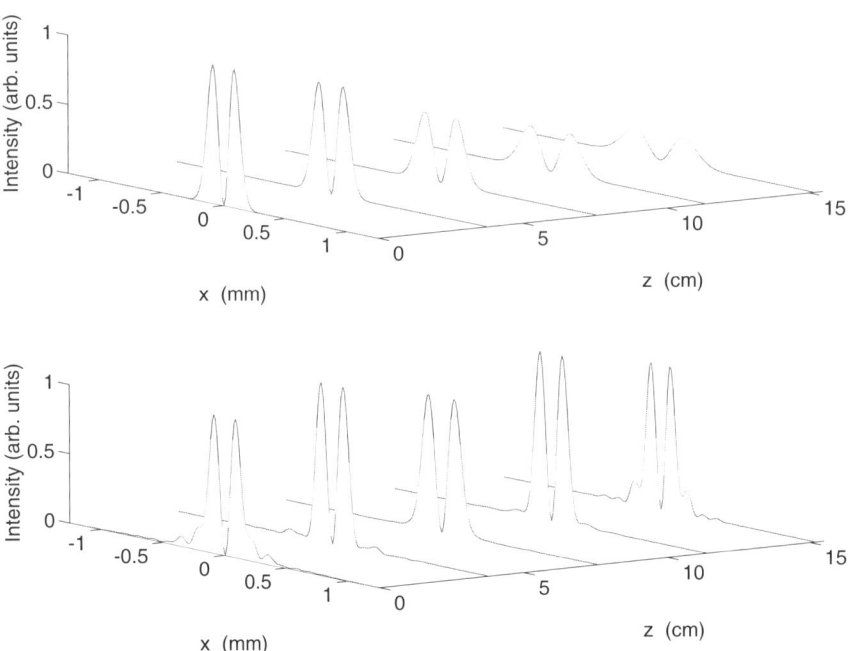

FIGURE 6.15. Comparison between dark region widening in (a) conventional "doughnut" mode and (b) synthesized nonexpanding dark beam with the same initial notch shape.

generated with low information-content, phase-only diffractive elements in on-axis configuration.

Using the present procedure to generate a dark beam, in a specific experiment, constraints were defined that keep a constant shape and size inside the dark region

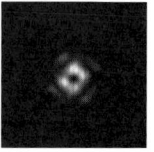

FIGURE 6.16. Nonexpanding dark beam: experimental transverse intensity distributions at distances corresponding to Fig. 6.15.

along an axial distance of 15 cm. A diffractive element was designed with phase-only modulation and low information content (128×128 pixels of size $100 \times 100 \mu m$), and an incident plane wave with $\lambda = 633$ nm. The result is shown in Fig. 6.15, where it is compared with a conventional Gaussian "doughnut" mode.

The experimentally obtained crossections are shown in Fig. 6.16.

6.8   Discussion and Future Prospects

This chapter addressed the issue of information storage from a 3D region of space onto a 2D surface. Fundamental and technological limitations were discussed as well as ways to mitigate them. An important question to ask is, are there other approaches that are able to relax the above limitations? Another question to be addressed is, can volume recording of information solve the problem? Well-known tomographic techniques, widely used for medical imaging and other applications, indicate a positive answer to these questions. However, the solution cannot be in a single reconstructed coherent wavefront as demonstrated in Sections 6.2 and 6.3. Even if that wave front is generated by a volume of stored information, it still is a 2D surface.

As a consequence of the above considerations, the generation of 3D light structures that overcome the limitations discussed in this chapter must rely on the generation of multiple wavefronts, preferably mutually incoherent. If the multiple wavefronts are mutually coherent, interference effects will reduce the amount of useful information, usually to an amount available on a single wavefront. Such multiple wavefronts can be generated in various ways. For example, one could record several DOEs and distribute them in space in a proper way. One such approach is practiced in cylindrical display holograms. Although coherently reconstructed, the observer views a virtual image, and therefore, the various wavefront components diverge and do not overlap to interfere. It would be extermely difficult to generate a real image by illuminating the hologram from outside. Thus, to reconstruct a real image from multiple DOEs, it is necessary to use mutually incohererent illuminating waves.

Multiple wavefronts can be also generated by a thick recording medium. The multiplexing capability of such a medium for holographic recording is well known[57] and the problem, for our purpose, is to properly superpose the various wavefronts. To avoid interference effects, each reconstruction must be implemented with a different source, probably different wavelength, or high-speed time multiplexing can be employed for applications such as displays.

Being a relatively new field of research, the applications of the paradigms described in this chapter are not yet fully realized. Nevertheless, various additional works have appeared,[58−70] and several interesting applications have already been implementeded. Among these applications we may note measurements,[71, 72] 3D lithography,[52] 3D display,[73] and unconventional optical elements with novel design of their point-spread function (PSF).[55, 74, 75]

Specially designed WFs can be utilized in a novel method for surface analysis with resolution beyond the diffraction limit. In this method, the surface is consecutively illuminated by a collection of WFs and the diffracted wave is measured. Adequate postprocessing of the information using space-frequency representations leads to the reconstruction of subwavelength surface features.

Another application worth mentioning here is that of special lithography. Conventional lithographic methods are based on contact printing or high-accuracy imaging. The applicability of these methods is strongly limited by diffraction effects such as resolution and depth of focus. An important example is the area of micro-electromechanics, in which relatively deep structures are fabricated. Among other things, these structures must be electrically connected and combined with electronic devices that require lithographic procedures on nonplanar surfaces. To implement such a lithographic process and mitigate the diffraction effects, light structures within a 3D region of space can be taylored.

The point-spread function of conventional optical systems and components is relatively simple and mainly determined by basic geometrical and material parameters. Employing the methods discussed in this chapter, complex point-spread functions can be engineered and adapted to sophisticated applications in communications, computing, and the high-technology industry.

In conclusion, we may state that 3D light sculpturing reached an advanced stage where various applications are already possible, but many more are expected to emerge as more advanced techniques are developed.

References

[1] D. Gabor, "A new microscopic principle," *Nature*, **161**, 777–778 (1948).

[2] D. Gabor, "Microscopy by reconstructed wavefronts," *Proc. Roy. Soc. A.* **197**, 454 (1949).

[3] D. Gabor, "Microscopy by reconstructed wavefronts: II," *Proc. Phys. Soc. B.* **64**, 449 (1951).

[4] Y.N. Denisyuk, "Photographic reconstruction of the optical properties of an object in its own scattered radiation field," *Sov. Phys. Dok.* **7**, 543 (1962).

[5] E.N. Leith and J. Upatnieks, "Reconstructed wavefronts and communication theory," *J. Opt. Soc. Am.* **52**, 1123 (1962).

[6] E.N. Leith and J. Upatnieks, "Wavefront reconstruction with continuous tone transparencies," *J. Opt. Soc. Am.* **53**, 522 (1963).

[7] R. Piestun, B. Spektor, and J. Shamir, "Wave fields in three dimensions: Analysis and synthesis," *J. Opt. Soc. Am. A* **13**, 1837–1848 (1996). See also R. Piestun, B. Spektor, and J. Shamir, CC PUB No. 107, Department of Electrical Engineering, Technion-I.I.T., Haifa, Israel, June 1995.

[8] O. Bryngdahl and F. Wyrowski, "Digital holography-Computer-generated holograms," in: Progress in Optics, E. Wolf, ed., (North-Holland, Amsterdam, 1990).

[9] H.P. Herzig, ed., *Micro-optics, elements, systems, and applications*, (Taylor & Francis, Hampshire, U.K., 1997).

[10] R. Piestun and J. Shamir, "Control of wave-front propagation with diffractive elements," *Opt. Lett.* **19**, 771–773 (1994).

[11] M. Born and E. Wolf, *Principles of Optics*, 6th ed., (Pergamon Press, Oxford, 1980).

[12] J.W. Goodman. *Introduction to Fourier Optics*, 2nd. ed, (McGraw-Hill, San Francisco, 1996).

[13] J. Shamir, *Optical Systems and Processes* (SPIE Press, New York, 1999).

[14] A.W. Lohmann, "Three-dimensional properties of wave-fields," *Optik* **51**, 105–117 (1978).

[15] A.W. Lohmann, D. Mendlovic, Z. Zalevsky, and G. Shabtay, "The use of Ewald's surfaces in triple correlation optics," *Opt. Comm.*, **144**, 170–172 (1997).

[16] D. Gabor, "Light and information" in *Progress in Optics*, E. Wolf, ed., (North-Holland, Amsterdam, 1971).

[17] J.R. Fineup, "Iterative method applied to image reconstruction and to computer-generated holograms," *Opt. Eng.* **19**, 297–305 (1980).

[18] G. Toraldo di Francia, "Degrees of freedom of an image," *J. Opt. Soc. Am.* **59**, 799–804 (1969).

[19] B.R. Frieden, "Evaluation, design, and extrapolation methods for optical signals, based on use of the prolate functions," in E. Wolf, ed., *Progress in Optics*, IX (North-Holland, Amsterdam, 1971), 311–407.

[20] R. Piestun and D.A.B. Miller, "Electromagnetic degrees of freedom of an optical system," *J. Opt. Soc. Am. A*. V. 17 #5, May 2000, 892–902.

[21] R. Piestun and D.A.B. Miller, "Electromagnetic degrees of freedom in diffractive optics," *European Optical Society Topical Meeting on Diffractive Optics*, EOS Topical Meetings Digest Series **22**, 4, Jena, Germany (August 1999).

[22] P.J.M. van Luarhoven and E.H.L. Aarts, *Simulated Annealing: Theory and Applications* (D. Reidel, Dordrecht, The Netherlands, 1987).

[23] D. Lawrence, *Genetic Algorithm and Simulated Annealing* (Morgan Kaufmann, Los Altos, California, 1987).

[24] U. Mahlab and J. Shamir, "Optical pattern recognition based on convex functions," *J. Opt. Soc. Am. A*, **8**, 1233–1239 (1991).

[25] U. Mahlab and J. Shamir, "Comparison of iterative optimization algorithms for filter generation in optical correlators,"*Appl. Opt.* **31**, 1117–1125, (1992).

[26] P.L. Combettes and H.J. Trussel, "Method of successive projections for finding a common point of sets in metric spaces," *J. Optimization Theory Applicat.* **67**, 487–507 (1990).

[27] D.C. Youla and H. Webb, "Image restoration by the method of convex projections: Part 1–Theory," *IEEE Trans. Med. Imaging* **TMI–1**, 81–94 (1982).

[28] Y. Censor and T. Elfving, "A multiprojection algorithm using Bregman projections in a product space,"*Numerical Algorithms* **8**, 221–239 (1994).

[29] T. Kotzer, N. Cohen and J. Shamir, "Image reconstruction by a novel parallel projection onto constraint set method,"*Opt. Lett.* **20**, 1172–1174 (1995)

[30] T. Kotzer, N. Cohen and J. Shamir, "A projection algorithm for consistent and inconsistent constraints" *SIAM J. Optimization* **7**, 527–546 (1997).

[31] D. Lyszyk and J. Shamir, "Signal processing under uncertain conditions by parallel projections onto fuzzy sets," *JOSA A* **16**, 1602–1611 (1999).

[32] M. Nazarathy and J. Shamir, "Fourier optics described by operator algebra," *J. Opt. Soc. Am. A* **70**, 150–151 (1980).

[33] M. Nazarathy and J. Shamir, "First-order optics-A canonical operator representation-lossless systems," *J. Opt. Soc. Am.* **72**, 356–364 (1982).

[34] M. Nazarathy and J. Shamir, "First-order optics-operator representation for systems with loss or gain," *J. Opt. Soc. Am.* **72**, 1398–1408 (1982).

[35] A.E. Siegman, *Lasers* (University Science Books, Mill Valley, California, 1986).

[36] P.L. Combettes, "The foundations of set theoretic estimation," *Proc. IEEE* **81**, 182–208 (1993).

[37] R. Aharoni and Y. Censor, "Block-iterative projection methods for parallel computation of solutions to convex feasibility problems," *Linear Algebra its Applications* **120**, 165–175 (1989).

[38] L.G. Gubin, B.T. Polyak, and E.V. Raik, "The method of projections for finding the common point of convex sets," *USSR Comput. Math. Math. Phys.* **7**, 1–24 (1967).

[39] R.W. Gerchberg and W.O. Saxton, "A practical algorithm for the determination of phase from image and diffraction plane pictures," *Optik* **35**, 237–246 (1972).

[40] R. Gerchberg, "Super resolution through error energy reduction," *Optica Acta* **21**, 709–720 (1974).

[41] A. Papoulis, "A new algorithm in spectral analysis and bandlimited extrapolation," *IEEE Trans. Circuits Syst.* **CAS-22**, 735–742 (1975).

[42] A.W. Lohmann, and D.P. Paris, "Binary Fraunhofer holograms generated by computer," *Appl. Opt.* **6**, 1739–1774 (1967).

[43] J. Durnin, "Exact solutions for nondiffracting beams. I. The scalar theory," *J. Opt. Soc. Am. A* **4**, 651–654 (1987).

[44] P. Szwaykowski and J. Ojeda-Castaneda, "Nondiffracting beams and the self-imaging phenomenon," *Opt. Comm.* **83**, 1–4 (1991).

[45] G. Indebetouw, "Nondiffracting optical fields: some remarks on their analysis and synthesis," *J. Opt. Soc. Am. A* **6**, 150–152 (1989).

[46] A. Vasara, J. Turunen, and A.T. Friberg, "Realization of general nondiffracting beams with computer-generated holograms," *J. Opt. Soc. Am. A* **6**, 150–152 (1989).

[47] V.P. Koronkevitch and I.G. Palchikova, "Kinoforms with increased depth of focus," *Optik*, **87**, 91–93 (1991).

[48] N. Davidson, A.A. Friesem, and E. Hasman, "Holographic axilens: high resolution and long focal depth," *Opt. Lett.* **16**, 523–525 (1991).

[49] N. Davidson, A.A. Friesem, and E. Hasman, "Efficient formation of nondiffracting beams with uniform intensity along the propagation direction," *Opt. Comm.* **88**, 326–330 (1992).

[50] J. Rosen, B. Salik, and A. Yariv, "Pseudo non-diffracting beam generated by radial harmonic function," *J. Opt. Soc. Am. A* **12**, 2446 (1995).

[51] R. Piestun and J. Shamir, "Generalized propagation invariant wave-fields," *J. Opt. Soc. Am. A*, **15**, 3039–3044 (1998).

[52] R. Piestun, B. Spektor, and J. Shamir, "Pattern generation with extended focal depth," *Appl. Opt.* **37**, 5394–5398 (1998).

[53] R.M. Herman and T.A. Wiggins, "Production and uses of diffractionless beams," *J. Opt. Soc. Am. A* **8**, 932–942 (1991).

[54] R. Piestun, B. Spektor, and J. Shamir, "Diffractive optics for unconventional light distributions," *Photonics West: Optoelectronics and Microoptical Devices*, Feb. 4–10, 1995, San Jose. *Proc. SPIE* **2404**.

[55] R. Piestun, B. Spektor, and J. Shamir, "Unconventional Light Distributions in 3–D domains," *J. Mod. Opt* **33**, 1495–1507 (199).

[56] J. Rosen, and A. Yariv, "Snake beam: aparaxial arbitrary focal line," *Opt. Lett.* **20**, 2042–2044 (1995).

[57] R.J. Collier, C.B. Burckhardt, and L.H. Lin, *Optical Holography* (Academic Press, New York, 1971).

[58] J. Rosen and A. Yariv, "Synthesis of an arbitrary axial field profile by computer-generated holograms," *Opt. Lett.* **19**, 845–847 (1994).

[59] V.V. Kotlyar, S.N. Khonina, and V.A. Soifer, "Iterative calculation of diffractive optical elements focusing into a three-dimensional domain and onto the surface of the body of rotation," *J. Mod. Opt.* **43**, 1509–1524 (1996).

[60] V. Soifer, V. Kotlyar, and L. Doskolovich, *Iterative Methods for Diffractive Optical Elements Computation* (Taylor & Francis, London, 1997).

[61] B. Salik, J. Rosen, and A. Yariv, "One-dimensional beam shaping," *J. Opt. Soc. Am. A* **12**, 1702 (1995).

[62] B.Z. Dong, G.Z. Yang, B.Y. By, and O.K. Ersoy, "Iterative optimization approach for designing an axicon with a long focal depth and high transverse resolution," *J. Opt. Soc. Am. A* **13**, 97–103 (1996).

[63] Y.Y. Schechner, R. Piestun, and J. Shamir, "Wave propagation with rotating intensity distributions," *Phys. Rev. E* 54, R50–R53 (1996).

[64] J.A. Davis, E. Carcole, and D.M. Cottrell, "Nondiffracting interference patterns generated with programmable spatial light modulators," *Appl. Opt.* **35**, 599–602 (1996).

[65] T. Haist, M. Schonleber, and H.J. Tiziani, "Computer-generated holograms from 3D-objects written on twisted-nematic liquid crystal displays," *Opt. Comm.* 299–308 (1997).

[66] R. Liu, B.Z. Dong, G.Z. Yang, and B.Y. By, "Generation of pseudo-nondiffracting beams with use of diffractive phase elements designed by the conjugate-gradient method," *J. Opt. Soc. Am. A* **15**, 144–151 (1998).

[67] R. Tommasini, F. Lowenthal, J.E. Balmer, and H.P. Weber, "Iterative method for phase-amplitude retrieval and its application to the problem of beam-shaping and apodization," *Opt. Comm.* **153**, 339–346 (1998).

[68] H. Stark, Y.Y. Yang, and D. Gurkan, "Factors affecting convergence in the design of diffractive optics by iterative vector-space methods," *J. Opt. Soc. Am.* **16**, 149–159 (1999),

[69] N. Guerineau and J. Primot, "Nondiffracting array generation using an N-wave interferometer," **16**, 293–298 (1999).

[70] U. Levy, D. Mendlovic, Z. Zalevsky, G. Shabtay, and E. Marom, "Iterative algorithm for determining optimal beam profiles in a three-dimensional space,"*Appl. Opt.* **38**, 6732–6736 (1999).

[71] R. Piestun, M. Friedmann, B. Spektor, and J. Shamir, Diffractive elements for surface investigation with light structures," *European Optical Society Topical Meeting on Diffractive Optics (Finland)*, EOS Topical Meetings Digest Series **12**, 82 (1997).

[72] M. Friedmann, L. Piestun, E. Paquet, and J. Shamir, "Surface analysis using multiple coherent beams," *The 19th Convention of the IEEE in Israel, IEEE Proc.*, 537–540 (November 1996).

[73] R. Piestun, J. Shamir, B. Wesskamp, and O. Bryngdahl, "On-axis computer generated holograms for 3–D display," *Opt. Lett*, **122**, 922–924 (1997).

[74] R. Piestun, B. Spektor, and J. Shamir, "On-axis binary-amplitude computer generated holograms," *Opt. Comm.* **136**, 85–92 (1997). See also: R. Piestun, and J. Shamir, CC PUB No. 106, Department of Electrical Engineering, Technion-I.I.T., Haifa, Israel (June 1995).

[75] B. Spektor, R. Piestun, and J. Shamir, "Dark beams with constant notch," *Opt. Lett.* **21**, 456–458 (1996).

Part IV

Applications

7
Solar Holography

Juanita R. Riccobono and Jacques E. Ludman

7.1   Introduction

Solar holograms are those holograms that are designed and used to take solar energy and divert, concentrate, disperse, reflect, or otherwise modify it for a variety of purposes. Most of the applications discussed here will concern high-efficiency holograms used for heat and power generation, lighting, rejection of certain wavelengths, and space heating.

One of the fundamental features of any solar hologram is substantial dispersion, because the bandwidth range is from 0.4 microns to 0.7 microns for the visible portion of the spectrum, from 0.4 microns to 1.1 microns for photovoltaic applications, or out to 2.0 microns for infrared (IR) utilization. A typical solar hologram diverts incoming solar energy through a substantial angle ($20°$ to $150°$). In addition, the hologram may be designed to not only divert, but also to concentrate the light, or to disperse it over an area. Because of the demand for broadband operation, relatively high-efficiency materials are used so that the hologram may be extremely thin and still have high efficiency. For a bandwidth that covers the visible spectrum, from 0.4 microns to 0.7 microns, the film thickness must be greater than 1 micron or else the holograms are not sufficiently broadband. There are materials that are capable of the appropriate performance with thin layers, and they include dichromated gelatin (DCG) and photopolymers such as those made by DuPont (DuPont Holographic Materials, Wilmington, DE) or Polaroid (Cambridge, MA).

There are a variety of applications for such solar holograms, and they include:

(1) **Solar Power Generation**: There are a variety of different types. Northeast Photosciences (Hollis, NH) pioneered the use of concentrated solar holograms to

deliver focused light that is chromatically dispersed so that two or more photovoltaic cells can be placed at appropriate locations such that their bandgap is matched to the appropriate region of the spectrum. Other work includes the design of solar holograms to match quantum dot solar cell technology so that each portion of the dispersed and concentrated solar spectrum is matched to the ideal photocell bandgap.

(2) **Solar Lighting**: This is accomplished by designing a window or aperture holographic system that diverts more than 50% of the incident solar illumination in from the window and up onto the interior ceiling. If the ceiling is white and reflecting, the light will be distributed within the interior space and provide an alternative source of lighting rather than the use of electric lighting, which introduces significant unwanted heat.

(3) **Solar Heating**: Solar holograms can be designed so that substantial portions of the incoming infrared energy are diverted away from an interior living or office area to heat-storing devices. Such holograms do not impede vision through the aperture and allow the interior to store heat for the night. In addition, holograms can be designed so that they perform both a lighting function, as well as diverting infrared energy into a room. Such holograms can be designed so that depending on the altitude of the sun, they divert the heat inside or outside the building or structure. Typically, during the height of the sun (summer), the diversion will be to the exterior, and when the sun is at a lower angle, the diversion of the infrared will be into the interior where it is needed.

(4) **Solar Rejection of Certain Wavelengths**: Holograms can be designed to reflect IR or ultraviolet (UV) radiation all or part of the day or year. In this way, heat and damaging UV radiation can be kept out of an interior office or living space. Alternatively, the heat or UV radiation can be concentrated and diverted elsewhere for use in infrared heating systems or UV chemistry and detoxification reactions.

(5) **Reduction of Reflection Losses of Certain Wavelengths**: Holograms can be used to reduce reflection losses of high refractive index materials like solar cells and solar absorbers so that more useful energy is available to these devices.

7.2   Fabrication

There are many materials that may be used to fabricate holograms. DCG and photopolymers (made by DuPont[1-3] and Polaroid[4-6]) are typically used in solar holography because these materials can have high efficiency in thin layers. DCG can be fabricated according to the method of Saxby[7] and McGrew,[8] whereas photopolymers are used "off-the-shelf." In either case, the material must be exposed to the appropriate wavelength and power. After exposure, the material must be processed. DCG requires wet processing similar to that used in photography, whereas photopolymers typically require baking followed by UV-cure.

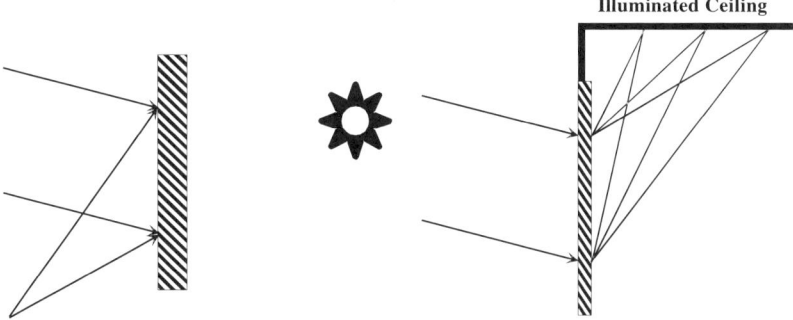

FIGURE 7.1. (a) Fabrication arrangement and (b) playback of holograms used for lighting applications.

The optical setup used to record the hologram depends on the end-use of the hologram. The hologram can be designed to redirect, concentrate, or block incident light or energy. The light or energy may be directed to a position horizontal or vertical to the plane of the hologram. It may also be separated by wavelength such that a range of wavelengths goes to one position while another range goes to another position. The diffracted light may remain diffuse for lighting applications or be focused to a spot. In addition, the hologram may reflect the incident light or infrared heating component.

Typical recording geometries for solar holograms place two beams at some angle less than 45° to the normal of the holographic plate. For lighting applications, the hologram may be designed to passively track the sun from morning to evening, summer to winter. In this case (Fig. 7.1), two diverging beams are incident to the unexposed holographic plate, one at an angle of 40° to the normal while the other beam is 10° to the normal. The point source 10° from the normal is equivalent to the sun and must be kept at an "infinite" distance from the unexposed holographic plate. The second point source, at 40° to the normal, is placed at a distance approximately equal to 1.5 times the length of the holographic plate.

Holograms may also be designed to concentrate sunlight onto a surface. In this case (Fig. 7.2), one of the beams incident to the unexposed holographic plate is collimated, whereas the other beam diverges. The angle between these two beams varies according to the application. For example, the hologram can be used to concentrate sunlight onto solar cells. Here, the collimated beam that represents the sun is incident normal to the unexposed holographic plate. The diverging beam is positioned at some convenient angle to the normal and at a convenient distance from the unexposed plate. The placement of the diverging beam is important because it determines where the solar cells will be placed in the final system.

Holograms can also be designed to block or reflect certain wavelengths. Typically, these holograms reflect incident heat in summer, but divert it into an architectural space during winter (Fig. 7.3). This relies on the position of the sun because it is at a high altitude in the summer and a low altitude in winter. To accomplish this, a reflection setup is used where the unexposed holographic plate is

FIGURE 7.2. (a) Fabrication arrangement and (b) playback of concentrating holograms used for power generation.

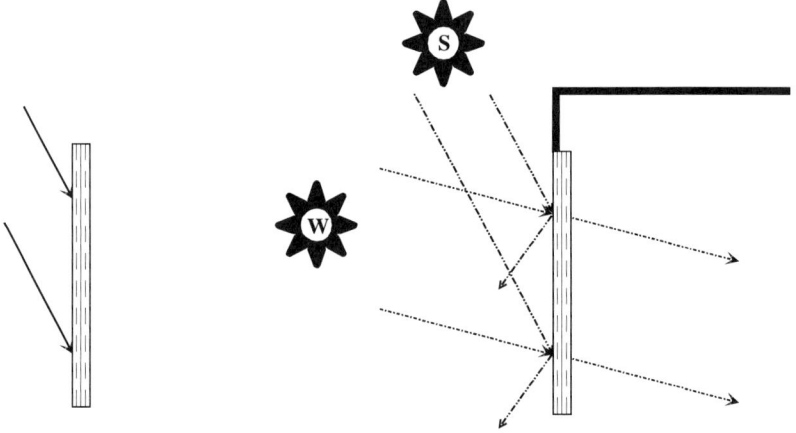

FIGURE 7.3. (a) Fabrication arrangement and (b) playback of thermally blocking holograms in summer and winter.

optically matched to a mirror. A collimated beam, incident to the holographic plate at a steep angle, reflects back on itself creating coplanar interference fringes. Now, during summer when the thermal energy of the sun is incident to the hologram at an angle greater than or equal to the angle at which it was written, the energy is reflected off the surface. During winter, the thermal energy passes through the hologram because it is incident at an angle less than the writing angle.

7.3   Applications

Solar holograms are used in numerous applications. Light in the visible or IR spectral regions can be used actively for power generation or passively for daylighting and heating. In addition, the exclusion or reflection of certain wavelengths or regions of the solar spectrum is also very important. Indeed, holograms are well suited to this task.

Holographic concentrators can be used to concentrate useful solar radiation onto solar cells. The solar cells generate electrical energy from the light energy. These systems are found both in space and on earth. Thermal power generation may also be a consequence of holographic concentrators.

Daylighting holograms can be used to diffract sunlight up onto a ceiling, deep in a room, without diffracting the light at eye-level. These holograms are broadband and may be able to passively track the movement of the sun across the sky, throughout the day and year. Therefore, the diverted visible light remains relatively stationary on the ceiling as the angle of the sun changes with the hour and the season. In addition, quality daylighting holograms do not distort the view to the outside.

Thermal holograms may be used to concentrate infrared heat energy. This heat energy may be used directly for heating, stored for later use, or converted into electricity for power generation. This class of hologram may also be used on glazing systems, in conjunction with daylighting holograms, to bring infrared heat into an architectural space during winter and to reflect it during summer.

Another type of hologram can be used to perform chemical reactions and to detoxify water. These holograms reflect and concentrate portions of the high-energy ultraviolet spectrum. Light in this wavelength range causes changes in electronic states and results in photochemical reactions and damage to DNA and microorganisms.

Finally, antireflection holograms can be used to reduce reflection losses of portions of the solar spectrum off a surface. These holograms are typically periodic surface structures and can be made relatively broadband. They are used on high refractive index materials such as solar absorber materials and some types of solar cells.

These applications will be discussed in more detail in the following sections:

7.4. Photovoltaic Concentration

   7.4.1. Space
   7.4.2. Terrestrial

7.5. Daylighting

   7.5.1. Commercial and Residential
   7.5.2. Greenhouses
   7.5.3. Other Applications

7.6. Thermal

   7.6.1. Heat and Power Generation
   7.6.2. Architectural Applications
   7.6.3. Thermal Photovoltaic Power Generation

7.7. Solar Chemistry and Detoxification
7.8. Antireflection

7.4   Photovoltaic Concentration

Solar cells convert incident solar radiation to electrical power. Flat panel arrays absorb the radiation directly and convert it into electrical energy with an efficiency of 10% to 15%. Tracking the sun, concentrating the incoming radiation, and spectrally splitting the radiation increases solar cell efficiency. Although tracking increases the efficiency, it requires additional technology, at significant cost, for terrestrial applications but is inherent in orbiting satellites. Concentrating the incoming radiation increases the efficiency by 10%, for an overall efficiency of 20% to25%.[9, 10] Splitting the spectrum into its component wavelengths can add an additional 10%, for a total of 30% to 35%, to the efficiency.[9]

Holography has many advantages over conventional methods of concentration and spectral splitting. Holographic concentrators are diffractive elements that inherently disperse incident solar radiation. Thus, one holographic element can both concentrate and spectrally split incoming radiation onto appropriate solar cells. Conventional methods require two elements, typically, a Fresnel lens and a dichroic mirror, diffraction grating or prism, or a Fresnel lens and a multijunction solar cell; in these cases, overall system efficiency suffers high losses. Holographic solar concentrators are able to achieve high diffraction efficiency, are lightweight, and are easily replicated. The materials typically used for holographic concentrators are of high optical quality and have a large bandwidth so that they operate over the visible portion of the spectrum.

Thus, it is no wonder that groups throughout the world are working on holographic concentrating systems for photovoltaic power systems. Groups in Europe, Asia, and North America are developing systems for use in space and on earth.

Most groups use transmission holograms[11–31], although reflection holograms[11–13, 15, 32–34] can be used. Reflection holograms generally have high diffraction efficiency, although they are not very broadband, unless several centered at different wavelengths are multiplexed together. The geometry necessary for a reflection-based holographic concentrator–photovoltaic system is complex.[11, 12, 15, 32] In addition, the swelling and shrinkage of the holographic emulsion during processing greatly affects the reconstruction angle and bandwidth of reflection holograms.[15, 35, 36]

Transmission holograms also achieve high diffraction efficiency, and they are broadband.[11, 13, 19–27, 37, 38] Although they suffer from chromatic aberrations, these aberrations can be minimized with a suitable system-geometry. Some groups use an off-axis geometry[12, 13, 15, 19–24, 37, 38] (Fig. 7.4a) that focuses a spectrally dispersed line at some angle to the hologram, whereas other groups use an in-line geometry[11, 16–18, 25, 27, 28, 30, 31] (Fig. 7.4b and c) that focuses and spectrally disperses sunlight either along the optical axis of the hologram or parallel to the hologram. Although both types of systems operate over a large bandwidth, the in-line system reduces chromatic aberrations of transmission holograms, but has a lower efficiency than does the off-axis system.[25–28] The off-axis geometry isolates the solar cells from most of the infrared heat so that all incident light is confined to the appropriate cell, which converts light to electricity. Cooling is a requirement

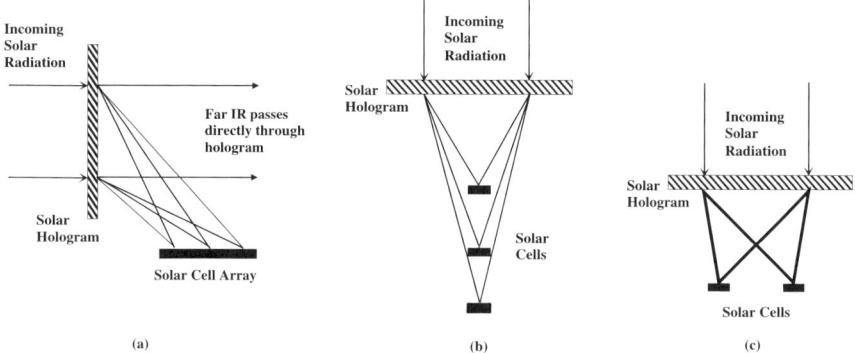

FIGURE 7.4. Holograms can be used to disperse the solar spectrum and focus it onto appropriate solar cells. An (a) off-axis or (b, c) in-line geometry with cells either positioned along the (b) optical axis or (c) parallel to the hologram can be used.

with the in-line geometry because as much as 50% of the incident sunlight is converted to heat.[17, 28, 30] When solar cells are arranged along the optical axis of an in-line system, individual cells are positioned accordingly or a multijunction cell is used. In either case, there is shading of successive photovoltaic cells. Neither the off-axis system nor the in-line system with cells parallel to the hologram suffers shading losses.

Virtually any holographic system that concentrates and disperses the spectrum requires tracking of the sun from morning to evening. The reason for this, of course, is that the incident beam radiation must remain perpendicular to the hologram, and the hologram and solar cells must maintain their relative orientation. Alignment of the solar cells must be precisely positioned along the dispersed spectral line.

Concentration values for these holographic systems vary. They are dependent on the geometry of the system, the tracking tolerance of the system, and the chromatic aberrations. Values from 1 to 200 have been calculated,[11, 13, 16, 19, 28, 30–32, 34] whereas experimentally determined values are 50 for different transmission systems[19, 28] and 65 for a reflection-based system that uses multiple holograms.[32]

Many different solar cell combinations are possible, of course. Silicon and gallium arsenide are typically the solar cells used in the laboratory because of their availability. However, they are not necessarily the optimum cells for these holographic concentrating systems because Si has a bandgap of 1110 nm and GaAs has a bandgap of 890 nm. Other cells of interest include AlGaAs, CdS, GaSb, InGaAs, GaInP, and the cascade cell GaAs- InGaAs. Theoretically, any number of different cells can be used; however, from a cost-benefit perspective, only two or three different cells achieve a substantial improvement in efficiency.

In the future, however, quantum dot solar cell technology[87, 88] may offer a high-efficiency, low-cost solution. Silicon quantum dots are fabricated by pulsed laser ablation and are assembled in a linear array with a size gradient parallel to the surface of the substrate that supports the quantum dots. This size-graded structure allows smaller quantum dots to absorb photons at shorter wavelengths, whereas comparatively larger dots absorb photons at both longer and shorter wavelengths. Thus, quantum dots operate over a large bandgap range, from 1.1 eV to 4.1 eV (300 nm to 1130 nm), and they are able to capture approximately 80% of the sun's emitted photons to generate carriers.

7.4.1 Space

There are numerous benefits to using these holographic–photovoltaic systems in space because they are lightweight and have high efficiency. In addition, they are particularly suited to orbiting satellites because satellites inherently track the sun. There are, of course, a number of considerations for space-based systems. The holograms, holographic substrate, adhesive, solar cells, and support structure or frame must all be able to survive the environment of space. They must also maintain their shape, maintain their optical properties, survive the stresses of launch, and remain operative for a long period of time without any maintenance. The entire structure must be carried within the spacecraft and deployed once in orbit.

The environments encountered in space vary widely, depending on the altitude and orbit of the spacecraft.[39] Spacecraft orbiting the earth at high altitudes may encounter radiation doses that exceed 10^7 rad/day; however, this dosage affects only the top microns of material. This is not important to the hologram because it is sealed between two sheets of the substrate material. If the hologram was exposed, these dose levels would be very important. There are materials, such as DCG and Polaroid DMP128, which have been tested and are considered space-resistant.[40, 41] Candidate substrate materials, such as Mylar and Kapton, have excellent radiation resistance and lifetime, as well.[39] Because damage by charged particles and UV radiation is generally limited to the surface layer of the support structure material, it does not affect bulk material properties. Support structures can be made from polyvinyl carbazole (PVK) and filled phenolics, as well as from thermoplastics such as PET and PBT.

Photovoltaic cells have been used in space for at least 30 years and have proven to be very dependable. They are the primary power source for most, if not all, U.S. Department of Defense and NASA satellites.[42] Consequently, there is much information on the radiation effects on solar cells and different types of cells used in space.[43–45]

Obviously, care must be taken in the choice of materials for space applications. There are, however, ample candidate materials with long lifetimes and high stability in space that are suitable for holographic materials, substrate, protective covering, solar cells, and support frame.

7.4.2 Terrestrial

There are a number of features of these holographic concentrating and spectral-splitting systems that make them very attractive for earth-based systems. These advantages include use of only one optical element to both concentrate and disperse the spectrum, decreased Fresnel reflection losses, high concentration and diffraction efficiency, and, in the case of the off-axis holographic systems, low resistance and shading losses on solar cells and avoidance of unwanted IR heating.

These holographic systems require solar tracking; however, the requirements for the off-axis systems are generally much less stringent in one dimension than for the in-line concentrating systems. Because the holograms focus to a long thin line, tracking in the direction that shifts the line along itself is only moderately critical: All of the light stays on the solar cells; only the location of the junction between the solar cells is slightly changed. Tracking tolerances in the other direction are as critical as for any concentrating system because light can miss the solar cells.

Conventional terrestrial systems are either flat plate collectors or use Fresnel lenses to concentrate the entire spectrum onto the solar cells. Currently, there are no commercial systems that both concentrate and disperse the spectrum. There are significant advantages to both splitting and concentrating the spectrum, as previously discussed. Economic and environmental considerations also point to the need for these systems because they could cost as little as 5.7 cents per Kilowatt hour, and show an improvement of 96.6% in efficiency over conventional Fresnel concentrating systems.[21]

7.5   Daylighting

Daylighting is an effective means of reducing both lighting and cooling costs in buildings. Most conventional daylighting techniques, such as louvers, blinds, mirrors, and prisms, obstruct the view to the outside, are heavy or bulky, and require maintenance and adjustment for optimal performance.[46-50] In addition, most are unable to track the sunlight or to deliver light deep into the interior of a room. Some, such as tracking mirror systems coupled with infrared rejection coatings, are expensive. Daylighting holograms, however, provide an efficient and effective method for the even distribution of sunlight within a room.[47, 51-58] Holograms may be used to diffract light up onto the ceiling, deep in a room, without diffracting the light at eye-level. The holograms have a broad bandwidth and are able to passively track the movement of the sun across the sky, throughout the day and year.[59]

Holographic daylighting systems have been developed to diffract sunlight in large quantities up onto the ceiling, deep in the interior of a room or building with high efficiency.[47, 51-53, 55-57, 60, 65] This diffracted light remains relatively stationary on the ceiling as the angle of the sun changes with the hour and the season. It is the visible portion of the sunlight that is diverted because the holograms possess a broad bandwidth. Such holographic daylighting systems have a very long life when made with appropriate materials. They may be used to retrofit existing windows or be contained in a glazing system.[51, 53, 55, 56, 58] Quality holograms that divert sunlight to the interior ceiling have an additional feature: The view through them to the outside is relatively undistorted, although it may be somewhat dimmer.[52]

7.5.1   Commercial and Residential

There is a significant amount of energy used in buildings in the United States. In 1950, 27% of all energy used was in buildings. In 1970, that percentage grew to 33%, and by 1990, it reached 36%. Currently, more than 30 quadrillion British Thermal Units (quads) of energy are used in buildings alone. It is estimated that this number will increase, assuming a moderate exponential growth rate, to about 42 by the year 2015. Sixty percent of all electricity and 40% of all natural gas used in the United States are in buildings, and 41% of the electricity used in commercial buildings is for lighting. Commercial buildings currently use about 13 quads of energy, in which lighting accounts for about 28% of the total energy use or 3.61 quads and space cooling accounts for about 16% or 2.1 quads.[61]

Large commercial buildings typically require lots of lights because daylight cannot penetrate deeply into the building's interior. This large lighting load produces heat that results in an increase in the cooling load. In a typical large office building in San Francisco, for example, 55% of the annual cooling load is due to the lighting.[61] Thus, if the lighting load was reduced, the cooling load would also be reduced.

Use of holographic daylighting technology can result in significant savings in energy; the lighting load alone can be reduced between 30% and 70% because a 50% efficient daylighting hologram delivers the equivalent of ten 75-W light bulbs or 12,000 lumens.[53] In addition, there are many ergonomic benefits to using natural sunlight, including increased concentration and performance and decreased incidence of seasonal-affected disorder.

Some of the key features of daylighting holograms are that they have a long lifetime, are inexpensive, and are transparent and deliver light deep into the interior of a building. Passive tracking daylighting holograms are preferred because they distribute the incident visible light onto the ceiling from morning to evening throughout the year. These holograms may be installed directly in a double-paned window glazing system or laminated to glass or plastic.

The shortcoming of daylighting holograms is the separation or dispersion of the different wavelengths or colors of sunlight, which creates a "rainbow effect" on the ceiling. This chromatic effect is visible only on the ceiling because as light is reflected and diffused by the white matte-finish of the ceiling, white light is transmitted into the workspace. Although this chromatic effect is inherent to the design of these daylighting holograms, several methods have been used to reduce these effects, including tiling[53, 54] and the use of multiple holograms or gratings.[29, 47, 52, 55–56, 58, 62] In each case, the result is the same. There is mixing of the light such that there is a large region of white light in the center of the ceiling with a reddish fringe near the window and a bluish fringe deep in the room.

7.5.2   Greenhouses

Holograms similar to daylighting holograms can be used in greenhouses in regions with a temperate climate. Greenhouses are not popular in northern regions during cold months because large amounts of heat, at significant expense, are needed to keep plants healthy. By insulating the north side of the greenhouse and applying holograms to the south-facing windows to divert light into the northern half of the greenhouse, heating costs can be cut in half.[63] This is clearly advantageous because locally grown produce can be distributed rapidly without sacrificing taste and quality.

In an experiment, holograms were installed on the south face of a greenhouse, whereas the north face was insulated.[63] The holograms were very efficient and effective at diffracting sunlight from the south side of the building to the north side, which allowed plants to be grown throughout the greenhouse. After three months of growing, from January 1 to March 30, heating costs for the holographic greenhouse with insulation were cut in half, compared with the heating costs for a traditional greenhouse without insulation. Plant yield was 90%, compared with the yield in the uninsulated greenhouse.

Although plants require a certain amount of sunlight every day, they do not necessarily require the entire visible spectrum. Plants may be illuminated by different wavelengths of light at different times of the day and by light of variable intensity throughout the day. The integrated total of the light received by the plants over the course of a day is the important factor. Thus, chromatic effects are not detrimental to plants so long as they receive sufficient amounts of red and blue wavelengths throughout the day.

7.5.3   Other Applications

In addition to daylighting holograms, holograms can also be used as light guides and as passive shading devices.[64–66] Holograms used in the renovation of a building guide light from a skylight down two and a half stories into the interior of the building.[64] They can also be used on skylights to distribute light more evenly in a room.[65] Stacks of reflection holograms are designed to transmit 80% of sunlight in winter but only 20% in summer.[29, 62]

Obviously, the potential applications of holograms for daylighting, greenhouses, solar rooms, and other architectural purposes are numerous and significant. The fact that use of daylighting holograms can lead to significant energy savings because both lighting and cooling loads are reduced, and that insulation of half a commercial greenhouse to R10 represents at least a 45% savings of energy costs, is noteworthy.[63]

7.6   Thermal

7.6.1   Heat and Power Generation

The heat from holographic solar concentrators can be used to generate electricity or heat a fluid. The holographic solar concentrators act like lenses or mirrors by focusing thermal energy onto an absorber or collector system with a working fluid.

The temperature difference can then be used to generate power or simply to heat the fluid.

Both transmission[17, 67] and reflection[68–70] holograms can be used. Obviously, the transmission holograms act like lenses and the reflection holograms act like mirrors, but the reflection holograms require a broad bandwidth to reflect sufficient amounts of thermal energy. Of course, stacks of holograms or multiple holograms may be used to achieve this. Transmission holograms, however, have the benefit of simply transmitting infrared energy to an absorption or collection system.

Stacks of reflection holograms operate most effectively when each hologram has a large, yet distinct, bandwidth that does not overlap that of the adjacent hologram because overlap leads to cross-talk. Stacks with three holograms covering the wavelength range from 425 to 1075 nm had an efficiency of 51%, with an 18% loss due to stacking.[68, 69] Stacks of five holograms covering the range from 350 to 1800 nm were proposed because they should yield an efficiency of better than 70%.[68, 69] These holographic stacks can be focused to a line for use with a parabolic trough or focused to a point for use with a parabolic dish.

A hybrid collector can also be used to generate electricity.[17] Here, holographic concentrators generate electricity via photovoltaic cells (Fig. 7.4c), as discussed above. The excess thermal energy that decreases efficiency of photovoltaic cells is collected with an absorber plate beneath the cells that heats a fluid to 100°C. This system has an electrical efficiency of 22% and a thermal efficiency of 35%.[17]

A similar system has been proposed for use in space on a satellite.[71] In this case, electricity is generated with an off-axis holographic concentrating system (Fig. 7.4a), as discussed earlier. Thermal energy that passes directly through the hologram heats a fluid contained in pipes or vessels. As the fluid is heated, it expands and performs work in the satellite. Because the satellite is mobile, it can be maneuvered so that specific concentrating units face the sun, thus, generating electricity and heating fluids, whereas other units are situated on the dark side, where electricity is not generated and the fluid remains cold and contracted and does not perform work.

7.6.2 Architectural Applications

The rejection of unwanted radiation is also important for solar applications. The heating of windows by infrared radiation is undesirable due to the additional heat load and subsequent cost, whereas exposure to UV radiation damages fabrics, upholstery, and carpeting. Holograms are an obvious choice to reflect portions of the spectrum in order to reduce the heat load or reduce UV damage. Ideally, only the visible portion of the solar spectrum should penetrate windows, at least in hot climates or in the summer, and it may be preferable to reduce the amount of transmitted visible light. Therefore, the holograms should be broadband to reflect large portions of the spectrum, and they should be sensitive to the angle of the sun. Several methods have been proposed to reflect unwanted radiation off windows. Each method uses reflection holograms made with DCG; however, each method is designed for a different region of the spectrum and for a different purpose.

Broadband Lippmann holograms have been fabricated that reflect 43.5% of the incident energy in the wavelength range 760–1160 nm.[72] A total of 85.3% of the incident visible spectrum is still transmitted through these holograms. These holograms are written in thin film DCG, but care must be taken during processing to get broadband holograms and to reduce unwanted harmonics that reduce transmission and result in tinting.

Broadband Bragg–Lippmann holograms that reflect radiation in the near-IR and near-UV regions have been proposed.[73, 74] In this case, three holograms, each with a bandwidth of less than 500 nm, are used together. One hologram reflects the near-UV, whereas the other two reflect portions of the near-IR. A Bragg–Lippmann hologram made in DCG with a bandwidth of 500 nm (600 nm to 1100 nm) was produced that reflected more than 80% of incident radiation.

A holographic system that reflects incident radiation in the visible and near-IR during the summer, when sunlight is incident at an angle of 50° to 60°, has also been proposed.[62] This system transmits 80% of incident radiation during the winter (incidence angle, 20°–30°) and blocks 80% during the summer. Two reflection holograms, one red and the other blue, are pasted onto glass slats. Several of these slats are put together to act as venetian blinds. The holographic slats are adjustable, as are blinds, with respect to the window and incoming radiation.

Another proposed system (discussed in Section 7.2) uses holograms to reflect infrared radiation off a window in the summer, when the altitude of the sun is high, and to transmit the infrared in the winter when the sun is at a low angle.[75] In this way, the need for cooling is reduced in the summer, but the sun's thermal rays still penetrate the window in the winter. These holograms are broadband reflection holograms, which operate from 800 to 1200 nm. They are fabricated by the interference of two collimated beams separated by a fixed angle. These holograms could be multiplexed to the daylighting holograms,[53] which operate in the visible region, described above.

7.6.3 Thermal Photovoltaic Power Generation

Long wavelength radiation from combustion, solar, nuclear, or chemical sources can be used to generate electricity directly with low bandgap photovoltaic cells such as GaSb. The thermal source emits photons with different amounts of energy. Those photons that are matched to the bandgap of the photovoltaic cells can be focused onto the cells with holograms.[76–78] Photons that are not matched to the cells' bandgap are directed elsewhere by the holograms. In this way, the holograms optimize the direction and distribution of energy bands onto the properly matched photovoltaic cell, thereby increasing the overall efficiency. These systems are designed to be lightweight and portable, for use in remote locations, disaster relief, military, and space missions.

A demonstration system was set up using a halogen lamp as the thermal source and Si, GaAs, and GaSb as photovoltaic cells.[76–78] A parabolic reflector was placed around the source with an off-axis holographic concentrator positioned at the opening of the parabola. In this way, the radiation incident to the hologram was collimated. Solar cells were placed in appropriate positions relative to the hologram.

The results of this experiment clearly indicate that it is possible to use a thermal source with holographic concentrators and long wavelength photovoltaic cells.

7.7   Solar Chemistry and Detoxification

Short-wavelength, high-energy UV light is used in photochemical reactions. Light in the wavelength range 200 to 400 nm causes changes in electronic states, particularly for double and triple bonds between carbon and between carbon and oxygen. These electronic transitions can result in the breaking of bonds, formation of radicals, formation of new chemical species, as well as damage to DNA and mircoorganisms. Use of a very narrow band of UV light can aid in the synthesis of fine chemicals. A high concentration of UV can kill microorganisms and detoxify water. Reflection holograms operating in the UV have been made for these applications.

Narrowband reflection holograms were made to operate in the UV.[79–82] Holograms with a bandwidth of 80 nm centered at about 425 nm were fabricated to behave as parabolic mirrors with a focal length of 100 cm.[79] Reflection holograms with a bandwidth of 60 nm centered at about 370 nm, and with a diffraction efficiency of 80%, were fabricated in DCG at 488 nm.[80]

A holographic concentrator was designed to detoxify water at an operating wavelength of about 380 nm.[81] A 1-m^2 point focus concentrator was built using 24 holograms. The holograms were each 20 cm × 20 cm, with a bandwidth of about 100 nm, and provided 2.1 W of power. The combined 24 holograms gave a 1.75-cm^2 spot with a peak temperature of 200°C. Holograms located along the edges of the system had non- negligible spherical aberrations. In addition, chromatic aberrations of the holograms reduced the concentration ratio. Therefore, future work includes reducing the bandwidth of the holograms to increase the concentration ratio of the system and developing a line focus concentrator.

Cylindrical holographic mirrors were designed for use with the sensitizer zinc tetraphenylporphorine, which operates at 420 nm.[82] Large format holographic mirrors, 20 cm × 50 cm, were recorded in DCG coated on flexible polyester. After development, the holograms were bonded with UV-curing polymer to a glass cylinder with a focal length of 22.5 cm. These cylindrical holographic mirrors reflected light to a line focus with a concentration ratio of about 15. They had a bandwidth of 40 nm centered at 420 nm and an effective efficiency of 75%. Processing of these holograms could be varied to shift their operating wavelength and their bandwidth.[82, 83]

7.8   Antireflection

Solar absorbers operate best when all incident radiation or radiation over a certain wavelength range is absorbed. Solar cells operate most effectively when the incident radiation matches the bandgap of the cells. Typically, these solar materials have a high refractive index, which leads to considerable reflection. Reduction of these reflection losses greatly increases the efficiency of the absorber or cell; however, antireflection coatings made by stacking thin films of differing refractive index are costly and difficult with rough surfaces. Several methods for producing holographic antireflection gratings have been tested for use with these materials.

Broadband antireflection coatings can be made by microstructuring the surface of an absorber or solar cell. Periodic holographic double gratings or moth eyes can be created in the surface of a rough-textured material by sputtering a photoresist film onto the surface. The photoresist must then be exposed twice in orthogonal positions to interference fringes generated by laser beams. The period of the interference fringes is varied with the laser source and the angle of the incident beams. Structures with heights of 0.1 microns to 0.4 microns and with periods of 0.3 microns and 0.7 microns were produced at a wavelength of 458 nm.[84] These structures were formed on seven different samples. Although all samples showed reduced reflection in the IR, six showed increased absorption in the visible. Surface roughness strongly affected results in the visible range.

Relief grating microstructures with a high spatial frequency and aspect ratio can be engraved on n-InP solar cells with photoelectrochemical etching (PEC). The interference pattern is etched directly onto the surface of the cells. The resulting holographic diffraction grating has a triangular or u-shaped geometry, depending on the orientation of the cell crystal. The microstructures formed have a high aspect ratio with the average period on the order of 1 micron, much smaller than the wavelength of light. Reflection losses can be reduced by a factor of 10 over the visible region of the spectrum.[85]

Surface-relief gratings for solar receiver covers can be replicated using holography. In this case, the holographic master is produced in photoresist. The master is then reproduced in nickel, to act as a stamp in the embossing process. The nickel stamp replicates the surface-relief grating on glass coated with a thick film. Transmittance values for these coated glass covers were 96% to 97% at normal incidence and ranged from 95.4% up to 98.7% ($\lambda = 575$ nm) in the visible.[86]

7.9 Conclusion and Future Prospects

Solar holograms can be made to effectively perform a wide variety of tasks. They can be fabricated to operate over a large spectral bandwidth with high efficiency. In these days of rising fuel oil prices and noticeable effects of greenhouse gases, alternative heat and power-generating and lighting technologies are important. The need for these technologies is even more apparent now that energy companies are looking into distributed power systems. In addition, there is an obvious need for solar detoxification and antireflection technologies, and the demand will only grow.

Thus, it is important to note a demonstration project at the International Garden Exhibition in Stuttgart, Germany.[65, 66] Here, three houses utilize light-directing holograms for both daylighting and photovoltaic power generation. These houses have large, sloped glazing systems that form the roof. The highest section of the roof glazing system is lined with strips of solar cells. Gaps between the solar cell strips allow transmission of 35% of the incident radiation into the house while the remainder is converted to electricity. The remainder of the roof is covered with large, moveable glazing units that have holograms attached to the inside of the outer glazing surface and strips of solar cells that cover about half of the inner glazing surface. The holograms direct about 94% of the incident beam radiation onto the solar cells, at a concentration ratio of about 2.5. Gaps between the solar cells allow for the passage of about 45% of the diffuse daylight and a view to the outside.

The holograms used in this project to provide light to solar cells do not spectrally disperse the incident radiation, and they have a low concentration ratio. Also, they are not specifically daylighting holograms in that they do not direct light deep into the interior of the room, rather they, in conjunction with the entire glazing system, act as shading devices. However, the estimated total solar electricity gain for the three buildings in this project is 5000 kWh/a, and the solar heat gain from direct radiation is reduced to 20% for each of the buildings.

This is a significant achievement and engineering feat. It also serves as a demonstration of what can be accomplished through solar holography. Use of high-efficiency holograms to concentrate and split the solar spectrum onto appropriate solar cells for power generation could provide even more power over a similar area. Daylighting holograms could be used to increase the visible light within a building and to decrease the need for artificial lighting. In addition, daylighting holograms could be combined with holograms that reflect heat in the summer and transmit heat in winter for further savings. These are just a few ideas to potentially improve power output and energy savings. Many solar holographic technologies exist that generate power, reduce energy losses, and perform important niche functions, and combinations of these technologies could further improve overall system efficiency and savings.

References

[1] W.K. Smothers, B.M. Monroe, A.M. Weber, and D.E. Keys, "Photopolymers for holography," in *Practical Holography IV*, S.A. Benton, ed., January 18–19, Los Angeles, California, *SPIE OE/Laser Conference Proceedings* **1212–03**, 20–29 (1990).

[2] A.M. Weber, W.K. Smothers, T.J. Trout, and D.J. Mickish, "Hologram recording in du Pont's new photopolymer materials," in *Practical Holography IV*, S.A. Benton, ed., January 18–19, Los Angeles, California, *SPIE OE/Laser Conference Proceedings* **1212–04**, 30–39 (1990).

[3] A.M. Weber, "DU PONT'S OMNIDEX™ HOLOGRAPHIC RECORDING MA-
     TERIALS," *OPTO 7/HOLOGRAPHICS 90 Conference*, October 16–18, Nuremberg,
     Germany (1990).

[4] R.T. Ingwall and M. Troll, "Mechanism of hologram formation in DMP-128
     photopolymer," *Opt. Eng.* **28**, 586–591 (1989).

[5] R.T. Ingwall and H.L. Fielding, "Hologram recording with a new Polaroid photopoly-
     mer system," in *Applications of Holography*, L. Huff, ed., January 21–23, Los Angeles,
     California, *Proc. SPIE* **523**, 306–312 (1985).

[6] R.T. Ingwall, M. Troll, and W.T. Vetterling, "Properties of reflection holograms
     recorded in Polarioid's DMP-128 photopolymer," in *Practical Holography II*, T.H.
     Jeong, ed., January 13–14, Los Angeles, California, *Proc. SPIE* **747**, 96–101 (1987).

[7] G. Saxby, *Practical Holography*, (Prentice Hall International (UK) Ltd, Great Britian,
     1988).

[8] S. McGrew, "Color control in dichromated gelatin reflection holograms," in *Recent
     Advances in Holography*, T.C. Lee and P.N. Tamura, eds., February 4–5, Los Angeles,
     California, *Proc. SPIE* **215**, 24–31 (1980).

[9] S.M. Sze, *Physics of Semiconductor Devices*, 2nd ed. (Wiley, New York, 1981).

[10] J.L. Stone, "Photovoltaics: Unlimited electric energy from the sun," Physics Today,
     22 (1993).

[11] A. Abdurakhmanov, R.A. Zakhidov, A.M. Ikramov, A.V. Khaidarov, and A.U. Borisov,
     "INVESTIGATION OF THE FEASIBILITY OF CREATING A HOLOGRAPHIC
     CONCENTRATOR," *Geliotekhnika* **22**, 17–20 (1986).

[12] N. Aebischer, C. Bainier, M. Guignard, and D. Courjon, "Holographic concentra-
     tor insuring a chromatic dispersion on a photovoltaic cell array," in *Industrial and
     Commercial Applications of Holography*, M. Chang, ed., August 24–25, San Diego,
     California, *Proc. SPIE* **353**, 61–67 (1982).

[13] V.V. Afyan and A.V. Vartanyan, "A new approach to developing selective solar
     radiation concentrators," *Geliotekhnika* **20**, 22–26 (1984).

[14] V.V. Afyan, A.V. Vartanyan, and D.S. Strebkov, "Selective concentrators based on
     holograms for photovoltaic modules," *Geliotekhnika* **22**, 24–26 (1986).

[15] C. Bainier, C. Hernandez, and D. Courjon, "SOLAR CONCENTRATING SYSTEMS
     USING HOLOGRAPHIC LENSES," *Solar Wind Technol.* **5**, 395–404 (1988).

[16] W.H. Bloss, M. Griesinger, and E.R. Reinhardt, "Dispersive concentrating systems
     based on transmission phase holograms for solar applications," *App Opt* **21**, 3739–3742
     (1982).

[17] K. Fröhlich, U. Wagemann, B. Frohn, J. Schulat, and C.G. Stojanoff, "Development
     and fabrication of a hybrid holographic solar concentrator for concurrent generation
     of electricity and thermal utilization," in *Optical Materials Technology for Energy
     Efficiency and Solar Energy Conversion XII*, C.M. Lampert, ed., July 13–14, San
     Diego, California, *Proc. SPIE* **2017**, 311–319 (1993).

[18] K. Fröhlich, U. Wagemann, J. Schulat, H. Schütte, and C.G. Stojanoff, "Fabrication and
     test of a holographic concentrator for two color PV-operation," in *Optical Materials
     Technology for Energy Efficiency and Solar Energy Conversion XIII*, V. Wittwer, C.G.
     Granqvist, and C.M. Lampert, eds., April 18–22, Freiburg, Germany, *Proc. SPIE* **2255**,
     812–821 (1994).

[19] J.E. Ludman, J.L. Sampson, R.A. Bradbury, J.G. Martín, J.R. Riccobono, G. Sliker,
     and E. Rallis, "Photovoltaic systems based on spectrally selective holographic con-
     centrators," in *Practical Holography VI*, S.A. Benton, ed., February 11–13, San José,
     California, **1667**, 182–189 (1992).

[20] J.E. Ludman, J.R. Riccobono, N.O. Reinhand, I.V. Semenova, J.G. Martín, W. Tai, X. Li, and G. Syphers, "Holographic solar concentrator for terrestrial photovoltaics," *24th IEEE Photovoltaic Specialists Conference Record*, December 5–9, Waikoloa, Hawaii, 1994 IEEE First World Conference on Photovoltaic Energy Conversion, 1208–1211 (1994).

[21] J.E. Ludman, J.R. Riccobono, I.V. Semenova, N.O. Reinhand, W. Tai, X.L. Li, G. Syphers, E. Rallis, G. Sliker, and J.G. Martín, "THE OPTIMIZATION OF A HOLO-GRAPHIC SYSTEM FOR SOLAR POWER GENERATION," *Solar Energy* **60**, 1–9 (1997).

[22] N.O. Reinhand, V.D. Rumyantsev, I.V. Semenova, J.E. Ludman, and J.R. Riccobono, "Development of holographic optical elements to be used in space as solar energy concentrators," *Acad. Sci. Russia*, A. F. Ioffe Physical Technical Institute **1623**, 1–27 (1994).

[23] N.O. Reinhand, I.V. Semenova, J.E. Ludman, and J.R. Riccobono, "Holographic solar-energy concentrators," *J. Opt. Technol.* **64**, 336–340 (1997).

[24] C. Shakher and V. Ramamurthy, "THICK TRANSMISSION PHASE HOLO-GRAMS FOR PHOTOVOLTAIC CONCENTRATOR APPLICATIONS," *Solar Energy Materials* **16**, 215–221 (1987).

[25] C.G. Stojanoff, H.D. Tholl, and R. Kubitzek, "Design and optimization of holographic solar concentrators," in *Optical Materials Technology for Energy Efficiency and Solar Energy Conversion VI*, C.M. Lampert, ed., August 18–19, San Diego, California, *Proc. SPIE* **823**, 166–173 (1987).

[26] C.G. Stojanoff, R. Kubitzek, and St. Tropartz, "Optimization procedure for a holo-graphic lens solar concentrator," in *Optical Materials Technology for Energy Efficiency and Solar Energy Conversion VII*, C.G. Granqvist and C.M. Lampert, eds., September 19–21, Hamburg, Germany, *Proc. European Congress on Optics* **1016**, 226–232 (1988).

[27] C.G. Stojanoff, R. Kubitzek, St. Tropartz, K. Fröhlich, and O. Brasseur, "Design, fabrication and integration of holographic dispersive solar concentrator for terrestrial applications," in *Optical Materials Technology for Energy Efficiency and Solar Energy Conversion X*, C.M. Lampert and C.G.Granqvist, eds., July 25–26, San Diego, California, *Proc. SPIE* **1536**, 206–214 (1991).

[28] C.G. Stojanoff, K. Fröhlich, U. Wagemann, H. Schütte, J. Schulat, and P. Fröning, "New developments in holographic solar concentrators: a review," in *Presented at Optical Materials Technology for Energy Efficiency and Solar Energy Conversion XIII*, V. Wittwer, C.G. Granqvist, and C.M. Lampert, eds., April 18–22, Freiburg, Germany, *Proc. SPIE*, (1994).

[29] C.G. Stojanoff, "Review of the technology for the manufacturing of large format DCG–holograms for technical applications," in *Practical Holography XI and Holographic Materials III*, S.A. Benton and T.J. Trout, eds., February 10–11, San José, California, *Proc. SPIE* **3011**, 267–278 (1997).

[30] E.U. Wagemann, K. Fröhlich, J. Schulat, H. Schütte, and C.G. Stojanoff, "Design and optimization of a holographic concentrator for two-color PV-operation," in *Optical Materials Technology for Energy Efficiency and Solar Energy Conversion XII*, C.M. Lampert, ed., July 13–14, San Diego, California, *Proc. SPIE* **2017**, 252–263 (1993).

[31] Y.W. Zhang, C.S. Ih, H.F. Yan, and M.J. Change, "Photovoltaic concentrator using a holographic optical element," *Appl. Opt.* **27**, 3556–3560 (1988).

[32] G. Anders, D. Corlatan, K. Herz, and D. Schmid, "HOLOGRAPHIC SOFT CON-CENTRATORS FOR PV APPLICATIONS," in *Tenth E. C. Photovoltaic Solar Energy*

Conference, A. Luque, G. Sala, W. Palz, G. Dos Santos, and P. Helm, eds., April 8–12, Lisbon, Portugal, *Proceedings of the International Conference*, (Kluwer Academic Publishers, Netherlands, 1991), 1229–1232.

[33] K. Herz, D. Corlatan, G. Anders, and M. Schäfer, "HOLOGRAPHIC OPTICAL ELEMENTS FOR THE CONSTRUCTION OF TANDEM CELLS," in *Tenth E.C. Photovoltaic Solar Energy Conference*, A. Luque, G. Sala, W. Palz, G. Dos Santos, and P. Helm, eds., April 8–12, Lisbon, Portugal, *Proc. Conf.*, (Kluwer Academic Publishers, Netherlands, 1991), 15–18.

[34] C. Linsen, "Design of holographic photovoltaic concentrator," *Guangxue Jishu/Optical Technique* **6**, 29–31 (1993).

[35] H. Schütte and C.G. Stojanoff, "Effects of process control and exposure energy upon the inner structure and the optical properties of volume holograms in dichromated gelatin films," in *Practical Holography XI and Holographic Materials III*, S.A. Benton, and T.J. Trout, eds., February 10–11, San José, California, *Proc. SPIE* **3011**, 255–266 (1997).

[36] C.G. Stojanoff, W. Windeln, and S. Tropatz, "Investigation of the properties of dichromated gelatin layers and their influence on the diffraction efficiency and on the bandwidth of holographic solar concentrators," in *Optical Materials Technology for Energy Efficiency and Solar Energy Conversion VI*, C.M. Lampert, ed., August 18–19, San Diego, California, *Proc. SPIE* **823**, 174–180 (1987).

[37] C. Shakher, V. Ramamurthy, and H.L. Yadav, "OPTIMISATION OF THICK PHASE TRANSMISSION HOLOGRAM PROCESSING PARAMETERS FOR PV CONCENTRATOR APPLICATIONS," in *Ninth E. C. Photovoltaic Solar Energy Conference*, W. Palz, G.T. Wrixon, and P. Helm, eds., September 25–29, Freiburg, West Germany, *Proc. of the International Conference*, (Kluwer Academic Publishers, Netherlands, 1989), 799–801.

[38] C. Shakher and H.L. Yadav, "Dependence of diffraction efficiency of holographic concentrators on angle of illumination, hologram-thickness and wavelength of illuminating light," *J. Opt.* (Paris) **21**, 267–272 (1990).

[39] A.R. Frederickson, D.B. Cotts, J.A. Wall, and F.D. Bouquet, "Spacecraft dielectric material properties and spacecraft charging," American Institute of Aeronautics and Astronautics, New York (1986).

[40] J.P. Golden, G.P. Summers, and W.H. Carter, "Resistance of holograms made in Polaroid DMP128 photopolymer to ionizing radiation damage," *Opt. Lett.* **13**, 949–951 (1988).

[41] A. McKay and J. White, "Effects of simulated space environments on dichromated gelatin holograms," in *Optomechanical Design of Laser Transmitters and Receivers*, B.D. Seery, ed., January 16–17, Los Angeles, California, *Proc. SPIE* **1044**, 269–275 (1989).

[42] R.W. Francis, W.A. Somerville, and D.J. Flood, "Issues and opportunities in space photovoltaics," *20th IEEE Photovoltaic Specialists Conference Record*, September 26–30, Las Vegas, Nevada, 8–20 (1988).

[43] J. Tracy and J. Wise, "Space solar cell performance for advanced GaAs and Si solar cells," *20th IEEE Photovoltaic Specialists Conference Record*, September 26–30, Las Vegas, Nevada, 841–847 (1988).

[44] D. Warfield and J. Silver, "An analysis of spaceflight performance of several types of silicon solar cells on the LIPS III satellite," *21st IEEE Photovoltaic Specialists Conference Record*, May 21–25, Kissimimee, Florida, 1164–1166 (1990).

[45] I. Weinberg, C.K. Swartz, R.E. Hart, Jr., and R.L. Statler, "Radiation and temperature effects in gallium arsenide, indium phosphide, and silicon solar cells," *19th IEEE Photovoltaic Specialists Conference Record*, May 4–8, New Orleans, Louisiana, 548–557 (1987).

[46] R. Hopkinson, P. Petherbridge, and J. Longmore, *Daylighting* (Heinemann, London 1966).

[47] R. Ian and E. King, "HOLOGRAPHIC GLAZING MATERIALS FOR MANAGING SUNLIGHT IN BUILDINGS," *J. Architectural Planning Res.* **4**, 1–6 (1987).

[48] N. Rogers, J. Ballinger, and C. Dunkerly, *Architectural Science Review* **22**, 44–49 (1979).

[49] J. Schuman, F. Rubinstein, K. Papamichael, L. Beltrán, E.S. Lee, and S. Selkowitz, "Technology reviews: Daylighting optical systems," D.O.E. Report under Contract DE-AC03–76SF00098 (September 1992).

[50] L. Whitehead, R. Nodwell, and F. Curzon, "New efficient lightguide for interior illimination", *Appl. Opt.* **21**, 2755–2761 (1982).

[51] H. Bryan and E. King, "An evaluation of a holographic diffractive glazing material for improved utilization of sunlight in buildings," in *Proceedings of: The 15th National Passive Solar Conference*, Solar 90, S.M. Burley and M.J. Coleman, eds., March 19–22, Austin, Texas, ASES, 181–187 (1990).

[52] H.J. Gerritsen, "Diffractive daylighting: ways to obtain wide angular range, large efficiency, near achromatic operation," in *Optical Materials Technology for Energy Efficiency and Solar Energy Conversion XII*, C.M. Lampert, ed., July 13–14, San Diego, California, *Proc. SPIE* **2017**, 377–388 (1993).

[53] J. Ludman, J. Riccobono, G. Savant, J. Jannson, G. Campbell, and R. Hall, "Holographic Daylighting," in *Application and Theory of Periodic Structures*, July 10–12, San Diego, California, *Proc. SPIE* **2532**, 436–446 (1995). Invited Paper.

[54] K. Papamichael, L. Beltrán, R. Furler, E.S. Lee, S. Selkowitz, and M. Rubin, "Simulating the Energy Performance of Holographic Glazings," in *Optical Materials Technology for Energy Efficiency and Solar Energy Conversion XIII*, V. Wittwer, C.G. Granqvist, and C.M. Lampert, eds., April 18–22, Freiburg, Germany, *Proc. SPIE* **2255**, 763–771 (1994).

[55] H.D. Tholl, "Efficiency and bandwidth analysis of holographic glazing materials in the conical diffraction configuration," in *Optical Materials Technology for Energy Efficiency and Solar Energy Conversion XIII*, V. Wittwer, C.G. Granqvist, and C.M. Lampert, eds., April 18–22, Freiburg, Germany, *Proc. SPIE* **2255**, 497–507 (1994).

[56] H.D. Tholl, R. Kubiza, and C.G. Stojanoff, "Stacked volume holograms as light directing elements," in *Optical Materials Technology for Energy Efficiency and Solar Energy Conversion XIII*, V. Wittwer, C.G. Granqvist, and C.M. Lampert, eds., April 18–22, Freiburg, Germany, *Proc. SPIE* **2255**, 486–496 (1994).

[57] H.D. Tholl, C.G. Stojanoff, R. Kubiza, and G. Willbold-Lohr, "Design Optimization and manufacturing of holographic windows for daylighting applications in buildings," in *Optical Materials Technology for Energy Efficiency and Solar Energy Conversion XII*, C.M. Lampert, ed., July 13–14, San Diego, California, *Proc. SPIE* **2017**, 35–45 (1993).

[58] S. Weber, E. King, and R. Ian, "Holography, architecture, intention," in *Practical Holography II*, T.H. Jeong, ed., January 13–14, Los Angeles, California, *Proc. SPIE* **747**, 96–101 (1987).

[59] J. Ludman, "Holographic solar concentrator," *Appl. Opt.* **21**, 3057–3058 (1982).

[60] K. Papamichael, L. Beltrán, R. Furler, E.S. Lee, S. Selkowitz, and M. Rubin, "The energy performance of prototype holographic glazings," D.O.E. Report under Contract DE-AC03–76SF00098 (February 1993).

[61] U.S. Congress, Office of Technology Assessment, Building Energy Efficiency, OTA-E-518 U.S. Government Printing Office, Washington, D.C., May 1992).

[62] C.G. Stojanoff, H. Schütte, J. Schulat, R. Kubiza, and P. Fröning, "Fabrication of large format holograms in dichromated gelatin films for sun control and solar concentrators," in *Diffractive and Holographic Device Technologies and Applications IV*, I. Cindrich and S.H. Lee, eds., February 12–13, San José, California, *Proc. SPIE* **3010**, 156–167 (1997).

[63] R. Bradbury, J. Ludman, and J. White, "Holographic lighting for energy efficient greenhouses," in *Practical Holography*, T.H. Jeong and J.E. Ludman, eds., January 21–22, Los Angeles, California, *Proc. SPIE* **615**, 104–111 (1986).

[64] J. Powell, "Image hologram as visual light guide," in *Practical Holography VIII*, S.A. Benton, ed., February 7–9, San José, California, *Proc. SPIE* **2176**, 162–165 (1994).

[65] H.F.O. Müller, "Application of holographic optical elements in buildings for various purposes like daylighting, solar shading and photovoltaic power generation," *Renewable Energy* **5**, 935–941 (1994).

[66] H.F.O. Müller, "Innovative use of light-directing holograms with solar cells," *Solar Architecture* (1994).

[67] W. Windeln and C.G. Stojanoff, "Development of high efficiency holographic solar concentrator," in *Optical Materials Technology for Energy Efficiency and Solar Energy Conversion IV*, C.M. Lampert, ed., Aug 20–22, San Diego, California, *Proc. SPIE* **562**, 67–74 (1985).

[68] J.L. Hull, J.P. Lauer and D.C. Broadbent, "Holographic solar concentrator," in *Materials and Optics for Solar Energy Conversion and Advanced Lighting Technology*, C.M. Lampert, ed., August 19–21, San Diego, California, *Proc. SPIE* **692**, 68–74 (1986).

[69] J. Hull, J. Lauer, and D. Broadbent, "HOLOGRAPHIC SOLAR CONCENTRATORS," *Energy* **12**, 209–215 (1987).

[70] H.D. Tholl and C.G. Stojanoff, "Performance and bandwidth analysis of holographic solar reflectors," in *Optical Materials Technology for Energy Efficiency and Solar Energy Conversion VII*, C.G. Granqvist and C.M. Lampert, eds., September 19–21, Hamburg, Germany,*Proc. European Congress on Optics* **1016**, 233–238 (1988).

[71] Conversations between Robb Frederickson and Northeast Photosciences (1992–1994).

[72] C. Rich and J. Peterson, "BROADBAND IR LIPPMANN HOLOGRAMS FOR SOLAR CONTROL APPLICATIONS," in *Practical Holography VI*, S.A. Benton, ed., February 11–13, San José, California, *Proc. SPIE* **1667**, 165–171 (1992).

[73] J. Jannson, T. Jannson, and K. Yu, "SOLAR CONTROL TUNABLE LIPPMANN HOLOWINDOWS," in *Optical Materials Technology for Energy Efficiency and Solar Energy Conversion IV*, C.M. Lampert, ed., August 20–22, San Diego, California, *Proc. SPIE* **562**, 75–82 (1985).

[74] J. Jannson, T. Jannson, and K. Yu, "SOLAR CONTROL TUNABLE LIPPMANN HOLOWINDOWS," *Solar Energy Materials* **14**, 289–297 (1986).

[75] D. Vukicevic, J. Riccobono, and J. Ludman, "Passive holograms for thermal control of building interiors," S.D.I. Proposal SDIO93–016 (January, 1993).

[76] T. Regan, J. Martín, J. Riccobono, and J. Ludman, "Holographically enhanced TPV conversion," *24th IEEE Photovoltaic Specialists Conference Record.* December 5–9, Waikoloa, Hawaii, *1994 IEEE First World Conference on Photovoltaic Energy Conversion*, 1930–1933 (1994).

[77] T. Regan, J. Riccobono, and J. Martín, "TPV conversion of nuclear energy for space applications," *First NREL Conference on Thermophotovoltaic Generation of Electricity, AIP Conf. Proc.* **321**, 322–330 (1995).

[78] T. Regan, J. Martín, J. Riccobono, and J. Ludman, "Multi-source thermophotovoltaic energy onversion," *Application and Theory of Periodic Structures*, July 10–12, San Diego, California, *Proc. SPIE* **2532**, 458–462 (1995).

[79] J.A. Quintana, P.G. Boj, A. Bonmatí, J. Crespo, M. Pardo, and C. Pastor, "Solar holoconcentrators in dichromated gelatin," in *Practical Holography III*, S.A. Benton, ed., January 17–18, Los Angeles, California, *Proc. SPIE* **1051**, 160–162 (1989).

[80] P.G. Boj, J. Crespo, M.A. Satorre, and J.A. Quintana, "Double-layer method for increasing the bandwidth of UV spectrally responsive holograms in dichromated gelatin," *Appl. Opt.* **33**, 2917–2920 (1994).

[81] M. Romero, E. Conejero, and M. Sanchez, "Recent experiences on reflectant module components for innovative heliostats," *Solar Energy Materials* **24**, 320–332 (1991).

[82] Ph. Fröning, J. Schulat, and C.G. Stojanoff, "Fabrication of high efficiency reflection holograms for use as a holographic concentrator in the blue spectral domain around 420 nm," *Holographic and Diffractive Technologies*, G.J. Dausmann, ed., October 9–11, Berlin, Germany, *Proc. SPIE* **2951**, 25–35 (1996).

[83] E.U. Wagemann, Ph. Fröning, and C.G. Stojanoff, "A new method for recording large size holograms of the relfective type with adjustable spectral characteristics in DCG," in *Holographic and Diffractive Technologies*, G.J. Dausmann, ed., October 9–11, Berlin, Germany, *Proc. SPIE* **2951**, 2–12 (1996).

[84] A. Borne, A. Gombert, W. Graf, M. Köhl, and V. Wittwer, "Regular and random surface micro–structures as anti-reflection coatings for solar absorbers," in *Optical Materials Technology for Energy Efficiency and Solar Energy Conversion XI*, A. Hugot-LeGoff, C.G. Granqvist, and C.M. Lampert, eds., May 18, Toulouse-Labège, France, *Proc. SPIE* **1727**, 140–148 (1992).

[85] D. Soltz, L. Cescato, and F. Decker, "Photoelectrochemical etching of n-InP producing antireflecting structures for solar cells," *Solar Energy Materials Solar Cells* **25**, 179–189 (1992).

[86] A. Gombert, K. Rose, A. Heinzel, W. Horbelt, C. Zanke, B. Bläsi, and V. Wittwer, "Antireflective submicrometer surface-relief gratings for solar applications," *Solar Energy Materials Solar Cells* **54**, 333–342 (1998).

[87] J.G. Zhu, C.W. White, S.P. Withrow, J.D. Budai, R. Mu, and D.O. Henderson, "Synthesis and physical properties of semiconductor nanocrystals formed by ion implantation," *Nanostructured Materials, ACS Symposium Series* **679**, 198 (1998).

[88] D.O. Henderson, A. Ueda, Y.S. Tung, R. Mu, C.W. White, R.A. Zuhr, and J.G. Zhu, "Surface phonon characterization of InP, GaP, and GaAs quantum dots confined in dielectric hosts," *J. Phys. D Appl. Phys.* **30**, 1432–1435 (1997).

8

Holographic Optical Memories

Philip Hemmer, Selim Shahriar, Jacques Ludman, and H. John Caulfield

8.1 Introduction

Holograms lend themselves to several kinds of optical storage. We concentrate here on something unique to holograms—the page-oriented holographic memory. This allows a change in some reference beam parameter (most commonly, angle of incidence or phase pattern) to change what 2D "page" of data falls on a detector array. Because the reference beam can often be switched rapidly with no moving parts, holographic memories offer random access speeds no other memory can approach. Other advantages are sometimes sought. The reference beam can be encoded so that only users with the proper codes can read its content. The storage density can be very great—the order of a few bits per cubic wavelength. As the readout apparatus is very large, this advantage is most helpful when we have many holograms that can be read one at a time by the same readout system.

The selectivity of these thick holograms comes about because a 3D pattern must be matched exactly for an efficient conversion of the input light into the desired output image to occur. For historical reasons, that matching condition is often called Bragg matching.

After describing these holographic memories in some detail, we will consider two important applications for which Bragg-selective holographic memories are especially suited. One is the rapid identification of images via optical correlation, and the other is secure storage of large quantities of data. Both of these applications will be discussed in more detail after the basics of page-oriented, Bragg-selective, holographic optical memories are reviewed.

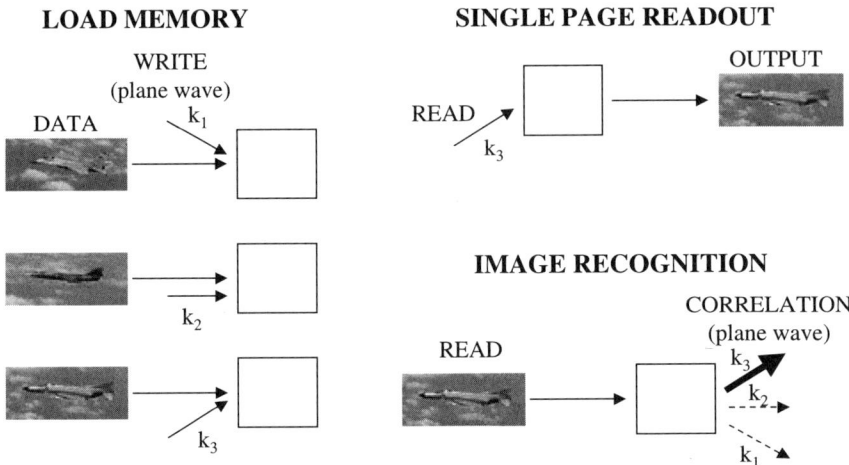

FIGURE 8.1. Operation of a Bragg-selective holographic optical memory. The memory is loaded by interfering an object laser beam, carrying page-oriented or image data, with a reference or write beam that is nominally a plane wave. Multiple images are loaded by using a reference beam having a unique propagation direction for each image. Readout of a single page is accomplished by inputting a laser read beam with the same angle as the write beam used to store the desired image. By replacing the plane-wave reference and read beams with an information-carrying beam, secure storage is accomplished. To perform image recognition, a test image is input to the memory using a laser beam. This generates a plane wave output in the same direction as the plane-wave used to store the matching image. Detection of the output beam direction identifies the test image.

8.2   Page-Oriented, Bragg-Selective, Holographic Optical Memories

Holographic optical memories are created by interfering two laser beams in a photosensitive medium. Typically, one of these laser beams carries information such as page-oriented data or an image, and the other is a reference or write beam that is nominally a plane wave, as shown in Fig. 8.1. To store multiple data pages, a different reference beam angle is used for each input page or image. The minimum difference between adjacent reference beam angles is determined by the Bragg angular selectivity, which depends inversely on the thickness of the hologram.[1] Typically, Bragg selectivity on the order of 1 mrad is achieved for a hologram thickness on the order of 1 mm. To selectively read out a single page of data, the hologram is illuminated with a read beam, which propagates in the same direction as the reference beam that was originally used to write the hologram of interest. If secure readout is required, the write and read beams can be replaced by an information-bearing beam that serves as a key. In this case, holograms can only be read out if the read beam is an exact replica of the write beam, where again the read beam angle determines which hologram is read out. Alternatively, the hologram can be read out by inputting an image-bearing laser beam from the same direction as the original data beams. This image-bearing read beam will produce a replica of the reference beam that was used to write the hologram of the matching image. By measuring the direction of the output reference beam replica, the test image is identified because the reference beam angle used to record each data hologram is known.

To determine the storage capacity of a Bragg-selective holographic memory, it is necessary to know the maximum number of holograms that can be written at a single spatial location. For thin samples, this will be determined by the Bragg angle selectivity because the range of possible reference beam input angles is limited. However, for thick holograms, on the order of 1 mm or larger, the maximum number of holograms is determined by the signal-to-noise ratio. This in turn is determined by the $M\#$ and the noise floor established by the scattering caused by surface and volume inhomogeneities. Consider first the role of the $M\#$, which determines the dynamic range and is defined as the maximum number of 100% efficient sinusoidal gratings that can be written at a given spatial location. If the number of multiplexed holograms in one location is M, then the efficiency of each hologram is given by:

$$\eta = \left(\frac{M\#}{M} \right)^2 .$$

The minimum efficiency is constrained by the fractional intensity of random scattering, η_S. A high-quality photosensitive material has low scattering, η_S less than 10^{-6}, and a large $M\#$ more than 5. For these numbers, we find the maximum number of holograms at a single spatial location to be at least 5000. Assuming a megapixel read/write head, this corresponds to about 1 Gbyte over the area of illumination, typically on the order of 1 cm^2.

8.3   Materials for Bragg-Selective Holographic Memories

The key to producing practical Bragg-selective holographic optical memories is identification of the optimum material. This is because the required Bragg selectivity of 1 mrad or better can only be achieved if the photosensitive material has dimensional stability on the order of 0.1% or better. Recently, much progress has been made in this area. For example, a number of impressive memory demonstrations[2] have been performed using photorefractive materials such as LiNbO$_3$. These materials were chosen because of their long storage times (in the dark) and the ability to erase and rewrite. Photorefractives can also be used for permanent storage by fixing. However, for permanent storage, polymer-based materials have advantages because they tend to have higher diffraction efficiencies, simplified exposure schedules, lower cost, and the potential for mass duplication.

To this end, there has been considerable research directed toward the development of superior holographic polymers in the former Soviet Union over the past few decades. This has led to a novel class of materials known collectively as PDA (photopolymer with diffusion amplification).[3] Recently, PDA has been adopted by a number of research groups, and some have improved on the original recipe.[4]

Aside from superior dimensional stability, the important properties of PDA are a high writing efficiency (about a factor of 10 better than lithium niobate) and very low scattering (10^{-6}). Briefly, the most commonly used material in this class consists of phenanthraquinone embedded in polymethylmethacrylate (PMMA). This medium uses the novel principle of diffusion amplification of holograms in polymeric recording media. Photoexposure of this system results in writing of two out-of-phase periodic structures that partially compensate each other. One of these is formed by a concentration distribution of chromatophore groups combined into macromolecules, whereas the other is formed by free chromatophore molecules. As a result of diffusion of the free molecules, their corresponding grating degrades, and the net diffraction efficiency is amplified *without additional processing*. The surviving grating is stored in macromolecules that do not diffuse.

To put this material in proper perspective, it is instructive to compare it with other photopolymeric systems contending for thick holography. Typically, the media studied previously used photochemical reactions of organic compounds in glassy polymeric matrices. One such medium is reoxan, whose action is based on varying the refractive index by sensitized photo-oxidation of anthracene compounds during exposure. The structure of the material with a hologram written on it consists of an out-of-phase concentration distribution of oxidized and unoxidized anthracene dissolved in the glassy, transparent polymer. With the passage of time (days at 80°), the hologram efficiency varies on account of the slow but real diffusion of anthracene molecules (both oxidized and unoxidized) in the polymer. Thus, both gratings deteriorate, which results in an overall reduction in the hologram efficiency. In contrast, in PDA, only one of the grating decays, whereas the other remains unchanged. Because the hologram efficiency is determined by the difference in the amplitudes of these gratings, diffusion in PDA leads to enhancement of the overall hologram efficiency. Experimental evidence shows that the long-term stability of the PDA hologram is orders of magnitude better than that of reoxan holograms. At room temperature, holograms written in this material survive many years without noticeable degradation. In addition, due to the diffusion-induced amplification, the hologram efficiency for PDA can be made much higher than for reoxan. With an exposure of 1 J/cm^2, holograms of near 100% efficiency are easily written in 1-mm-thick samples.

To demonstrate the quality of this PDA material, we have stored and recalled 10 images corresponding to different orientations of a U.S. Air Force resolution chart. These were written with reference beam angles 5 mrads apart, in a sample about 2 mm thick. Figure 8.2 shows one of these images, comparing the original to the reconstruction. Note that the reconstructed image is virtually identical to the original, which indicates potential for a very low bit error rate.

ORIGINAL IMAGE #1

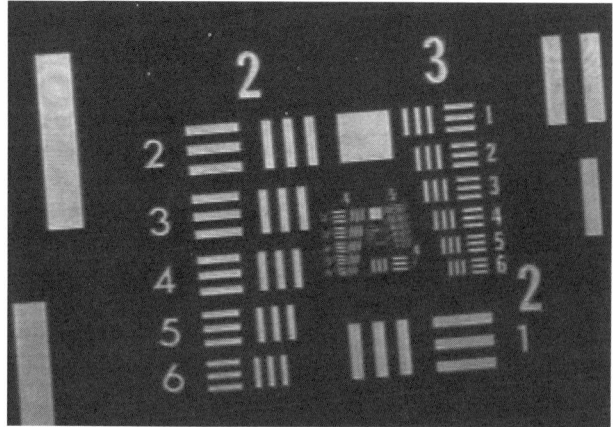

RECONSTRUCTED IMAGE #1

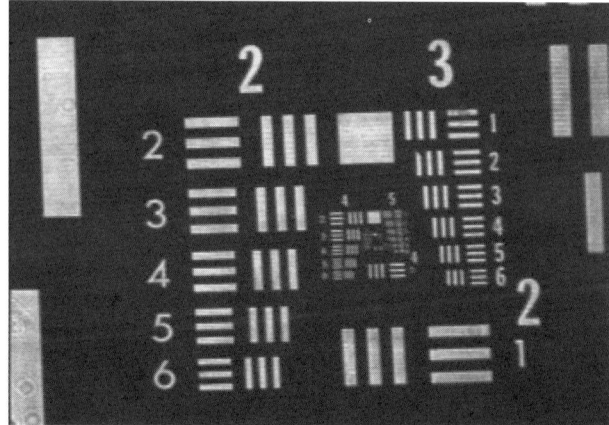

FIGURE 8.2. Comparison of the original image with the holographic reconstruction of one of the 10 images stored in a thick PDA sample, with reference beam angles 5 mrads apart.

8.4   Architecture for Writing Bragg-Selective Holographic Read-Only Memories

To make this discussion more concrete, it is useful to consider the details of a holographic optical memory demonstration unit that we plan to construct in the laboratory. In order to examine the details of the architecture, it is useful to consider first the constraints imposed by optics, detectors, and the holographic material. For the optical system, we choose to write the hologram in the image plane, and assume that the minimum resolvable spot size will be 4 microns, a goal easily achievable for a write wavelength of 514 nm and a read wavelength of 800 nm. For the readout detector array, we assume an integration time of 1 ms, during which at least 100 photons will be detected per pixel. Assuming the shot noise limit (achievable if the detector assembly is Peltier-cooled), this will correspond to a minimum signal-to-noise ratio of about 10. The holographic substrate for this memory unit will be a 5-mm-thick disk, with a diameter of 15 cm. It will be segmented into about 1000 square zones, each 4 mm × 4 mm in size, which corresponds to a fill factor of about 90%. For each zone, we will use 2D angle multiplexing, implemented with galvo-mounted mirrors.

Consider first the horizontal plane. The image beam will be incident at a 30° angle to the normal of the disk. The reference beam will be centered at a 60° angle with respect to the image beam. Around this center position, the reference beam will be scanned, using two scanning mirrors (M3 and M4 in Fig. 8.3), to avoid translational shifts, over an angular span of ±23° (about ±400 mrads), at a step size of about 1 mrad (larger than the Bragg selectivity, to avoid cross-talk). This corresponds to about 800 different Bragg-selected holograms. We will multiplex this further by a factor of 10 by tilting the reference beam in the vertical direction, using a third scanning mirror (M2 in Fig. 8.3). The total number of holograms written at one position will be 8000. The number of bits in a single page will be 1024 × 1024, so that the net storage capacity will be greater than 1 GByte per zone or 1 TByte total.

The beam expander (BE) and the lens assembly for 4f imaging are constructed to minimize aberrations, and they guarantee a minimum image resolution of 4 μm. To transfer the data from the SLM (spatial light modulator) to the interference plane, the image is reduced by about a factor of 2, because our SLM pixel size is about 7.6 μm. The waveplates and polarizers are oriented to maximize the image contrast.

A set of microcontrollers (such as the Motorola (Schaumburg, IL) M68HC11) in master–slave configurations will control the operation of the whole system. A memory buffer of about 100 MBytes (corresponding to 800 memory pages) will be used to hold the data temporarily before each horizontal angular scan. The microcontroller system will regulate the I/O handshake with the mainframe computer. It will also control the radial and angular positioning of the memory disk and scan the galvo-mounted mirrors, through a set of D/A converters and latches. The built-in A/D converters (each capable of monitoring up to eight channels) will be used to provide feedback stabilization of the galvos, with a bandwidth of up to 16 kHz. The microcontroller system will also regulate the shutters and allow for user interruption in-between each page in case of unexpected troubles. Electronically controlled attenuators will be used maintain the laser intensity at the desired level. The whole operation will be carried out on a vibration-isolated optical table.

An Argon laser operating at 514 nm, with a single-frequency, single-mode power exceeding 5 W will be used for writing the holograms. The beam will be expanded initially to a size of 3.5 cm (1/e) diameter to ensure good uniformity over the approximately 8-mm × 8-mm area of the SLM. To achieve the total exposure of 1 J/cm^2, necessary to write a single hologram with 100% efficiency, the exposure time needed for each page would be about 1 msec.

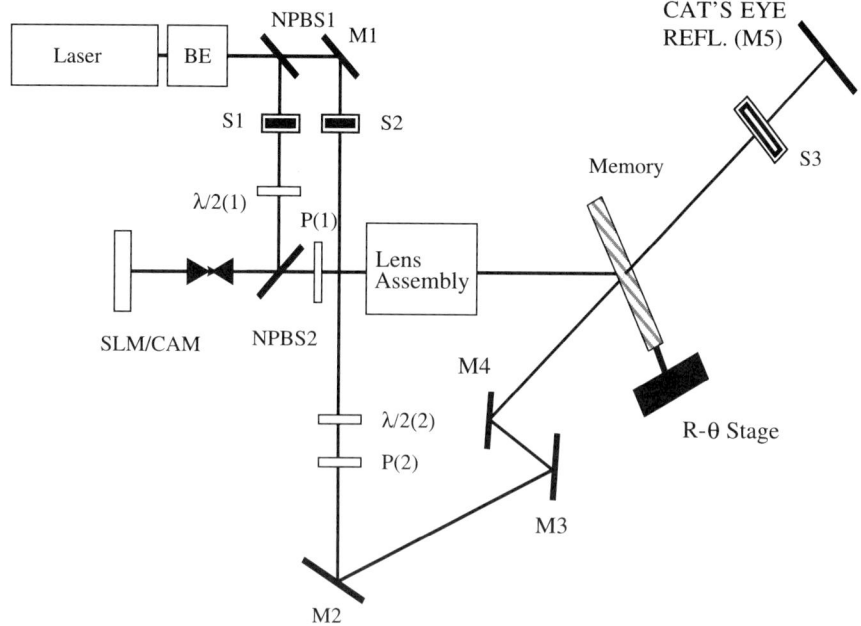

FIGURE 8.3. The basic schematic of the planned Bragg-selective holographic optical memory demonstration unit. The laser output is expanded by the beam expander (BE) and split into two beams controlled by shutters S1 and S2. The first half wave plate, $\lambda/2(1)$, and polarizer, P(1), will be used to optimize the contrast of the liquid crystal spatial light modulator (SLM) operating in reflection mode. The lens assembly will image this SLM onto the memory material. In the reference beam, half-wave plate and polarizer, $\lambda/2(2)$ and P(2), will be used to match the polarization of the information-bearing object beam. M3 and M4 are two horizontally deflecting galvos configured to compensate for laser beam translations. M2 is a vertical-deflecting galvo. The R-θ stage moves the memory disk to select the spot to be exposed. The cat's eye, shutter S3, and the pixel-matched camera (CAM) will be used for readout. This "phase-conjugate" readout geometry compensates for imperfections in the optics.

Prior to writing, the 1 Gbyte of data can first be encoded with the Reed–Solomon error correction algorithm. The level of overhead to be used is determined by the raw bit error rate (BER). For example, an overhead of 6 bits for every 9 bits of useful information can detect and correct for up to three random errors for each group of contiguous 60 bits of information stored, which corresponds to a raw BER of about 5%. In our system, however, the raw BER is not expected to be any higher than 10^{-4}, so that much lower overhead, of the order of 1%, would be enough.

During the writing stage, the shutters S1 and S2 will be open and S3 will be closed. Eight hundred data pages from a mainframe computer (with a memory bank large enough to store the 1 GByte of information to be written) will be loaded into the memory buffer of the SLM driver. The firmware in the SLM driver, in synchronization with the rest of the scanning and positioning control system, will upload each page of data to the SLM, and the data page will then be recorded in a hologram at the proper horizontal angle of the reference beam. As stated above, one page will be written every 1 msec. Once the 800 pages of data are recorded via horizontal angle multiplexing, the horizontal angle galvos will be reset to the starting position, and the vertical angle galvo will be updated to the next (vertical) angle. Then, the horizontal angle scanning sequence will resume until the next 800 pages are loaded. Once the 10 vertical angles have been multiplexed, the R-θ stage (see Fig. 8.3) will be moved electrically to choose the next spatial location. Without undue interruption, the whole writing process would take less than 3 hours. Eventually, it should be possible to run a few of these setups in parallel to increase the throughput.

8.5   Rapid Identification of Images via Optical Correlation

One very important application of holographic optical memories is for the rapid identification of an image. The importance of this application is clear in the context of military target or threat identification. For example, in a countermeasure system, a potential target must be identified as friendly or not, and it must be classified according to type to permit the correct countermeasures to be employed. Because hostile missiles generally travel at supersonic speeds, this entire process must be completed in a matter of seconds. To see how a holographic optical memory can be of value in such a system, it is necessary to consider the target identification process in more detail.

A typical target recognition system accepts a 2D image-like input from a synthetic aperture radar (SAR), a hyperspectral imager (HSI) system, or simply an infrared camera. Based on some criteria, a 2D region of interest (ROI) is identified. The task is then to identify the potential target in this ROI based on this image-like information. This identification process consists of first guessing what the potential target is, and then using what is known about the potential target and observation conditions to generate synthetic data that are compared with the input image. This image generation and comparison process is iterative, eventually converging to a solution that corresponds to accurate target identification.

Clearly, the speed and accuracy of the convergence are very sensitive to the quality of the initial guess. This initial guess is made by comparing the input image to a large number of stored, specially constructed, composite images using a correlation-based pattern recognition technique. Decomposing a 2D image into a bit-stream and digitally correlating to stored data is a time-consuming process, which limits the number of stored images to which the input can be compared. The best way, by far, to do correlation-based pattern recognition quickly and accurately is optically, by correlating the Fourier transform of the test image with a stored databank of appropriate transforms. Doing this optically is far superior to electronically because simple optics can make the transform of the test image and do the correlation in a matter of nanoseconds (i.e., at the speed of light). Fourier optical pattern recognition, as judged by papers published in the last 40 years, has been very successful at recognizing objects in noisy, cluttered scenes.

By using a Bragg-selective holographic memory to store the entire databank of appropriate Fourier transforms in a single holographic memory element, a high degree of parallelism can be achieved. Simply passing the test image through a lens makes the required Fourier transform. This transform is then optically correlated with all the stored Fourier transform pages at once. The entire process requires no computational steps. Thus, a large number of such correlations can be done in parallel, without a penalty in processing speed. This is not possible with present-day digital computers.

To visualize the implementation of an optical correlator with a Bragg-selective holographic memory, consider the architecture of our planned demonstration unit,

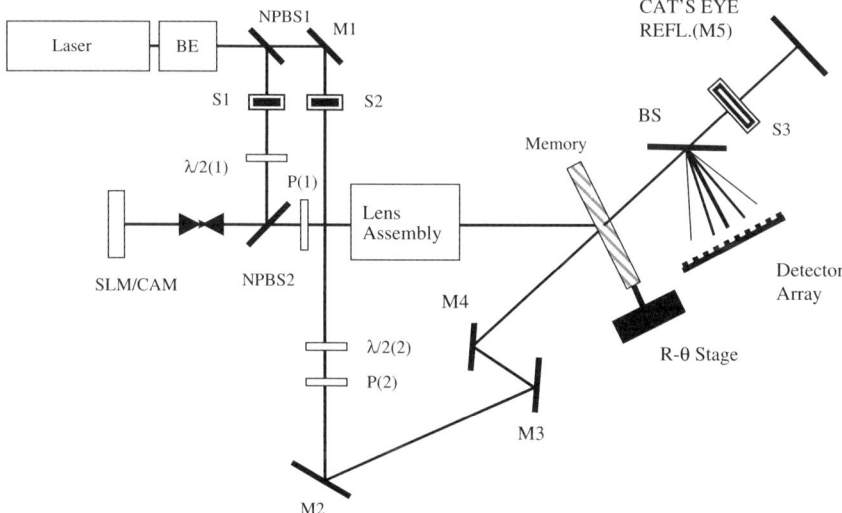

FIGURE 8.4. Basic architecture for the planned Bragg-selective holography optical correlator demonstration unit. The setup is nearly identical to that in Fig. 8.3 except with the addition of a bcamspfitter (BS) and dectector array in the readout path. To perform the target identification, image data from the appropriate region of interest are input into the spatial light modulator (SLM), where its information is impressed on a laser beam that in turn is used to interrogate the memory. The stored holograms generate replicas of their original reference beams, each of which propagates at a unique angle. The brightest reference beam corresponds to the best match and call be identified using the detector array. To use the holographic memory to recall the correct stored image, the output reference beam retroreflected by the cat's eye, M5, to produce an image on the camera (CAM) that is assumed to be pixel registered with the SLM.

which is shown in Fig. 8.4. The configuration is essentially identical to the recording configuration shown in Fig. 8.3, except that an additional beamsplitter and a detector array is inserted before the cat's eye reflector.

Consider a situation in which the object in a selected ROI, as identified by an appropriate algorithm in a SAR, HSI, or infrared (IR) imaging system. Our objective is to identify the object by finding the best possible match from among the thousands of composite images stored in the memory unit. First, the relevant test image is input to the SLM, and shutter 1 is opened, keeping the other shutters off. This produces an image-bearing beam, which when input into the memory unit will produce a set of output reference beam replicas, one corresponding to each image stored. The brightness of each of these output beams is proportional to the corresponding degree of similarity with the input test image. The detector array will identify the brightest spot and record the position of the relevant detector element. This position information will then be used in a look-up table to identify the image. Alternatively, the holographic memory can be used to recall the stored image by closing shutter 1, opening shutter 2, and scanning the reference beam until the correct detector element sees a bright signal. At this point, shutter 3 will be opened, and the reference beam will be reflected by the cat's eye mirror and project the best-match image onto the camera (CAM), which is assumed to be pixel-registered with the SLM.

Researchers at IBM (Armonk, NY) recently demonstrated a similar system for image search in a holographic memory, although on a very small scale. This technique has also been applied to problems such as robotic vision, and the simultaneous correlation of an input image with more than 10,000 stored images appears realistic.[5]

8.6   Secure Archival Storage of Large Quantities of Data

There is a need for massive, secure, read-only storage in a range of enterprises. For example, the detailed records of patients in a hospital can be stored in such a system. The law enforcement agencies can use this for storing criminal records. Artists, designers, and architects can have easy access to a wealth of pictures and drawings packaged in such a memory. Wide use of such a massive storage system would also help reduce the Internet traffic jam.

In addition to data storage, massive secure storage has potential application to video on demand. With the rapid increase in the bandwidth of fiber-optic telephone networks, it would soon be possible for consumers to receive movies over the phone line. The server system will upload a movie over the phone into a storage device connected to the viewer's video monitor. The server for such a system can use a few holographic optical memories to store all movies of interest. A variant of such a system can also widen the range of movies available to travelers in hotel rooms.

Massive, secure, read-only storage can be accomplished using the Bragg-selective holographic memory setup shown in Fig. 8.3. Selective readout of the desired page of data is accomplished by inputting a reference beam at the appropriate angle. To make this recall secure, the hologram is written and read out by a reference beam containing spatial information. Such information can be imposed on the relevant laser beam, using, for example, a second SLM, which is not shown. If a megapixel SLM is used for this purpose, the level of security will be high, possibly approaching the megabit regime.

Finally, there is a need for the robust, massive storage in space. The unique environment of orbiting satellites is such that radiation-hardened storage is needed. Holographic optical memories are well suited for such applications because the information can be stored so that each data bit is spread over the entire memory element. In this case, damage to a particular spatial location of the memory would result is a slight decrease in diffraction efficiency for all stored data, but no data would be lost completely.[6]

8.7   Summary

The potential applications of Bragg-selective holographic optical memories have been discussed. In particular, important applications were identified that rely on the unique properties of these memories, namely, their combined high storage density and capability for massively parallel input/output. These include image identification based on massively parallel optical correlations, which has application to military target recognition and commercial robotic vision, and high-density page-oriented storage, which has applications to secure databases and satellite information storage. To make the discussions of these applications concrete, we included details of our planned Bragg- selective holographic memory demonstration unit.

References

[1] H. Kogelnik, "Coupled wave theory for thick hologram gratings," *Bell Syst. Tech. J.* **48**, 2909 (1969).

[2] IBM Holographic Optical Storage Team, Laser Focus World, "Holographic storage promises high data density," (November, 1996), 81.

[3] A.V. Veniaminov, V.F. Goncharov, and A.P. Popov, "Hologram amplification by diffusion destruction of out-of-phase periodic structures," *Opt. Spectrosc. (USSR)* **70**, 505 (1992).

[4] S.H. Lin, K.Y. Hsu, W.Z. Chen, and W.T. Whang, "Phenanthrenequinone-doped poly(methyl methacrylate) photopolymer bulk for volume holographic data storage," *Opt. Lett.* **25**, 451 (2000).

[5] G.W. Burr, A. Xin, F.H. Mok, and D. Psaltis, "Large-scale rapid access holographic memory," *Proc. SPIE - Int. Soc. Opt. Eng. (USA)*, **2514**, 363–371 (1995).

[6] A.J. Hand, "Is holographic storage a viable alternative for space," *Photonics Spectra*, (June, 1998), 120.

9

Holography and Speckle Techniques Applied to Nondestructive Measurement and Testing

Pramod K. Rastogi

9.1   Introduction

Holographic and speckle techniques are important and exciting areas of research. They have established themselves as promising techniques in the field of optical metrology. Techniques within their folds for the measurement of static and dynamic displacements, topographic contours, and flow fields have been developed and demonstrated with success in a wide range of problems. The aim of this chapter is to provide a review of the holographic and speckle techniques applied to experimental mechanics and nondestructive testing. Numerous books have been published that treat these fields in detail.[1−18]

9.2   Holographic Interferometry

The development of holographic interferometry has redefined our perception of interferometry. A selected compilation of forerunner papers in the field is included in Sirohi and Hinsch.[19] The unique property of holographic interferometry to bring a wavefront, generated at some earlier time, stored in a hologram and released at a later time, to interfere with a comparison wavefront has made possible the interferometric comparison of a rough surface, which is subject to stress, with its normal state. Assuming the surface deformations to be very small, the interference of the two speckled wavefronts forms a set of interference fringes that are indicative of the amount of displacement and deformation undergone by the diffuse object.

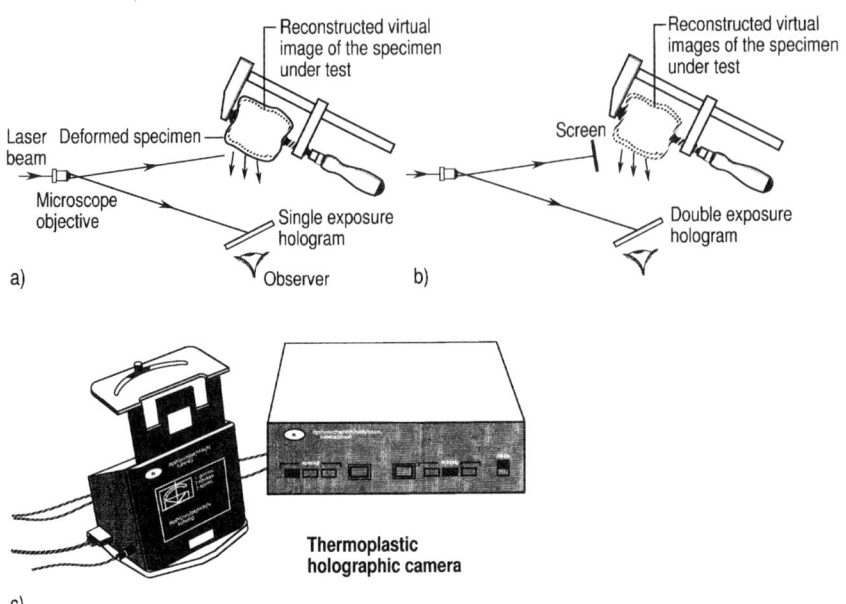

FIGURE 9.1. Schematic for the observation of interference patterns in (a) real-time and (b) double-exposure holographic interferometry. (c) Example of a commercial unit for implementing holographic interferometry.

There are several schemes that have been implemented to obtain the interferometric comparison of the wavefronts. Most important of these are briefly described below.

9.2.1   Schemes for Recording and Reconstructing Holographic Interference

In the real-time form of holographic interferometry, a hologram is made of an arbitrarily shaped rough surface. This is schematized in Fig. 9.1a. After development, the hologram is placed back in exactly the same position. Upon reconstruction, the hologram produces the original wavefront. A person looking through the hologram sees a superposition of the original object and its reconstructed image. The object wave interferes with the reconstructed wave to produce a dark field due to destructive interference. If the object is now slightly deformed, interference fringes are produced that are related to the change in shape of the object. A dark fringe is produced, whenever

$$\varphi(x, y) - \varphi'(x, y) = 2n\pi, \qquad n = 0, 1, 2, \ldots, \tag{1}$$

where, n is the fringe order and φ and φ' are the phases of the waves from the object in its original and deformed states, respectively. The method is very useful for determining the direction of the object displacement, and for compensating on the interferogram the influence of the rigid body motion of the object when subjected to a stress field.

Double exposure is another way to implement holographic interferometry. In this scheme shown in Fig. 9.1b, two exposures are made on the holographic plate: one exposure with the object in its original state and a second exposure with the object in a deformed state. On readout, two reconstruction fields are produced: one corresponding to the object in its original state and the other corresponding to the object in its stressed state. The reconstructed image is modulated by a set of interference fringes that contain information about how and how much the object was distorted between the two exposures. The fringe pattern is permanently stored on the holographic plate. The interference fringes denote the loci of points that have undergone an equal change of optical path between the light source and the observer. A bright fringe is produced whenever

$$\varphi(x, y) - \varphi'(x, y) = 2n\pi, \qquad n = 0, 1, 2, \ldots. \qquad (2)$$

The use of thermoplastic plates as recording material has gained in importance in recent years for implementing real-time and double-exposure holographic inter-ferometry setups, as it can be processed rapidly in situ using electrical and thermal processing. The plate is also cost effective as it is erasable and reusable at least several hundred times. It has brought to a virtual halt the use of silver-halide pho-tographic plates and the associated wet processing in holographic interferometry. In thermoplastic recording, the information is recorded as thickness variation cor-responding to a charge intensity pattern deposited on the thermoplastic layer. A unit commercialized by Newport Corporation (Newport, Corp, Irvine, California) until last year is shown in Fig. 9.1c. Its production has since been discontinued. Self-developing or photorefractive crystals are another interesting alternative to the use of film as the recording medium. A useful review on holographic recording materials can be found in Hariharan,[20] Troth and Dainty,[21] and Hafiz et al.[22] A view of a holographic setup in the author's interferometric nondestructive testing laboratory at the Swiss Federal Institute of Technology Lausanne is shown in Fig. 9.2. The experimental setup in the photograph was implemented for investigating the mechanical behavior of a glued laminate timber using real-time holographic interferometry. Iso in-plane displacement holographic moiré contours on the timber joint specimen are also shown in the photograph.

The use of holographic interferometry is often limited by its high sensitivity to rigid body movements. These movements cause the fringe pattern to alter com-pletely or lose their contrast to the extent of even vanishing from the object surface. Fringe control techniques have long been developed that in combination with a real-time procedure provide a highly satisfactory performance in deformation ap-plications. Fringe control can also be achieved in a holographic setup containing two reference beams.[23] The technique works by moving one reconstructed object image in relation to the other. Sandwich holography has also been a useful tool in overcoming the influence of rigid body movements. Reflection holographic in-terferometry is still another alternative used to obtain interference fringes in the presence of extraneous large object motions. In this mode of functioning, the holo-

FIGURE 9.2. An interferometric nondestructive testing laboratory at the Swiss Federal Institute of Technology Lausanne. The holographic moiré arrangement built around a tensile loading device, fitted with a fingertip joint wood specimen, is sensitive to the measurement of the displacement component, u, along x-axis. The fringe map on the lower right half of the figure displays in-plane displacement contours on the wood specimen.

graphic plate is rigidly attached to the object surface. The beam that illuminates the object-plate tandem serves both as object and as reference waves; the reference wave is the one that falls directly on the back of the holographic plate, whereas the object illumination beam is the one that is transmitted by the plate. Examples of interference images using conventional real-time and double-exposure holographic interferometry are shown in Figs. 9.3a,c–h and b, respectively. The picture in Fig. 9.3a shows the live monitoring of the development of the hatching behavior of a chicken embryo on a TV screen; the egg is placed inside an incubator.

Digital recording and numerical reconstruction of holograms is a relatively new development in holographic interferometry.[24–27] It consists of recording two holograms corresponding to undeformed and deformed states of an object using a charge-coupled display (CCD) camera. These images are stored in a digital image processing system. The two holograms are then reconstructed separately by numerical methods from the digitally stored holograms. The diffraction of a plane wave at the hologram is described by the Fresnel–Kirchoff integral. The Fresnel approximation is used because the dimension of the CCD chip is very small in comparison to the distance between the hologram and the reconstructed image. The Fresnel approximation allows for calculating the complex amplitude of the diffracted wave in the plane of the reconstructed image. The advantage of digi-

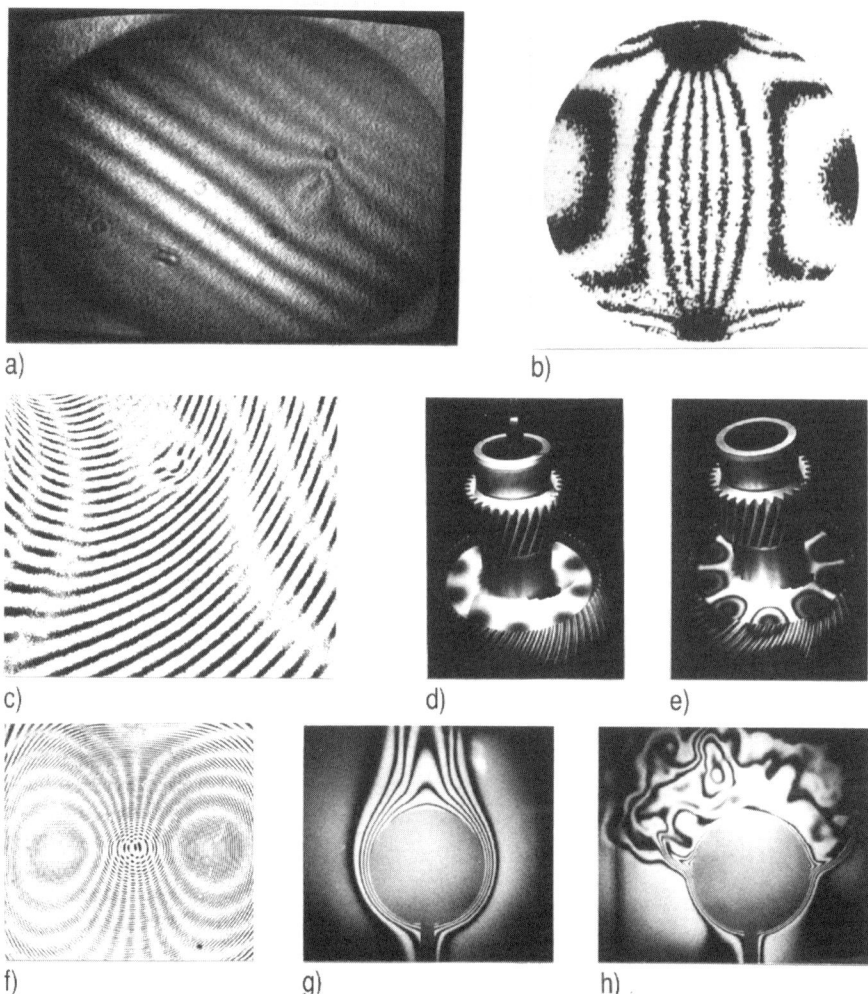

FIGURE 9.3. Examples of the application of holographic interferometry using conventional recording techniques: (a) live observation of the hatching behavior of a chick embryo; (b) in-plane displacement contours overlaid on a disk under diametral compression; (c) detection of flaw in a rubber plate using comparative holography; (d, e) vibration analysis of a helicoptor component at 16,153 Hz under increasing loads; the resonance mode shape hardly visible in (d) is well defined in (e); (f) fringe contours of the first derivative of out-of-plane deformation on a square aluminum plate clamped along its edges and subjected to a point load at its center; (g) visualization of a free convective thermal flow around a heated sphere; (h) visualization of a convective thermal flow around a heated rotating sphere; the sphere is rotating at the speed of 300 rounds per minute. (d, e: Courtesy of L.M. Rizzi, ENEL Research; g,h: Courtesy of S.M. Tieng).

tal holography is that it allows for computing both the intensity and the phase of a holographically stored wavefront during the numerical reconstruction. The reconstruction can also be performed by the convolution approach. The contours of constant optical phase change due to deformation are then obtained by subtracting the reconstructed phases of the undeformed from the deformed object wavefield. A hologram can be recorded on a CCD as long as the sampling theorem is fulfilled; that is to say, each period of the spatial variation of the hologram intensity is sampled by at least two pixels of the CCD array. The condition imposes a limit on the angle that object beam can make with the reference beam, which must necessarily be small given that the resolution of CCDs is relatively low. The applications of the method are thus limited mainly by the pixel size in CCD array. The maximum spatial frequency that must be resolved by the recording medium is determined by the maximum angle between the reference and the object wave

$$f_{max} = \frac{2}{\lambda} \sin \frac{\Delta\theta_{max}}{2}, \tag{3}$$

where λ is the wavelength of the laser source. Because the CCD cameras have resolutions of around 100 lines/millimeter, the maximum angle between the object and reference waves is limited to a few degrees. Approaches have been described to provide the possibility of working with larger angles between interference waves, meaning thereby that larger object sizes can also be accommodated by the method. Examples of phase distribution images obtained with digital holographic interferometry are shown in Fig. 9.4.

9.2.2   Fringe Localization and Decorrelation Aspects in Holographic Interferometry

The two wavefronts released for interference in double-exposure holographic interferometry are exactly identical both in their broad structure and in their detail. The wave scattered by a rough surface shows rapid variations of phase, which have little correlation across the wavefront. The perfect identity of the wavefronts in their minute-detail structure is an essential condition for the two waves to interfere and give rise to a phase difference, at an observation point O. Because the condition of the identity of the minute-detail structure of the waves is fulfilled only by the waves originating from the corresponding points on the two object surfaces, undeformed and deformed, any pair of waves scattered from the regions other than the identical points do not contribute to the fringe formation. The phase difference varies randomly over the ray directions contained within the aperture of the viewing system that would normally prevent the observation of fringes with large aperture optics near such points. However, a surface may exist at a distance from the object at which the value of phase difference is stationary over the cone of ray pairs defined by the viewing aperture. Interference fringes are localized in the region where the variation in phase difference is minimum over the range of viewing directions. This approach has been widely used to compute fringe localization

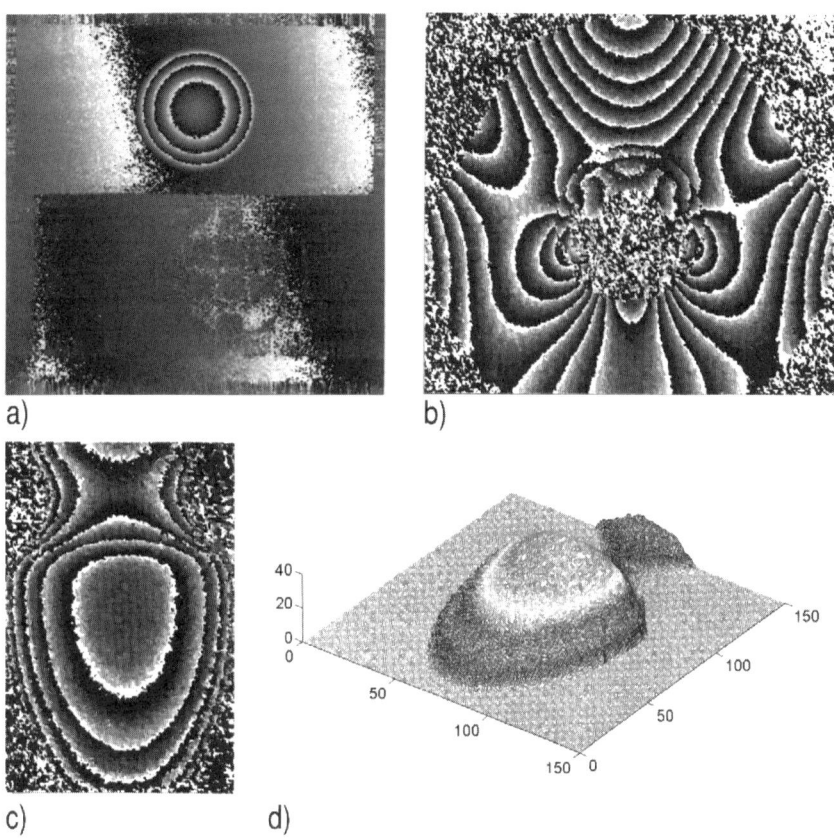

FIGURE 9.4. Examples of the application of digital holographic interferometry: (a) phase distribution obtained by the change in refractive index method; (b) phase map of a vibrating object; (c) shape measurement of a vase; contour interval in the phase map corresponds to 7.1 mm; (d) 3D representation of the object in (c); scale is shown in millimeters. (a: Courtesy of T. Kreis; b–d: Courtesy of G. Pedrini).

for any object displacement and for any arbitrary illumination and observation geometry.

A factor leading to the loss of fringe visibility arises from the relative displacement of the interfering rays at the observation plane. In the presence of a lateral displacement component at the image plane, the fringes tend to move away from the object surface. The fringe visibility, equal to the normalized autocorrelation function of the diffracted waves, is given by the Fourier transform of the pupil function limiting the spectral extent of the object. Another factor that adds to a loss of fringe visibility arises from the relative displacement of the interfering wavefronts at the pupil plane (Fig. 9.5a, b). This displacement produces a decorrelation of the elements of the wavefronts that contribute to the image formation. Certain portions of the wavefront that were within the aperture before loading are

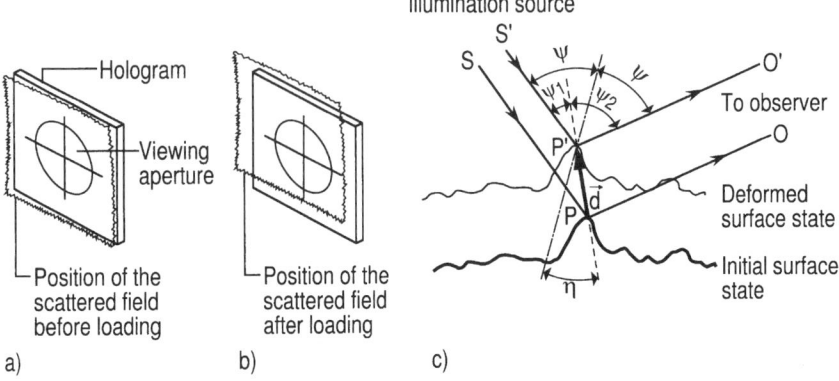

FIGURE 9.5. In the real-time holographic interferometry, configuration (a) shows the position of the recorded wavefront in front of the hologram, with the object in its state of rest, and (b) shows the position of the reconstructed wave-front in front of the hologram, after the loading of the object. (c) Optical geometry to calculate the optical phase change caused by an arbitrary displacement of a surface point P.

no longer within it after loading. Inversely, portions of the wavefronts not going through the aperture originally will pass through after deformation. Because, due to the random nature of the local detail of the two wavefronts, only identical regions can interfere, a loss of correlation between the two wavefronts is observed by way of a decrease in the fringe visibility of the interference pattern.

9.2.3 Quantitative Measurement of Wavefront Phase

A diffusely reflecting object is deformed such that a point P on the surface moves to P′ as shown in Fig. 9.5c. In double-exposure or real-time holographic interferometry, the two wavefronts related to the states of a surface before and after deformation are reconstructed simultaneously and compared. The surface is illuminated by light incident in the direction SP and is viewed along the direction PO. The image intensity of a holographic interferogram is

$$I(x, y) = I_0(x, y)\{1 + V(x, y)\cos \Delta\varphi(x, y)\}, \tag{4}$$

where $I_0(x, y)$ is the background intensity and $V(x, y)$ is the fringe contrast. The change of optical path length due to displacement d is

$$\Delta\varphi = \frac{2\pi d}{\lambda}(\cos \Psi_1 + \cos \Psi_2), \tag{5}$$

where Ψ_1 and Ψ_2 are the angles that the illumination and observation waves make with respect to the direction of displacement; (5) can be written in the form

$$\Delta\varphi = \frac{2\pi d}{\lambda}\cos \eta \cos \Psi, \tag{6}$$

where Ψ is the angle of bisection between the optical paths $S'P'$ and $P'O'$ and η is the angle that the bisector makes with the direction of displacement. The term $d\cos\eta$ that is the resolved part of the displacement PP' in the direction of the bisector implies that the fringe pattern provides the measure of the displacement component along the bisector of the angle between the incident and viewing directions.

Given the exponential growth in the power of digital computers and image processing techniques, the last decade has seen a rapid development of techniques for the automatic and precise reconstruction of phases from fringe patterns. These are based on the concepts of fringe tracking, Fourier transform, carrier frequency, and phase shift interferometry.[28−30] Quantitative phase data for object displacements, for example, can be generated in a few seconds using phase-shift interferometry techniques. With this technique for the quantitative measurement of the wavefront phase, one records a series of holograms by introducing artificially known steps of phase differences in the interference image in (4)

$$I_i(x, y) = I_0(x, y)\{1 + V(x, y)\cos(\Delta\Psi_i(x, y) + \Delta\Psi_i)\}, \qquad i = 0, 1, 2, \ldots,$$
(7)

where $\Delta\Psi$ is the phase shift that is produced, for example, by shifting the phase of the reference wave. A minimum of three intensity patterns are required to calculate the phase at each point on the object surface. Known phase shifts are commonly introduced using a piezoelectric transducer, which serves to shorten or elongate the reference beam path by a fraction of wavelength; the phase-shifter is placed in one arm of the interferometer. Numerous phase-shifting techniques have been developed and can be incorporated to holographic interferometry to generate a phase map, $\Delta\varphi(x, y)$, which corresponds to the object displacement. The calculated phase $\Delta\varphi(x, y)$ is independent of the terms $I_0(x, y)$ and $V(x, y)$, which considerably reduces the dependence of the accuracy of measurements on the fringe quality. The introduction of phase-shift interferometry not only provides accurate phase measurements, but also eliminates the problem of phase sign ambiguity of the interference fringes. The phase-shifting techniques have also found use in the quantitative determination of object deformations, vibration modes, surface shapes, and flow patterns.[28−34] Examples of phase contours obtained with phase-shifting holographic interferometry are shown in Fig. 9.6. Dynamic phase-shifting techniques allow for producing time sequences of deformation maps for studying objects subjected to time-varying loads.[35−36] The concept is based on considering the normal phase variations caused by a dynamic phenomena as equivalent to the phase shifting that is introduced in a holographic interferometer to perform the phase evaluation task. An example of the result obtained by applying dynamic phase shifting is shown in Fig. 9.7. This method extends the possibility of applying the method to continuous deformation measurements.

9.2.4 Measurement of Static Displacements

Equation (6) shows that observation along a single direction yields information only about the resolved part of the surface displacement in one particular direction.

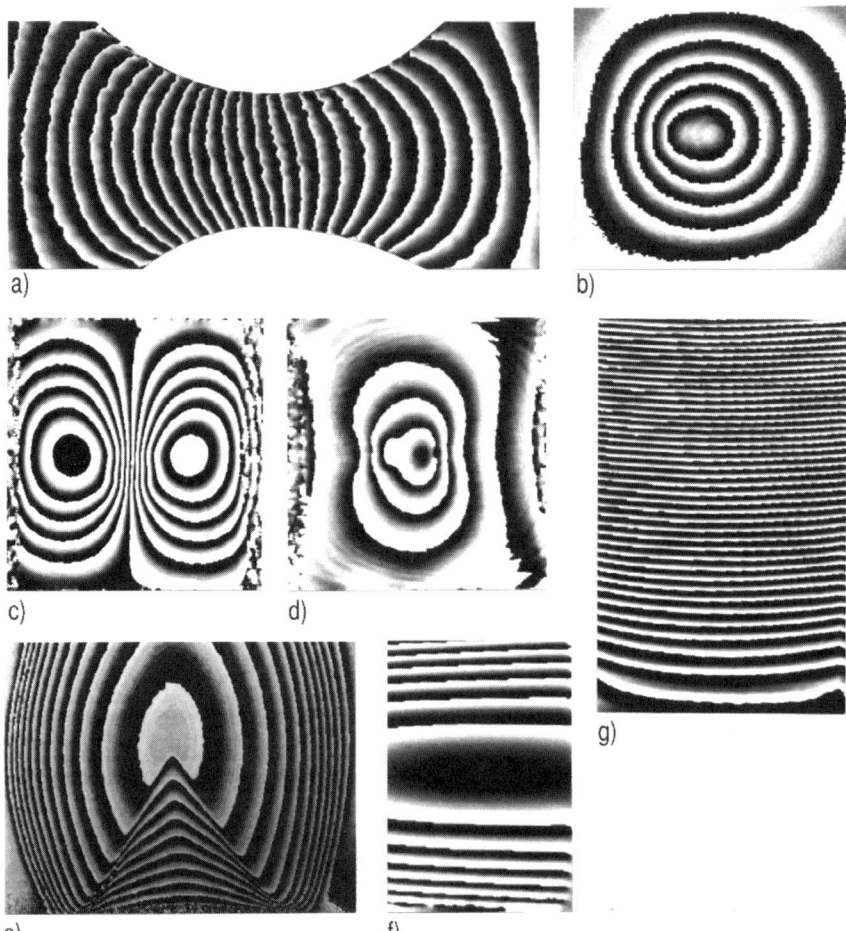

FIGURE 9.6. Examples of holographic interferometry implemented with phase-shifting techniques: (a) in-plane phase contours of an aluminum sheet under tension; (b) difference displacement contours reveal the presence of a defect in an aluminum plate; (c) phase contours relative to slope change in an aluminum plate clamped along its edges and submitted to a load applied at its center; (d) phase contours relative to curvature or the second derivative of displacement along x-direction reveal the presence of a flaw in an aluminum plate; (e) phase distribution of a test flame; (f) topographic phase contours of a 3D object; (g) phase distribution corresponding to a large out-of-plane displacement; the distance between two adjacent contour planes is 16.5 μm. (e: Courtesy of S.M. Tieng).

The approach to measuring 3D vector displacement would be to record holograms from three different directions to obtain three independent resolved components of displacement. The computation of the resulting system of equations would then allow for determining the complete displacement vector.

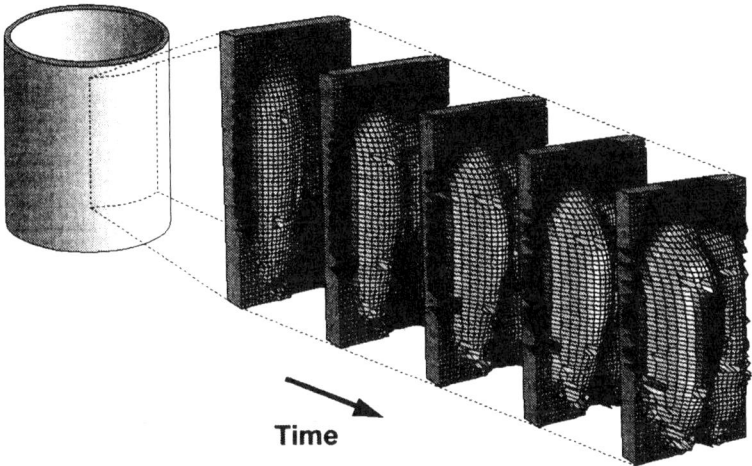

FIGURE 9.7. Deformation states of a composite spatial-telescope body submitted to localized thermal loading. The fringe pattern is recorded at a camera rate of 10 Hz. A time-frequency analysis based on wavelet transforms is used to reconstruct the "movie" of the deformation. The five images above span a time interval of 40 seconds. (Courtesy of X. Colonna de Lega).

Because the object surface is usually illuminated and viewed in a near-normal direction, the fringe patterns related to the line of sight component of displacement are simple to obtain. The corresponding fringe equation is given by

$$w = \frac{n\lambda}{2},$$

(8)

where w is the out-of-plane component of displacement and n is the fringe number.

On the other hand, displacements orthogonal to the line of sight can be obtained by illuminating the surface at near-grazing incidence and viewing it in a direction close to the illumination beam. The procedure, however, has the drawback that one dimension of the surface as viewed would be foreshortened to zero. A number of optical configurations have been devised to overcome this problem. The in-plane component of displacement can be measured by illuminating the object along two directions symmetric to the surface normal and observing along a common direction.[37] The in-plane displacements, in a direction containing the two beams, are mapped as a moiré between the interferograms due to each illumination beam. The moiré appears as a family of fringes

$$u = \frac{n\lambda}{2 \sin \theta_e},$$

(9)

where u is the displacement component in the x-direction and θ_e is the angle that the illumination beams make with the surface normal. Figure 9.3b shows the u pattern corresponding to a disk under diametral compression. Of the other methods of interest for measuring in-plane displacements, Rastogi[38] proposes recovering the in-plane displacement information by reconstructing the reference wave of a double-beam illuminated object, Pirodda[39] proposes the use of conjugate wave holographic interferometry to obtain fringes corresponding to in-plane displacements, and more recently, Rastogi and Denarie[40] have extended the use of the phase-shifting technique to facilitate the quantitative analysis of fringe patterns obtained with holographic moiré. The phase map in Fig. 9.6a illustrates the type of result that one can obtain with phase-shifting holographic interference. Examples of application of the method to study crack propagation and creep behavior in wood are shown in Figs. 9.8a, b. The fringe contours in Fig. 9.8a and b correspond to the out-of-plane, w, and in-plane, u, components of displacement, respectively. The method has also been used to investigate the formation of the fracture process zone in a fiber-reinforced concrete specimen subjected to a wedge-splitting test. An example of the type of results obtained in an in-plane displacement-sensitive interferometer is shown in Fig. 9.8c.

Another technique of significant interest in nondestructive testing is comparative holography that provides the contours of path variations related to the difference in displacements or shapes of two objects. The possibility of instantaneously comparing two objects is of considerable interest in nondestructive inspection. A number of approaches have been developed to compare two macroscopically similar but physically different surfaces.[41−45] These methods provide information about the resolved part of the difference in displacement vector along the direction bisecting the illumination and observation rays. In the case of illumination and observation normal to the object surface, the fringe equation becomes

$$\Delta w = \frac{n\lambda}{2},\tag{10}$$

where Δw is the difference in displacement component along z-direction. The technique provides a tool to compare the mechanical responses of two nominally identical specimens subjected to same loading and as well the shapes of two nominally identical specimens. These features are useful in detecting anomalies in a test specimen with respect to the flaw-free master specimen. Figures 9.3c and 9.6b illustrate fringe contours corresponding to Δw.

9.2.5 Holographic Shearing Interferometry

Holographic shearing interferometry provides directly the patterns of slope change contours by laterally shearing wavefronts diffracted from the object surface. There are numerous ways to achieve shearing, such as, by introducing a Michelson-type configuration, an inclined glass-plate device or a split-lens assembly in a holographic interferometer. The role of the shearing device is to enable the observation of a point on the object along two distinct neighboring directions. As a result of lateral wavefront shearing in the interferometer, a point in the image plane receives contributions from two different points on the object. Assuming that the illumi-

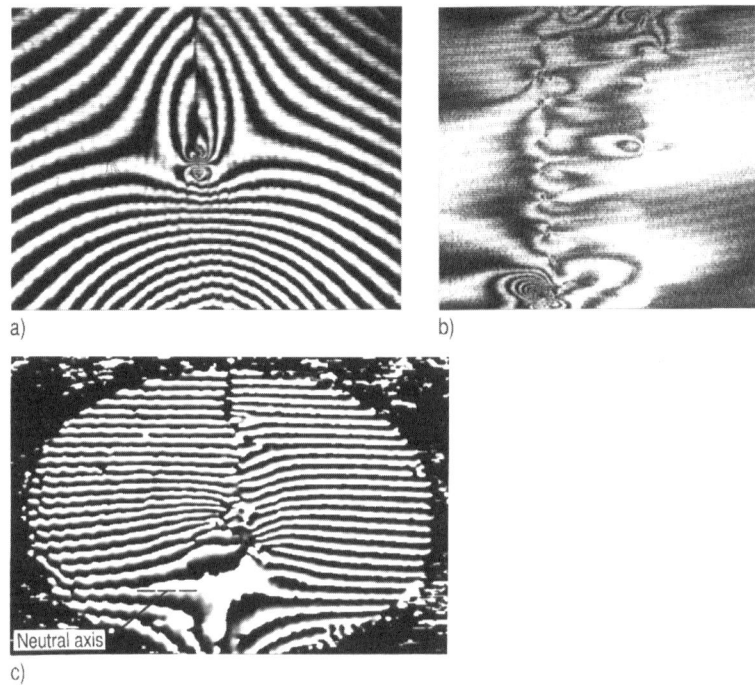

FIGURE 9.8. Examples of applications of holographic interferometry in materials testing: (a) phase map corresponding to the out-of-plane displacement component, w, in a wood specimen subjected to a wedge-splitting test; (b) interferogram corresponding to the in-plane displacement component, u, in a fingertip joint wood specimen under tensile load; the phase contours depict the creep behavior in the wood assembly; (c) phase map corresponding to the in-plane displacement component, u, in a fiber-reinforced concrete specimen submitted to a wedge-splitting test. The interferograms also illustrate the process of crack propagation and formation of fracture process zone in the two tested materials.

nation beam lies in the x-z plane and makes an angle θ_e with the z-axis, and the observation is carried along the direction normal to the object surface, the change of optical phase due to deformation is

$$\Delta\varphi = \frac{2\pi}{\lambda}\left[\frac{\partial u}{\partial x}\sin\theta_e + \frac{\partial w}{\partial x}(1 + \cos\theta_e)\right]\Delta x, \qquad (11)$$

where Δx is the object shear along the x-direction. For $\theta_e = 0$, (11) reduces to

$$\frac{\partial w}{\partial x} = \frac{n\lambda}{2\Delta x}. \qquad (12)$$

Equation (12) displays a family of holographic moiré fringes corresponding to the contours of constant slope change. Examples of related fringe patterns are shown in Figs. 9.3f and 9.6c, which illustrate the case of a centrally loaded square aluminium plate clamped along its boundary. A dual image-shearing device has been used to obtain whole-field phase maps corresponding to curvature, Fig. 9.6d, and twist distributions.[46]

9.2.6   Vibration Measurement

Holographic interferometry has considerably simplified the investigation of resonance frequencies and the mode shape of vibration of specimens.[31, 32, 47–49] Consider that a hologram is recorded of an object vibrating sinusoidally in a direction normal to its surface. The exposure time is supposed to be much longer than the period of vibration. The specimen is illuminated at an angle θ_e and viewed at an angle θ_o to its surface normal. The complex amplitude of the wave reconstructed by the hologram is proportional to the time average of the complex amplitude of light in the hologram plane over the exposure interval T. The squaring of the reconstructed complex amplitude leads to the expression for the irradiance distribution in the image

$$I(x, y) = I_0(x, y)J_o^2\left[\frac{2\pi}{\lambda}d(x, y)(\cos\theta_e + \cos\theta_0)\right],\tag{13}$$

where $d(x, y)$ is the amplitude of vibration, ω is the circular frequency of vibration, and J_o is the zero-order Bessel function of the first kind. The virtual image is modulated by the $J_o^2(\xi)$ function. The dark fringes correspond to the zeros of the function $J_o^2(\xi)$. The fringe pattern is characterized by a very bright zero-order fringe, which corresponds to the nodes, a decreasing irradiance, and an unequal spacing between the successive zeros. The method does not provide the user with information on the relative phase of vibration between different object points nor on the mode in which the surface is vibrating until the hologram has been recorded and reconstructed. Holographic time average fringes shown in Fig. 9.4d and e correspond to the vibration analysis of an Agusta helicopter component.

The possibilities of studying the response of a vibrating object in real time extends the usefulness of the technique to identify the resonance states of the object. A single exposure is made of the object in its state of rest. The plate is processed, returned to its original condition, and reconstructed. The observer looking through the hologram at the sinusoidally vibrating object sees the time-averaged intensity

$$I(x, y) = I_0(x, y)\left[1 - J_0\left\{\frac{2\pi}{\lambda}d(x, y)(\cos\theta_e + \cos\theta_o)\right\}\right].\tag{14}$$

The method has half the sensitivity of the time-average holography. The fringe contrast is much lower than the time-average fringes.

Stroboscopic holography is an alternative approach to study vibrations. The hologram is recorded by exposing the photographic plate twice for short time intervals during a vibration cycle. The pulsed exposures are synchronized with the vibrating surface. This being equivalent to making the surface virtually stationary during the recording yields cosinusoidal fringes on reconstruction. The constant displacement fringes are related to the change in object position between the two exposures. In the real-time variant of stroboscopic holography, a hologram of a nonvibrating object is first recorded. If the vibrating object is illuminated strobo-scopically and viewed through the hologram, the reconstructed image from the hologram interferes directly with the light scattered from the object to generate live fringes.

Temporally modulated holography represents a more generalized form of vibra-tion analysis in which the object or reference waves are time dependent. Temporal modulation techniques in their different variants have been employed to produce useful functions such as enhancing or reducing the sensitivity of measurements, or to map the relative phase as well as the amplitude of a vibrating surface. The variants most often employed are frequency-translated holography, amplitude-modulated holography, and phase-modulated holography. Double-pulse holography has be-come a routine technique for vibration analysis of nonrotating objects. The double pulse freezes the object at two points in the vibration cycle. The study of the vibra-tion of rotating structures would also require some form of rotation compensation to ensure correlation between the wavefields scattered from the two states of the vibrating and rotating object.[50] An example of the application of holographic inter-ferometry in a nonlaboratory environment, for example, in the vibration analysis of a railway bridge is shown in Fig. 9.9. The knowledge of amplitude distribution of vibrations is necessary to optimize the design characteristics of a bridge with regard to the noise emission.

9.2.7 Flow Measurement

Holographic interferometry is a useful technique in the observation of the spatial variation of refractive index, density and temperature distributions, and parameters of paramount importance in areas such as aerodynamics, plasma diagnostics, and heat transfer.[33, 34, 51–53] In a flow-visualization-specific holographic interferome-ter, two consecutive exposures are made, usually the first exposure without flow and the second in the presence of a flow field. The motion of the fluid flow inside the test section causes density variations. Double-pulsed holography is used if the flow field is changing rapidly. The change of optical phase difference due to flow between the exposures is

$$\Delta\varphi(x, y) = \frac{2\pi}{\lambda} \int_0^L \{n(x, y, z) - n_0\}dz, \tag{15}$$

where $n(x, y, z)$ is the refractive index distribution during the second exposure, n_o is the uniform refractive index during the first exposure, and L is the length of the test section. Using the Gladstone–Dale equation, the expression for phase difference becomes

$$\Delta\varphi(x, y) = K \int_0^L \rho(x, y, z)dz - KL\rho_0, \tag{16}$$

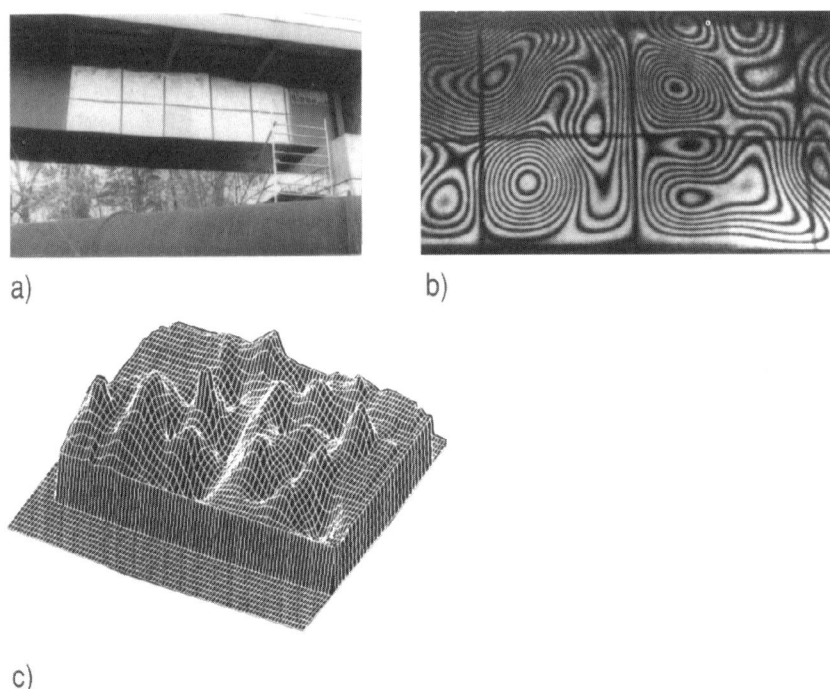

a)   b)

c)

FIGURE 9.9. Example of the application of holographic interferometry to the investigation of large structures in nonlaboratory conditions: (b) and (c) display results obtained from vibration analysis of a railway bridge in (a). A running train excites the vibrations in the bridge. (Courtesy of H. Steinbichler, Labor Dr. Steinbichler GmbH).

where K is the Gladstone–Dale constant and ρ_o is the density in the no-flow situation. Assuming that the properties of the flow are constant in the z-direction,

$$\Delta\varphi(x, y) = KL\{\rho(x, y) - \rho_0\}. \tag{17}$$

The interference pattern contours the change in the density field of the flow. The change in density per fringe is given by λ/KL. Figures 9.3g, and h and 9.6e illustrate interferograms of flow fields.

Holographic interferometry has found use in a wide range of applications in experimental mechanics and nondestructive testing. A glimpse of some of these applications is found in Fig. 9.10.

9.2.8   Contouring of Three-Dimensional Objects

Holographic interferometry encompasses a group of techniques that provide information on the relief variations of a 3D object surface.[8, 54–57] A practical way to display this information consists of obtaining contour maps that show the intersection of an object with a set of equidistant planes perpendicular to the line of sight. The wavelength change and immersion methods apart, holographic tech-

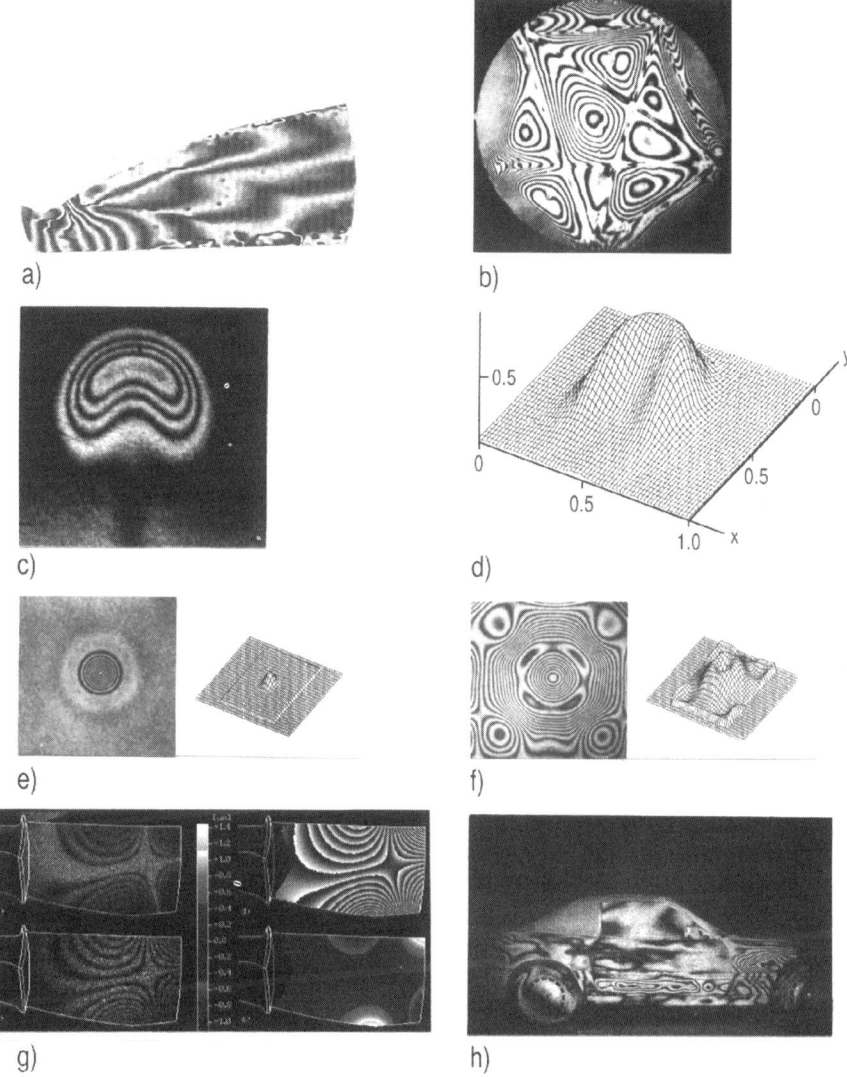

FIGURE 9.10. Examples of the application of holographic interferometry in diverse areas to illustrate the method's usefulness and its ever widening acceptability in nondestructive measurement and testing: (a) visualization of the flow field of a high-speed jet; (b) measurement of thermal deformations of a parabolic antenna under space-simulated conditions; (c, d) visualization of the temperature field in a liquid and the 3D plot of the field; (e, f) display of amplitude distributions produced on a metal plate after (e) 90 μs and (f) 500 μs of the application of an impact load on the plate; (g) vibration analysis of a turbine blade in an out-of-plane displacement configuration (*see color insert*); (h) vibration analysis of a car body excited in a wind tunnel. (Courtesy of H. Steinbichler, Labor Dr. Steinbichler GmbH).

niques for contouring applications are, in general, the fruit of modifications in the illumination and observation directions.

In the two-wavelength method, an on-axis collimated beam illuminates the object. The holographic recording of the object is made on a photographic plate. A light of slightly different wavelength $\lambda_2 (\lambda_1 \neq \lambda_2)$ now illuminates the object and the hologram. This results in the lateral and longitudinal displacements of the reconstructed image. Tilting the collimated reference beam by an appropriate amount eliminates the lateral displacement. The interference fringes arise due to the longitudinal displacement of the reconstructed image with respect to the real position of the object. The contour sensitivity per fringe is given by

$$\Delta z = \frac{\lambda_1 \lambda_2}{2(\lambda_1 - \lambda_2)} \tag{18}$$

The two lines 477 nm and 488 nm from an argon ion laser would, for example, give a contour interval of the order of 10 μm.

In the immersion method, the specimen is immersed in a transparent liquid of refractive index n_1. A hologram of the object is recorded, and the liquid is next changed to one of refractive index n_2. The interference fringes arise due to the change of optical path in the light rays traversing the two liquids. The successive fringes in the interference pattern correspond to increments of

$$\Delta_z = \frac{\lambda}{2|n_2 - n_1|}. \tag{19}$$

In the dual-beam multiple sources contouring, the specimen is illuminated by two beams incident at equal angles on both sides of the optical axis. A hologram of the object is recorded. The illumination beams are then tilted by the same amount, $\Delta\theta_e$, and in the same direction around an axis perpendicular to the plane of the paper. An observer looking through the hologram sees a family of moiré fringes. The moiré results from the interaction of the fringe patterns generated by the two beams. Moiré fringes with an incremental height interval of Δz are seen to intersect the object, where

$$\Delta z = \frac{\lambda}{2 \sin \theta_e \sin \Delta\theta_e}. \tag{20}$$

In the multiple sources contouring method, illuminating the specimen from two different directions generates the contour fringe pattern. The waves scattered from an object illuminated by a collimated beam at an angle are recorded on a holographic plate, and the object and its reconstructed image are viewed through the hologram. A set of parallel equidistant fringes is projected onto the object surface by introducing a small tilt, $\Delta\theta_e$, to the illumination beam. Providing an appropriate reference beam rotation and a holographic plate translation generates contouring surfaces normal to the line of sight. The contour sensitivity per fringe is given by

$$\Delta z = \frac{\lambda}{\sin \theta_e \sin \Delta\theta_e}. \tag{21}$$

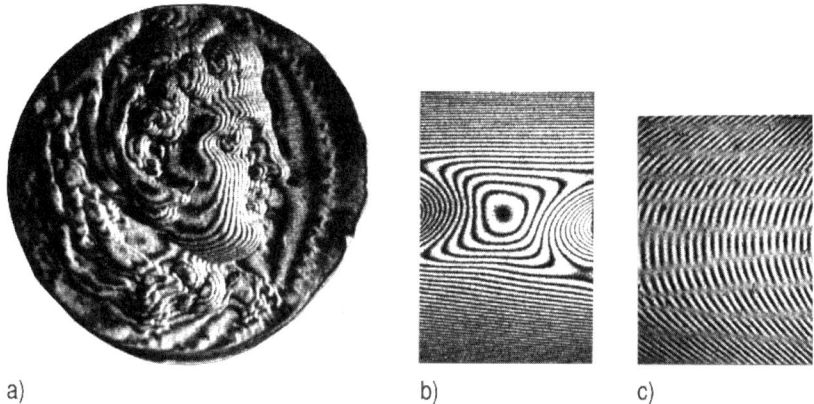

a)   b)   c)

FIGURE 9.11. Examples of the application of holographic interferometry to shape measurement: (a, b) topographic contours with sensitivities of 170 μm and 135 μm, respectively; (c) fringe contours corresponding to slope change variations of a curved surface.

This approach has been applied to the measurement of large out-of-plane deformations undergone by a deformed object.[58] Figure 9.6g displays phase distribution corresponding to the out-of-plane displacement of a cantilever beam subjected to a large load at its free end. The distance between two contour planes is 16.5 μm. On the other hand, examples of shape measurement using holographic interferometry are shown in Figs. 9.6f and 9.11a, and b.

Holographic shearing interferometry has been extended to measure the variation of slopes on a 3D surface. The configuration[59] uses the principle of wavefront shearing to bring to coincidence in the observation plane the waves scattered by two neighboring points on the specimen surface. The surface is illuminated by a plane wave parallel to the x-z plane and making an angle θ_e to the surface normal. In the real-time recording scheme, the initial sheared wavefields from the two scattering points are recorded on a holographic plate. The illumination beam is next tilted by a small angle $\Delta\theta_e$. The interference between the live and the reconstructed wavefields gives rise to a family of moiré fringes. The distance between two adjacent contour planes corresponds to

$$\Delta\frac{\partial z}{\partial x} = \frac{\lambda}{\Delta x \sin\theta_e \sin\Delta\theta_e},$$
(22)

where Δ_x is the lateral shift along the x-axis. The sensitivities of the described methods can be tuned in a wide range. An example of topographic slope contouring using holographic moiré is shown in Fig. 9.11c.

9.3   Speckle Techniques

When an optically rough surface is illuminated with a coherent light, the rays are scattered in all directions and at all distances from the surface. The scattered wave distribution displays a random spatial variation of intensity called a speckle pattern. The speckled appearance is caused by the interference of the elementary waves emanating from the surface. A wide range of speckle techniques, has emerged, which are classified into three broad categories: speckle photography, speckle interferometry, and speckle shearing interferometry. Background knowledge and a strong insight in the field can be developed by reading the series of books and book chapters in Jones and Wykes,[6] Francon,[10] Erf,[11] Dainty,[12] Sirohi,[13] Malacara,[14] Williams,[15] Cloud,[16] and Rastogi,[17, 18]. A selected list of trend setting papers in the area can be found in Sirohi[60] and Meinlschmidt et al.[61] Recently, two entire journal issues have appeared on the subject.[62, 63]

9.3.1   Speckle Photography

The bulk motion of subjective speckle patterns is used as an information carrier of a variety of mechanical quantities, such as deformations, strains, and shapes. A subjective speckle pattern of a specimen is recorded on a photographic plate. The specimen is deformed, and a second exposure is made on the same plate. At an arbitrary point on the object surface, the components of speckle shifts $\Omega_x(x, y)$ and $\Omega_y(x, y)$ along x- and y- directions are, respectively, given by

$$\Omega_x(x, y) = u(x, y) + \frac{x}{d_o} w(x, y), \tag{23}$$

$$\Omega_y(x, y) = v(x, y) + \frac{y}{d_o} w(x, y), \tag{24}$$

where v is the in-plane component of displacement in the y-direction. The influence of the out-of-plane displacement component on the in-plane component of speckle shift becomes progressively important as the observation point on the object surface shifts away from the optical axis.

The photographic plate records two laterally displaced speckle patterns that are relatively shifted. The shift between speckles can be obtained by pointwise filtering or by whole-field filtering. The point-by-point approach consists of illuminating the recorded specklegram by a narrow laser beam. The laser beam diffracted by the speckles lying within the beam area gives rise to a diffraction halo whose distribution is given by the autocorrelation of the aperture function. The halo is modulated by an equidistant system of fringes arising from the interference of two identical but displaced speckles. The direction of these fringes is perpendicular to the direction of displacement. The magnitude of displacement, inversely proportional to the fringe spacing p, is given by

$$d_\xi = m\sqrt{\Omega_x^2 + \Omega_y^2} = \frac{\lambda s}{p}, \tag{25}$$

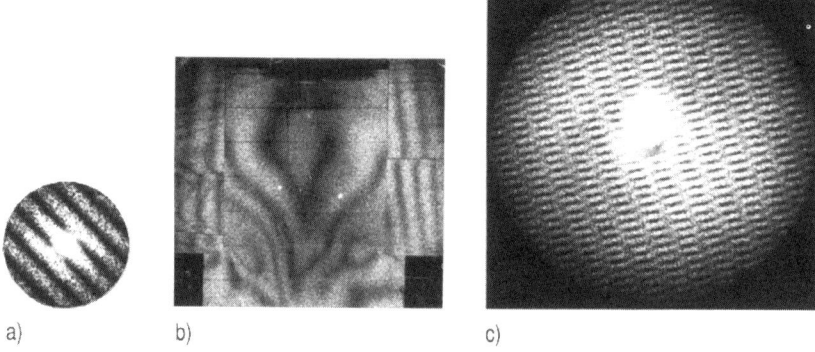

FIGURE 9.12. Examples of (a) Young's fringe pattern obtained by pointwise analysis of a specklegram; (b) in-plane displacement, v, contours on a wooden beam structure subjected to deformation obtained by whole-field filtering of a specklegram; (c) Young's moiré obtained by pointwise analysis of a speckle-shearogram; moiré fringes are directly related to the in-plane strain undergone by the interrogated object point.

where d_ξ is the displacement in the image plane, m is the magnification factor of the speckle recording imaging system, and s is the distance between the specklegram and the observation screen. The halo modulated by an equidistant system of fringes or the so-called Young's fringes is shown in Fig. 9.12a.

The reconstruction of a whole-field interference image requires placing the recorded specklegram in a Fourier filtering arrangement. An observer, looking through a small offset aperture placed in the focal plane of the Fourier transform lens, sees contours of speckle displacement along the direction defined by the position of the aperture. The fringe contours are described as the loci of points, where

$$d_\xi = \frac{n\lambda f}{\zeta_\xi}, \qquad (26)$$

f is the focal length of the Fourier transform lens, ξ denotes the azimuth, and ζ_ξ is the radial distance from the position of the zero diffraction order. The in-plane displacement contours, relative to u, obtained on a wooden beam structure by whole field filtering are shown in Fig. 9.12b.

Very recently, a novel technique based on the principles of speckle photography and wave shearing, appropriately termed as speckle shearing photography, has emerged for direct measurement of in-plane surface strains.[64] The strain information stored at each point on the object surface is read out on a pointwise basis. The halo modulated by an equidistant system of moiré fringes or the so-called Young's moiré fringes is shown in Fig. 9.12c.

High-resolution photographic film is generally used to register the speckle patterns in speckle photography. The rapid development of solid-state detectors and computers during the last decade has, however, made video recording of speckle patterns and computer processing of the images an interesting alternative. Electronic speckle photography analyzes the motion of a characteristic random pattern before and after displacement on a finite subimage basis. Fast and accurate algorithms have been developed that perform the cross-correlation calculations in the frequency domain via fast Fourier transform algorithms. Electronic speckle photography uses the peak position of a 2D cross-correlation function to quantitatively represent the speckle displacement, the measurement accuracy being proportional to the width of the correlation peak. Despite the lower resolution offered by solid-state detectors, the electronic speckle photography has the advantage of being easy to use, and free of any sign of ambiguity in displacement direction measurement because all images are stored on different frames and the digital processing of images offers a greater flexibility in the choice of the evaluation algorithm.

9.3.2   Speckle and Speckle Shearing Interferometry

Speckle interferometry is an extremely versatile tool for the measurement of out-of-plane and in-plane components of displacements as well as for nondestructive inspection. There is an ever-increasing interest in the application of speckle interferometry to measure the shape of 3D objects. Methods have also been investigated to obtain directly the spatial derivatives of the surface topography of an arbitrarily shaped object. Speckle shearing interferometry has been widely used for measuring derivatives of surface displacement of objects undergoing deformation. The later methods make use of a shearing device in the imaging system, which results in two superimposed images of the object on the CCD chip of an electronic camera. Of sensitivity comparable to that of holographic interferometry, these methods are based on the coherent addition of speckle fields diffracted by object and reference fields.

A useful method of generating speckle correlation fringes is based on the electronic subtraction of intensities, before and after object deformation. Unlike traditional photographic film-based methods, the digital extensions of speckle interferometry or speckle shearing interferometry allow quantitative phase data to be obtained quickly and in electronic form. The scope of this section will be limited to summarizing the basic principles of phase contour map formation in digital speckle pattern interferometry (DSPI) and digital speckle shearing pattern interferometry (DSSPI). Because most of the basic optical techniques used to measure surface deformations and surface topography in holographic interferometry have found extension and have been configured to DSPI and DSSPI, no attempt will be made in this section to deal with specific optical configurations used in different measurement tasks.

Digital Speckle Pattern and Digital Speckle Pattern Shearing Interferometry

Figure 9.13 shows a schematic of the DSPI arrangement for out-of-plane component measurement. The outcoming beam from a laser source is divided into an object beam and a reference beam. A reference beam derived from the same laser source as the object beam is added to the object beam on the CCD array. The

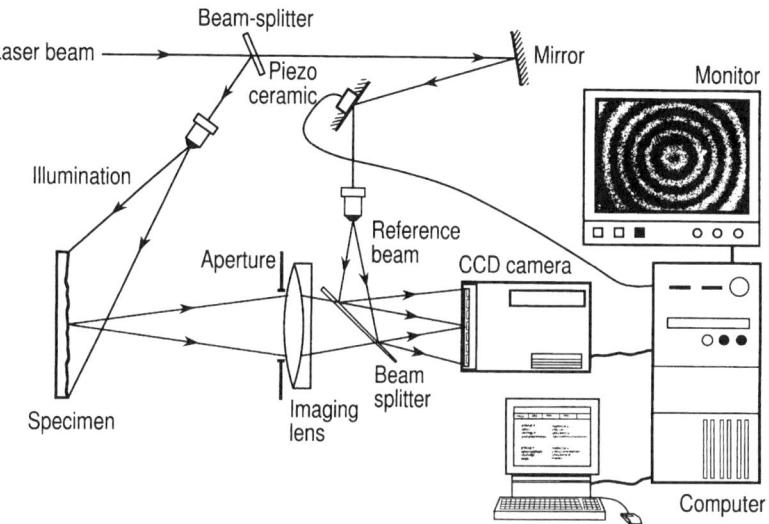

FIGURE 9.13. Schematic of a digital speckle pattern interferometer in an out-of-plane displacement-sensitive configuration.

reference beam is either a plane wave or another speckle field not necessarily originating from the object surface under study. The addition of a reference beam to the object wave introduces a significant change in the behavior of the resultant speckle pattern when the object is deformed. Because the resultant speckle pattern is formed by the interference of two coherent fields, the intensity in the pattern would naturally depend on the relative phase distributions of the added fields. Writing the complex amplitudes of the object, A_1, and reference, A_2, beams at any point (x, y) on the CCD array as

$$A_1(x, y) = a_1(x, y) \exp(i\varphi_1(x, y)), \tag{27}$$
$$A_2(x, y) = a_2(x, y) \exp(i\varphi_2(x, y)), \tag{28}$$

where $a_1(x, y)$ is the amplitude and $\varphi_1(x, y)$ is the phase of the light scattered from the specimen surface in its initial state. The irradiance at any point (x, y) on the image plane is

$$I_i(x, y) = |A_1(x, y)|^2 + |A_2(x, y)|^2 + 2|A_1(x, y)| |A_2(x, y)| \cos(\varphi_R(x, y)), \tag{29}$$

where the quantity $\varphi_R(x, y)$ represents the random phase. Assuming the output camera signal to be proportional to the irradiance, the video signal is

$$V_i \propto I_i(x, y). \tag{30}$$

The analog video signal v_i from the CCD camera is sent to an analog-to-digital converter, which samples the video signal at the TV rate and records it as a digital frame, say, of size 512×512 pixels and 256 intensity gradations, in the memory of a computer for subsequent processing.

If the object is deformed, the relative phase of the two fields will change, causing a variation in the irradiance of the speckle pattern. The displacements and displacement gradients applied on the object surface are assumed to be small. Following the application of the load, the complex amplitude of the wavefront scattered from the object surface onto the CCD array is

$$A_1'(x, y) = A_1(x, y) \exp(i \Delta\varphi(x, y)), \tag{31}$$

where $\Delta\varphi(x, y)$ is the phase difference introduced in the original object beam due to object deformation.

The irradiance distribution after object deformation becomes

$$I_f(x, y) = |A_1(x, y)|^2 + |A_2(x, y)|^2$$
$$+ 2|A_1(x, y)| |A_2(x, y)| \cos(\varphi_R(x, y) - \Delta\varphi(x, y)). \tag{32}$$

A common method to produce speckle correlation fringes is based on the direct electronic subtraction of the intensity of the displaced surface from that of the initial surface state. The video signal v_f is digitized and subtracted pixel by pixel from the digital frame corresponding to v_i. The subtracted video signal v is given by

$$v = 4a_1a_2 \sin\left(\varphi_R + \frac{\Delta\varphi}{2}\right) \sin\frac{\Delta\varphi}{2}. \tag{33}$$

Because the video signal in (33) also has negative going values, it must be rectified. The full-wave rectification suppresses negative values. Consequently, the brightness on the monitor is proportional to $|v|$,

$$B = C|\sqrt{I_1 I_2} \sin\left(\varphi_R + \frac{\Delta\varphi}{2}\right) \sin\frac{\Delta\varphi}{2}|, \tag{34}$$

where, C is a proportionality constant, $\sqrt{I_1 I_2}$ gives the background illumination, and $\sin\left(\varphi_R + \frac{\Delta\varphi}{2}\right)$ represents the speckle noise that varies randomly between 0 to 1 across the object image; (34) describes the modulation of the high-frequency noise by a low-frequency interference pattern related to the phase difference term, $\Delta\varphi$.

The brightness will be minimum whenever

$$\Delta\varphi = 2n\pi, \qquad n = 0, 1, 2, \ldots. \tag{35}$$

Thus, all of those regions where the speckle patterns, before and after deformation, are correlated will appear dark. On the other hand, the condition for maximum brightness is

$$\Delta\varphi = (2n + 1)\pi, \qquad n = 0, 1, 2, \ldots. \tag{36}$$

This implies that all of those regions where the speckle patterns, before and after deformation, are uncorrelated will appear bright. Consequently, the monitor will display correlation fringes representing contours of constant $\Delta\varphi$. The computer controls the data acquisition, stores the digitized image—a number of image buffers of 512×512 pixels resolution are available—performs image processing, and interfaces with graphical peripheral for the display and printout of results. The high sampling rate means that 25 speckle patterns, by the European TV standard, or 30 speckle patterns, by the American TV standard, become available on the TV monitor each second. The developments in digital cameras have gone through a period of boom in these last few years, with emphasis on high resolution, i.e., ever more pixels packed in CCD sensors, and high frame speeds. Several approaches have been described for calculating the phase maps in DSPI. One such approach termed as "phase of the differences" for the generation of visible correlation fringes is described below.[65]

In digital speckle interferometry, the path length of the reference beam is varied in a controlled manner. This is done in Fig. 9.13 by displacing a mirror mounted on a piezoelectric transducer and placed in the reference arm of the interferometer. Of the schemes available for introducing phase shifts, one may cite the use of piezoelectric transducer, polarization-based devices, liquid crystal phase shifters, source wavelength modulation, and so on. DSPI allows for the measurement of phase values at individual pixels by the phase-stepping technique.

Assuming the object to be in its state of rest, this procedure requires measurements of the intensities $I_{i,1}(x, y)$, $I_{i,2}(x, y)$ and $I_{i,3}(x, y)$ at each pixel point (x, y) for three positions of the reference beam mirror. Between each recording, the phase is stepped by 0, $2\pi/3$ and $4\pi/3$, respectively. The corresponding speckle patterns may be written as

$$I_{i,k+1}(x, y) = I_1 + I_2 + 2\sqrt{I_1 I_2} \cos\left(\varphi_R + \frac{2k\pi}{3}\right), \qquad k = 0, 1, 2. \quad (37)$$

These intensity distributions are recorded and digitised. The speckle phase at each pixel can be calculated from

$$\varphi_R(x, y) = \tan^{-1} \frac{\sqrt{3}(I_{i,3} - I_{i,2})}{2I_{i,1} - I_{i,2} - I_{i,3}}. \quad (38)$$

The object is deformed. The corresponding speckle pattern becomes

$$I_{f,k+1}(x, y) = I_1 + I_2 + 2\sqrt{I_1 I_2} \cos\left(\varphi_R + \Delta\varphi + \frac{2k\pi}{3}\right), \qquad k = 0, 1, 2. \quad (39)$$

The computation of new speckle phase at each pixel gives

$$\varphi_R(x, y) + \Delta\varphi(x, y) = \tan^{-1} \frac{\sqrt{3}(I_{f,3} - I_{f,2})}{2I_{f,1} - I_{f,2} - I_{f,3}}. \quad (40)$$

Subtraction of (38) from (40) gives the value of the phase difference $\Delta\varphi$ arising from object deformation at each pixel point. A phase measurement example, experimental analogs of (38), (40), and $\Delta\varphi$, is shown in Figs 9.14a, b, and c, respectively. Other popular algorithms for phase calculations require measurement of the intensity of four, five, or more images. DSPI has found a wide range of applications in experimental mechanics and nondestructive testing. Some of these applications are shown in Fig. 9.15.

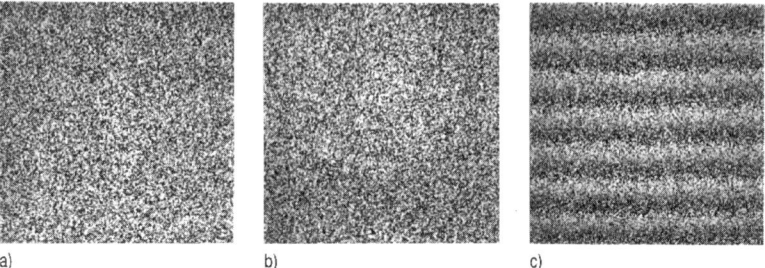

FIGURE 9.14. Display of (a) a reference phase map $\varphi_R(x, y)$; (b) a deformed phase map $\varphi_R(x, y) + \Delta\varphi(x, y)$; and (c) the difference between the phase maps in (a) and (b), $\Delta\varphi(x, y)$. (Courtesy of A. Moore).

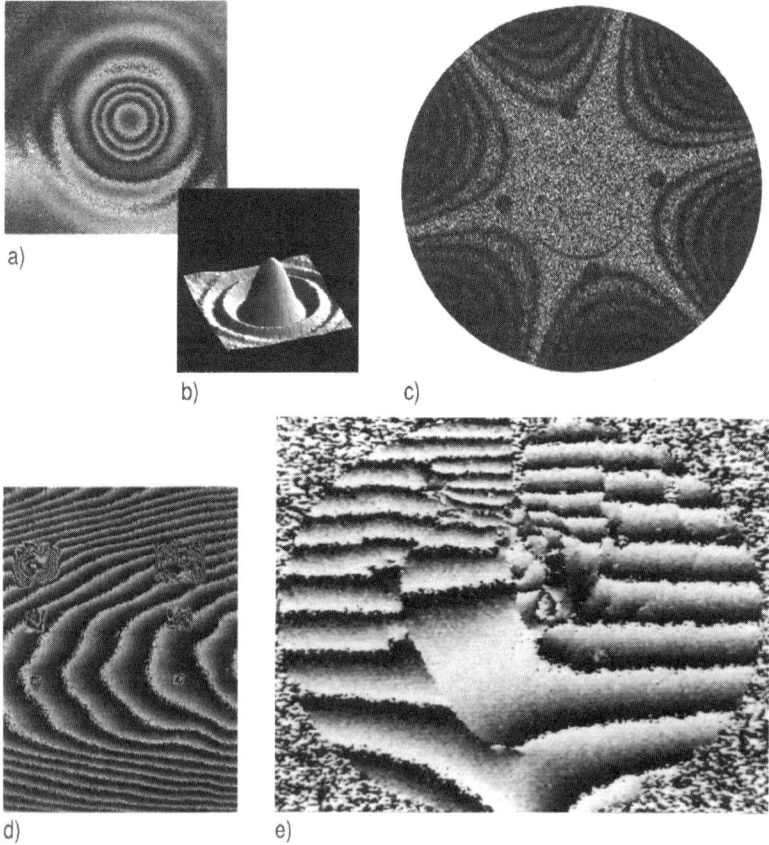

FIGURE 9.15. Examples of DSPI in the evaluation of (a, b) shock propagation on a plate; (c) resonances of a brake disk; (d) defects in composite materials; and (e) the development of a fracture process zone in a fiber-reinforced reactive powder concrete. (a–d: Courtesy of H. Steinbichler, Labor Dr. Steinbichler GmbH).

FIGURE 9.16. (a, b) Examples of DSSPI in the measurement of in-plane strains on a thin aluminum sheet subjected to tension: Phase contours in (a) and (b) correspond to $\partial u/\partial x$ and $\partial u/\partial y$, respectively. (c–e) Examples of DSSPI in the measurement of the derivatives of out-of-plane deformations: Phase contours in (c), (d), and (e) correspond to $\partial w/\partial x$, $\partial^2 w/\partial x^2$, and $\partial^2 w/\partial x \partial y$, respectively.

In digital speckle shearing pattern interferometry, the shearing of the speckle wavefront originating from the same object point renders the combined pattern sensitive to the rate of change of phase across it. In all other aspects, the method is similar in functioning to digital speckle interferometry. If an image frame captured after deformation is subtracted from a reference frame captured before deformation, the resulting brightness distribution is as for the case of DSSPI given by (34), with the only difference that the phase term, $\Delta\varphi(x, y)$, in the equation is now related to the Cartesian components of gradient of displacement. Examples of DSSPI in the quantitative measurement of the derivatives of displacements are shown in Fig. 9.16. The figure shows phase contours of (a) $\partial u/\partial x$, (b) $\partial u/\partial y$, (c) $\partial w/\partial x$, (d) $\partial^2 w/\partial x^2$, and (e) $\partial^2 w/\partial x \partial y$. DSSPI has also been applied to a wide variety of applications in experimental mechanics and nondestructive testing. Two such applications are displayed in Fig. 9.17.

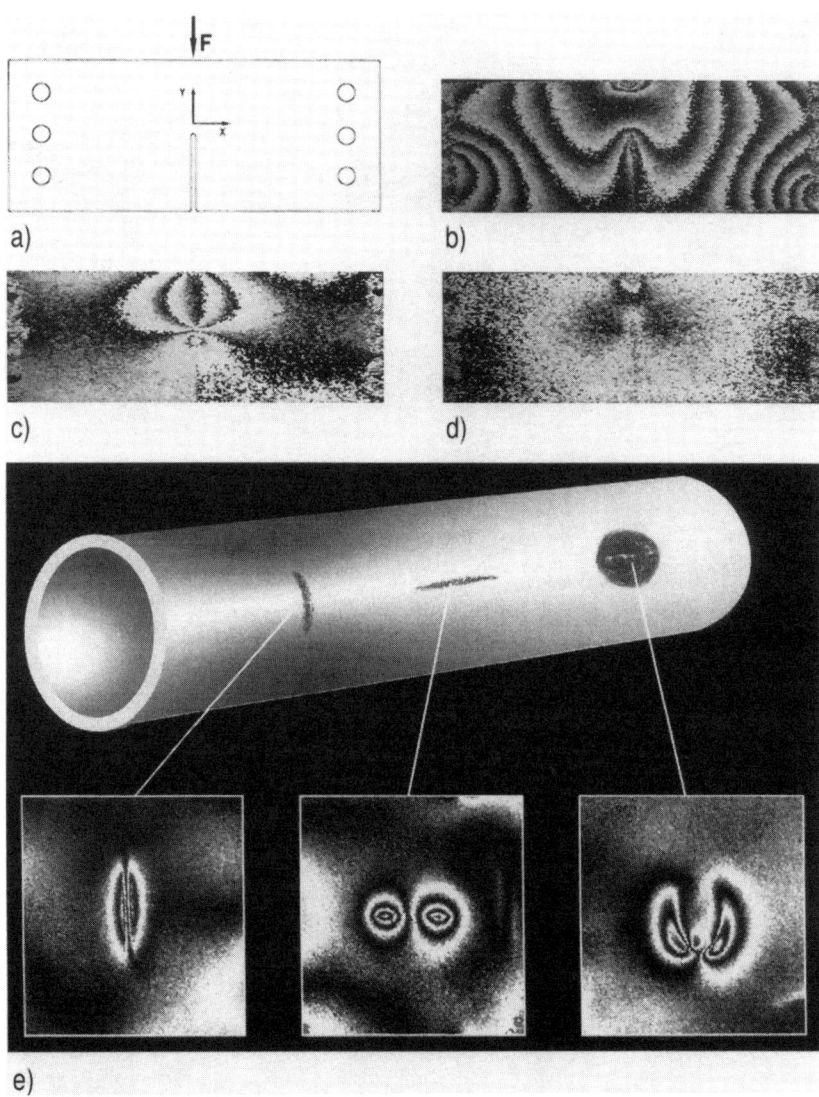

FIGURE 9.17. Examples of applications of DSSPI in experimental mechanics and nonde-structive testing: (a-d) measurement of strains on a fixed notched metal plate submitted to load F. Phase contours in (b), (c), and (d) correspond to $\partial w/\partial x$, $\partial u/\partial x$, and $\partial v/\partial y$, re-spectively. (e) Testing of a steel pipe: Flaws, inhomogeneities, and rusty spots are detected. (Courtesy of H. Steinbichler, Labor Dr. Steinbichler GmbH).

9.4   Conclusions

Holographic interferometry is a valuable tool for strain and vibration analysis and for defect detection. Several optical configurations for achieving holographic interference have been discussed with special emphasis on various approaches in which the basic technique can be implemented. The powerful potential of holographic interferometry should be apparent to the reader from the few selected examples of measurements and applications of the technique in engineering and sciences.

The description of holographic interferometry has been appended with a brief discussion on speckle photography, digital speckle pattern interferometry, and digital speckle shearing pattern, including examples of measurements and applications of these techniques.

Unprecedented growth in computer technology and related digital processing techniques have provided an undeniable advantage to holographic and speckle techniques in the way of their simplicity of use, and rapid and real-time display of whole-field phase maps accompanied by fast quantitative evaluation of these contours. Given these powerful attributes, we can confidently expect holographic and speckle techniques to not only continue to grow and develop, but also to have a major impact in tasks related to measurements, inspection, and verification of computer codes in a wide range of applications in engineering and scientific fields.

References

[1] R.K. Erf, *Holographic Non-Destructive Testing*, (Academic, New York, 1974).

[2] C.M. Vest, *Holographic Interferometry*, (Interscience, New York, 1979).

[3] Y.I. Ostrovsky, M.M. Butusov, and G.V. Ostrovskaya, *Interferometry by Holography*, (Springer-Verlag, Berlin, 1980).

[4] W. Schumann, and M. Dubas, *Holographic Interferometry*, (Springer-Verlag, Berlin, 1979).

[5] W. Schumann, J.-P. Zürcher, and D. Cuche, *Holography and Deformation Analysis*, (Springer-Verlag, Berlin, 1985).

[6] R. Jones and C. Wykes *Holographic and Speckle Interferometry*, 2nd ed, (Cambridge University Press, Cambridge, 1989).

[7] Y.I. Ostrovsky, V.P. Schepinov, and V.V. Yakovlev, *Holographic Interferometry in Experimental Analysis*, (Springer Verlag, Berlin, 1991).

[8] P.K. Rastogi, *Holographic Interferometry—Principles and Methods*, (Springer-Verlag, Berlin, 1994).

[9] T. Kreis, *Holographic Interferometry—Principles and Methods*, (Akademie Verlag, Berlin: 1996).

[10] M. Francon, *Laser Speckle and Application to Optics*, Academic Press, (New York, 1979).

[11] R.K. Erf, *Speckle Metrology*, Academic Press, (New York, 1978).

[12] J.C. Dainty, *Laser Speckle and Related Phenomenon*, 2nd ed. (Springer Verlag, New York, 1984).

[13] R.S. Sirohi, *Speckle Metrology*, (Marcel Dekker, New York, 1993).

[14] D. Malacara, *Optical Shop Testing*, (Wiley, New York, 1992).

[15] D.C. Williams, *Optical Methods in Engineering Metrology*, (Chapman and Hall, London, 1993).

[16] G. Cloud, *Optical Methods of Engineering Analysis*, (Cambridge University Press, London, 1994).

[17] P. K Rastogi, *Optical Measurement Techniques and Applications*, (Artech House, London, 1997).

[18] P.K. Rastogi, *Photomechanics*, Topics in Applied Physics, Vol. 77 (Springer-Verlag, Berlin, 2000).

[19] R.S. Sirohi and K.D. Hinsch, Selected Papers on Holographic Interferometry—Principles and Techniques, *SPIE Milestone Series* V **MS 144** (1998).

[20] P. Hariharan, *Optical Holography*, (Cambridge University Press, London, 1985), Chap. 7.

[21] R.C. Troth, and J.C. Dainty, "Holographic Interferometry using Anisotropic Self-Diffraction in $Bi_{12}SiO_{20}$," Opt. Lett. **16**, 53–55 (1991).

[22] A. Hafiz, R. Magnusson, J.S. Bagby, D.R. Wilson, and T.D. Black, "Visualization of aerodynamic flow fields using photorefractive crystals," Appl. Opt. **28**, 1521–1524 (1989).

[23] R. Dändliker, "Two-reference-beam holographic interferometry," in *Holographic Interferometry—Principles and Methods*, P.K. Rastogi, ed., (Springer-Verlag, Berlin 1994), Chap. 4.

[24] U. Schnars and W. Jüptner "Direct recording of holograms by a CCD target and numerical reconstruction," *Appl. Opt.*, **33**, 179–181 (1994).

[25] G. Pedrini, Y.L. Zou, and H.J. Tiziani, "Digital double pulse holographic interferometry for vibration analysis," *J. Modern Opt.* **42**, 367–374 (1995).

[26] E. Marquardt and J. Richter, "Digital image holography," *Opt. Eng.*, **37**, 1514–1519 (1998).

[27] T.M. Kreis, W.P.O. Juptner, and J. Geldmacher, "Digital holography: Methods and applications," *Proc SPIE* **3407**, 169–177 (1998).

[28] K. Creath, "Phase-measurement interferometry techniques," in *Progress in Optics*, XXVI, E. Wolf, ed., (North-Holland, Amsterdam, 1988), Chap. V, 349–393.

[29] J.E. Greivenkamp and J.H. Bruning, "Phase shifting interferometry," in *Optical Shop Testing*, 2nd ed., D. Malacara, ed., (Wiley, New York, 1992).

[30] D.W. Robinson and G.T. Reid, *Interferogram Analysis; Digital Fringe Pattern Measurement Techniques*, (IOP publications, Bristol, U.K., 1993).

[31] K.A. Stetson and W.R. Brohinsky, "Fringe shifting technique for numerical analysis of time average holograms of vibrating object," *J. Opt. Soc. Am. A*, **5**, 1472 (1988).

[32] K.A. Stetson and W.R. Brohinsky, "Electro-optic holography and its application to hologram interferometry," App. Opt., **24**, 3631–3637 (1985).

[33] S.M. Tieng and W.Z. Lai, "Temperature measurement of reacting flowfield by phase-shifting holographic interferometry," *J. Thermophys. Heat Transfer*, **6**, 445–451 (1992).

[34] T.A.W.M. Lanen, "Digital holographic interferometry in flow research," *Opt. Comm.* **79**, 386–396 (1990).

[35] X. Colonna de Lega, "Continuous deformation measurement using dynamic phase-shifting and wavelet transforms," *Proc. Conf. App. Opt. Optoelectron.*, K.T.V. Grattan, ed., Institute of Physics, 261–267 (1996).

[36] P. Jacquot, M. Lehmann, X. Colonna de Lega, "Deformation analysis of a communication telescope structure under non-uniform heating using holographic interferometry," *Proc. SPIE*, **3293**, 102–113 (1998).

[37] C.A. Sciammarella, P.K. Rastogi, P. Jacquot, and R. Narayanan, "Holographic-moiré in real time," *Experimental Mechanics*, **22**, 52–63 (1982).

[38] P.K. Rastogi, "Holographic in-plane measurement using reference-wave reconstruction: Phase stepping and application to a deformation problem," *Appl. Opt.* **34**, 7194–7196 (1995).

[39] L. Pirodda, "Conjugate wave holographic interferometry for the measurement of in-plane deformations," *Appl. Opt.* **26**, 1842–1844 (1989).

[40] P.K. Rastogi, and E. Denarié, "Visualization of in-plane displacement fields by using phase-shifting holographic moiré: Application to crack detection and propagation," *Appl. Opt.*, **31**, 2402–2404 (1992).

[41] D.B. Neumann, "Comparative holography: A technique for eliminating background fringes in holographic interferometry," *Opt. Eng.* **24**, 625–627 (1985).

[42] Z. Füzessy, and F. Gyimesi, "Difference holographic interferometry: Technique for optical comparison," *Opt. Eng.* **32**, 2548–2556 (1993).

[43] P.K. Rastogi, "Comparative holographic interferometry: A nondestructive inspection system for detection of flaws," *Experimental Mechanics* **25**, 325–337 (1985),

[44] P.K. Rastogi, "Comparative phase shifting holographic interferometry," *Appl. Opt.*, **30**, 722–728 (1991).

[45] P.K. Rastogi, "Direct and real-time holographic monitoring of relative changes in two random rough surfaces," *Physical Review A*, **50**, 1906–1908 (1994).

[46] P.K. Rastogi, "Visualization and measurement of slope and curvature fields using holographic interferometry: An application to flaw detection," *J. Modern Opt.*, **38**, 1251–1263 (1991).

[47] C.S. Vikram, "Study of vibrations," in *Holographic Interferometry—Principles and Methods*, P.K. Rastogi, ed., (Springer-Verlag, Berlin, 1994), Chap. 8.

[48] K.A. Stetson, and W.R. Brohinsky, "Electro-optic holography system for vibration analysis and nondestructive testing," *Opt. Eng.* **26**, 1234–1239 (1987).

[49] R.J. Pryputniewicz, and K.A. Stetson, "Measurement of vibration patterns using electro-optic holography," *Proc. SPIE*, **1162**, 456–467 (1989).

[50] M.A. Beek, "Pulsed holographic vibration analysis on high-speed rotating objects: Fringe formation, recording techniques, and practical applications," *Opt. Eng.*, **31**, 553–561 (1992).

[51] W. Merzkirch, *Flow Visualization*, (Academic Press, New York, 1987).

[52] R.J. Parker and D.G. Jones, "The use of holographic Interferometry for turbomachinery fan evaluation during rotating tests," *J. Turbomachinery*, **110**, 393–399 (1988).

[53] S.P. Sharma and S.M. Ruffin, "Density measurements in an expanding flow using holographic interferometry," *J. Thermophysics Heat Transfer*, **7**, 261–268 (1993).

[54] P.K. Rastogi and L. Pflug, "A fresh approach of phase management to obtain customized contouring of diffuse object surfaces of broadly varying depths using real-time holographic interferometry," *J. Modern Opt.* **37**, 1233–1246 (1990).

[55] P.K. Rastogi and L. Pflug, "Novel concept for obtaining continuously variable topographic contour mapping using holographic interferometry," *Appl. Opt.*, **29**, 4392–4402 (1990).

[56] P.K. Rastogi and L. Pflug, "A holographic technique featuring broad range sensitivity to contour diffuse objects," *J. Modern Opt.* **38**, 1673–1683 (1991).

[57] P. Carelli, D. Paoletti, and G.S. Spagnolo, "Holographic contouring method: Application to automatic measurements of surface defects in artwork," *Opt. Eng.* **30**, 1294–1298 (1991).

[58] P.K. Rastogi and L. Pflug, "Measurement of large out-of-plane displacements using two source holographic interferometry," *J. Modern Opt.* **41**, 589–594 (1994).

[59] P.K. Rastogi, "A multiple sources holographic technique for the measurement of slope change of a three-dimensional object," *J. Modern Opt.* **40**, 2389–2397 (1993).

[60] R.S. Sirohi, *Selected Papers on Speckle Metrology*, SPIE Milestone Series, MS 35, (SPIE Optical Engineering Press, Washington, 1991).

[61] P. Meinlschmidt, K.D. Hinsch, and R.S. Sirohi, *Selected papers on Speckle Pattern Interferometry—Principles and Practice*, SPIE Milestone series, MS 132, (SPIE Optical Engineering Press, Washington, 1996).

[62] P.K. Rastogi, ed., Special Issue on Speckle and Speckle Shearing Interferometry—1, *Opt. Lasers Eng.* **26**, 83–278 (1997).

[63] P.K. Rastogi, Guest Editor, Special Issue on Speckle and Speckle Shearing Interferometry—2, *Optics & Lasers in Engineering*, **26**, Nos. 4-5, p. 279–460, 1997.

[64] P.K. Rastogi, "Speckle shearing photography—A tool for direct measurement of surface strains," *Appl. Opt.*, **37**, 1292–1298 (1998).

[65] A.J. Moore, J.R. Tyrer, and F.M. Santoyo, "Phase extraction from electronic speckle pattern interferometry," *Appl. Opt.* **33**, 7312–7320 (1994).

10

Diffuser Display Screen

Gajendra Savant, Tomasz Jannson,
and Joanna Jannson

10.1   Introduction

Conventional diffusers are objects, materials, or films that diffuse incident light
from a point source. Almost every day we experience diffusion. For example, when
the sun is blocked by haze, clouds, or smoke, we do not see direct sunlight but
its diffused rays. Cold cathode fluorescent bulbs diffuse light through translucent
glass. Sun roofs, smoked or hazed glass or plastics, and various kinds of architectural glasses exhibit light diffusion. Screens for film, overhead, and slide projectors
diffuse light for display. Diffusers are now commonly employed in laptop computers, cockpit displays, display signboards, Hollywood movie production, and
toys. Diffusers scatter light so that an object in the path of the diffused light is illuminated from a number of directions. Typical state-of-the-art diffusers are made
from ground glass, photographic emulsion, or plastic. Opal glass, opaque plastic,
chemically etched plastic, machined plastics, cloth/resin, paper/nylon composites,
and paint surfaces are typical conventional diffusers. Another typical diffuser is a
frosted light bulb or ground glass plate or other rough surface placed near a light
source. Ground or milky glass diffusers scatter light uniformly in all directions
and have the effect of averaging noise.[1] When light is transmitted through or reflected from a medium, three phenomena ideally take place. Figure 10.1 illustrates
the cases of light being deflected, transmitted, or absorbed. These phenomena are
qualified by diffusion as illustrated in Fig. 10.2.[2]

Conventional diffusion is an effect of scattering from solid, liquid, or gaseous
particles. Uncontrolled scattering diffuses light in all directions, so that little of
the incident light is sent out at any one angle. As can be inferred from Fig. 10.2,

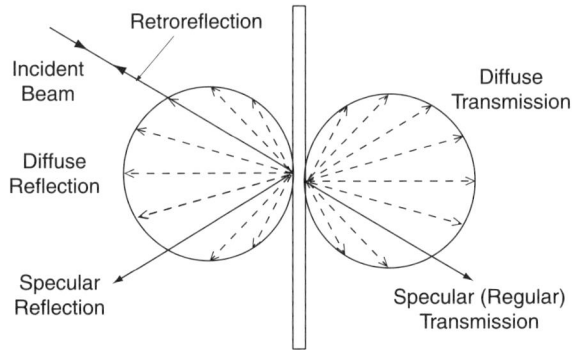

FIGURE 10.1. Idealized reflection and transmission.[2]

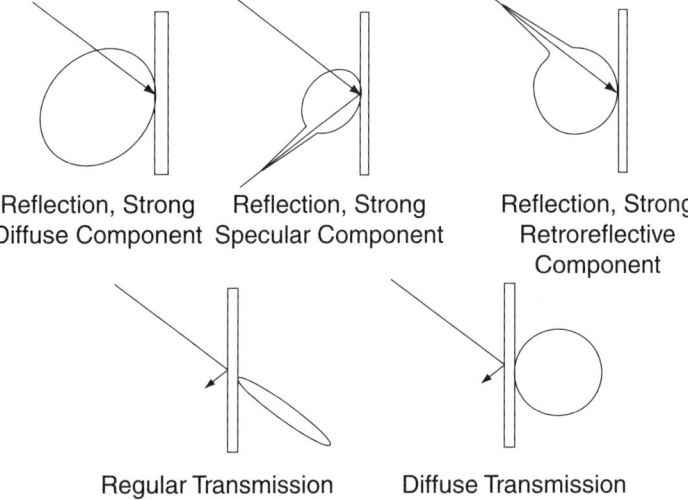

FIGURE 10.2. Reflection, transmission, and diffusion of light.[2]

diffused light cannot be controlled or its intensity increased, using conventional materials.

To control and enhance intensity and brightness, Physical Optics Corporation (POC, Torrance, CA) over the past 15 years has developed light-shaping diffusion (LSD) technology and products based on it. POC's LSDs are now mass produced holographically.

10.2   Historical Background

It is well known that coherent light incident on an optically rough surface such as a ground glass diffuser forms a random intensity pattern called speckle. Speckle is present anywhere beyond a diffuser illuminated with coherent light as discussed by Dainty.[3] A number of authors, including Dainty, have studied the statistics of speckle created by an illuminated diffuse object. The pioneering analysis was that of L.I. Goldfischer,[4] who describes the general structure of light scattered by diffusers, and the characteristic speckle pattern that resulted when he exposed photographic film directly to backscattered radiation from a diffuse surface illuminated by a coherent monochromatic source.

Lowenthal and Arsenault[5] describe the statistics of the image produced by an optical system with a coherent diffuse object: Although speckles are related to the random structure of a diffuser illuminated by coherent light, the spatial coherence of the light is related to the random structure of the source radiation.[6] Carter and Wolf[6] studied spatial coherence and quasi-homogeneous sources generally. Miller et al.[7] discussed the statistics of laser speckle patterns in a plane some distance from a coherently illuminated object. Kowalczyk[8] sets out the theory relating to thin phase diffusers in coherent imaging systems.

A specific form of computer-generated diffuser called a kinoform has also been studied. A kinoform is a computer-generated wavefront reconstruction device that, like a hologram, displays a 2D image. In contrast to a hologram, however, a kinoform yields a single diffraction order, so that reference beams and image separation calculations are unnecessary. A kinoform is purely phase, because it is based on the assumption that only phase information in a scattered wavefront is required for time construction of the image of the scattering object. Kinoforms are discussed by Lesem[9] and H.J. Caulfield.[10] Caulfield presents a general description of kinoform phase diffusers and shows that the angular spectrum of light scattered from a diffuser increases with the angular size of the aperture of the diffuser. Caulfield tested this by varying the distance between a diffuser mask consisting of a ground glass and a kinoform made of bleached silver halide. The textbook by J.W. Goodman, *Statistical Optics*,[14] is a useful reference in studying the statistics of diffusers, speckles, and partial coherence of light as well as the spatial coherence of light and scattering by moving diffusers. Collier et al.'s study is also helpful.[11]

10.3   Lambertian Diffuser

Conventional diffusers such as ordinary frosted glass, ground glass, milky glass, opaque plastics, machine/chemically etched plastic, and metal or cloth/resin composites that incorporate lenticular diffusers, are only capable of scattering light into a hemispherical pattern, and they achieve their diffusion by means of surface roughness or microscale integrity, which is difficult to control. Therefore, such diffusers scatter light nondirectionally, with rays randomly turned into a wide set of directions. This is distinct from the redirection at the surface of a mirror or lens by reflection or refraction, which does not increase the entropy or disorder of a light beam.

In nature, we see many degrees of scattering, such as with sunlight as it varies from the dazzle of a clear day, with only some blue-light scattering that colors the

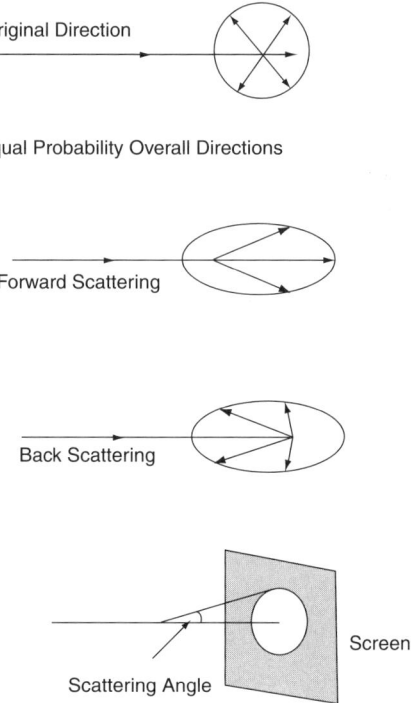

FIGURE 10.3. Natural scattering of light. At all scattering angles, there is circular symmetry.

sky, to an overcast day with a uniformly gray sky. Intermediate degrees produce a fuzzy glare in the general direction of the sun. This is volume scattering, the cumulative effect of passage through an extended gradually scattering medium, in this case, the earth's atmosphere. Some of the atmosphere's volume scattering processes send light in all directions uniformly, whereas some have concentrations in the forward or backward directions, depending on the nature of the scattering centers, whether molecular or particulate. In nature, scattering along any angle from the original direction of the light is uniform around a conical pattern centered on that original direction: There are no azimuthal variations in the scattering (see Fig. 10.3). We shall see that holographic diffusers are not subject to this limitation.

Artificial diffusers have become common in this century of electric illumination, with frosted and textured glass as examples. Like their natural counterparts, however, these exhibit highly diffuse scattering. Although not all light may be scattered, as when the surface of a piece of glass is only slightly frosted, that part of the light that is scattered goes into a wide, almost hemispheric, range of angles away from the original direction of light. If the cosine function of Lambert's law is satisfied,[12] the brightness function is angle independent.[12] Such a hemispheric scattering is termed "Lambertian" in the parlance of photometry, after one of its founders. These diffusers commonly rely on surface roughness, a microscale irregularity that consists of multiple random refractors that spread light out, away from its original incoming direction. The statistical characteristics of the irregularity are difficult to control, so that a designer has little ability to specify a particular scattering pattern. If he could do so, such scattering specificity would be a third method of directional control, in addition to reflection and refraction.

10.4   Principle of Diffusion

An illustration of light diffusion or scattering from a mask diffuser, which scatters light in numerous directions, and the recorded holographic diffuser is shown in Fig. 10.4. Each light beam scattered from the structure or diffuser satisfies (1).

$$\mu = |\mu|^{e^{i\theta}} = \mathrm{Re}[\mu] + \mathrm{Im}[\mu] = R + iI, \tag{1}$$

where i is the square root of -1. This phase representation of scattered light takes into account the amplitude and phase of the scattered beam for the coherent light case. If the light is coherent, the phase component must be considered because all waves interfere when light is coherent. A typical coherent light source is a laser.

The unit vector of observation $\vec{s}$ describes the input coordinates of the output intensity $f(sx, sy)$; θ is the angle between the z-axis and $\vec{s}$. Thus, using spherical coordinates, we obtain,

$$s_z = \cos\theta, \tag{2}$$
$$s_x = \sin\theta\cos\alpha, \tag{3}$$
$$s_y = \sin\theta\sin\theta, \tag{4}$$
$$s_x^2 + s_y^2 + s_z^2 = 1. \tag{5}$$

The horizontal (s_x, s_y) projections define the area in (s_x, s_y) space, analogous to the Fourier area in coherent optical processing. This "pseudo-Fourier" area characterizes the profile of the non-Lambertian diffuser spectrum. Typical examples of

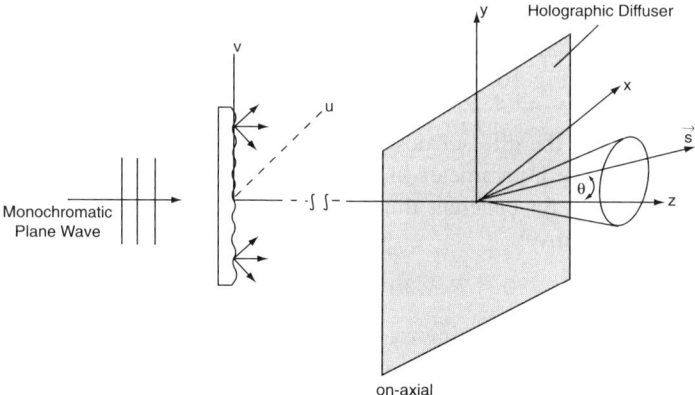

FIGURE 10.4. Illustration of monochromatic light beam incident on axis onto a mask diffuser to record a holographic diffuser.

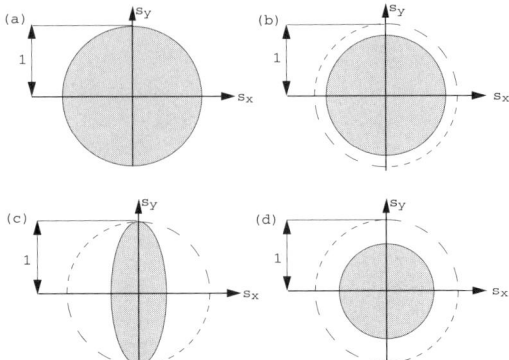

FIGURE 10.5. Typical examples of "pseudo-Fourier" areas for various on-axis diffusers, including (a) Lambertian diffuser, (b) limited Lambertian diffuser, (c) non-Lambertian (elliptical) diffuser, and (d) non-Lambertian cylindrical diffuser. Here, $s_x^2 + s_y^2 \leq 1$. The broader circle defines the boundaries of the homogeneous wave spectrum.

"pseudo-Fourier" areas are shown in Fig. 10.5. In this case, according to (5), we have

$$s_x^2 + s_y^2 \leq 1, \tag{6}$$

which defines the boundaries of a homogenous (propagating) wave spectrum.

The conventional formula for light scattered on a diffuser, derived by Goldfischer,[4] includes a number of constant terms that generate specular beams, reducing the diffuser's optical efficiency. In the holographic graded-index (GRIN) diffusers developed by the authors, however, these constant terms are attenuated to zero by multiscattering. Therefore, the following formula describes the monochromatic plane wave response of the diffuser:

$$J(s_x, s_y) = D \cos \theta \iint du\, dv\, P(u, v) \times P(u + s_x \times h, v + s_y \times h), \tag{7}$$

where D is proportionally constant, θ is the angle between the z-axis and $\vec{s}$-unit vector, illustrated in Fig. 10.4, and (s_x, s_y) are horizontal projections of the $\vec{s}$-vector, as defined by (2).

In the context of the diffuser design, the coordinates (u, v) can be treated as auxiliary, as can the h-parameter in length units. The pupil function P can in general be complex for coherent illumination. For incoherent illumination, it is always real and positive:

$$0 \leq P(u, v) \leq 1, \text{ for all } (u, v). \tag{8}$$

The first step in computer coding for the on-axis H1-diffuser and the monochromatic plane incident wave is simply a computation of (6) for various values of the h-parameter and various profiles of the pupil function $P(u, v)$.

As a first step in modeling the holographic diffuser, the authors developed a model for the broadband performance of a non-Lambertian on-axis diffuser. The diffuser properties are related to the size of the aperture (see Fig. 10.4), which limits the angular extent of a Lambertian recording wave. For monochromatic illumination of the non-Lambertian diffuser (NLD), the angular dependence of the intensity of the transmitted beam is described by an autocorrelation of the aperture function.

$$J(\vec{s}) = A \cos \theta \iint P(u, v) P \left[\left(u - \frac{\lambda_R}{\lambda} h s_x \right), \left(v - \frac{\lambda_R}{\lambda} h s_y \right) \right] du \, dv, \quad (9)$$

where θ is an angle relative to the diffuser normal, λ_R is the recording wavelength, λ is the replay wavelength, $P(u, v)$ is the aperture function, and $[s_x, s_y, s_z]$ are the direction cosines of a vector directed from the center of the H1-diffuser in the direction of observation.

To calculate the polychromatic properties, $J(\vec{s})$ can be integrated over the spectral band. For a Gaussian spectrum, as shown in Fig. 10.6, the diffuse intensity has an approximately conic distribution (see Figs. 10.7 to 10.11). The AM-2 atmospheric transmission model was used to weight the solar spectrum so that the diffusion properties under solar illumination could be calculated.

10.5   Non-Lambertian Light Shaping Diffuser

A light shaping diffuser is a non-Lambertian diffuser in the sense that it scatters light not in all directions but in a well-controled and well-defined solid cone. Whether the beam is coherent, pseudocoherent, or white light, the output light can always be well defined. The non-Lambertian diffuser is based on a new method of producing controlled scattering by means of a holographic diffuser, which consists of a thin ($<$ 1-mm) volume of spatially fluctuating refractive indices.

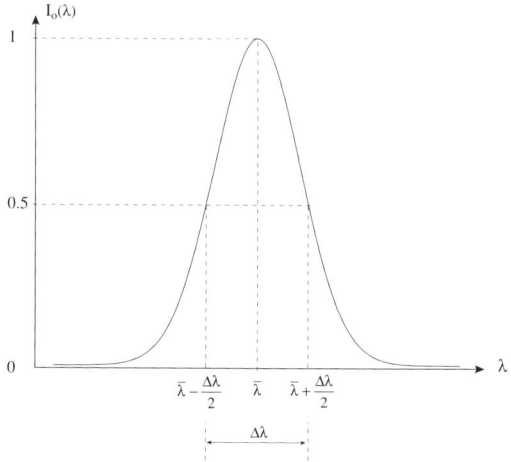

FIGURE 10.6. Gaussian illumination spectrum.

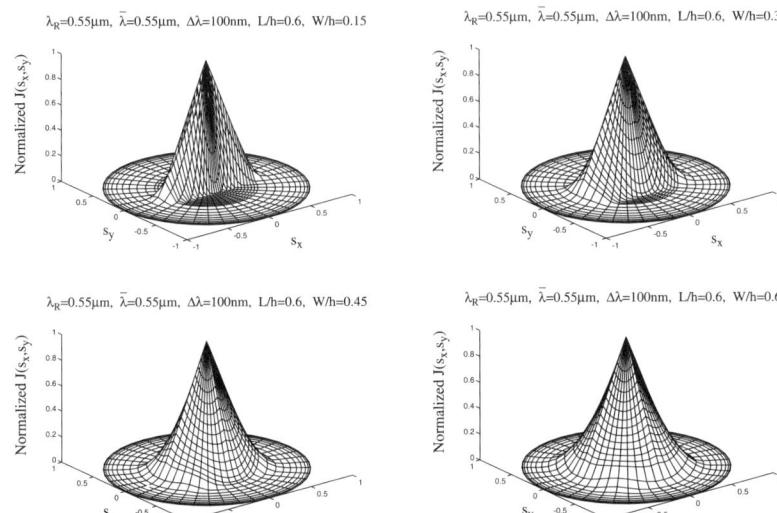

FIGURE 10.7. For square aperture and polychromatic wave with $\lambda_R = 0.55$ μm, $\bar{\lambda} = 0.55$ μm, $\Delta\lambda = 100$ nm, $L/h = 0.6$, (a) $W/h = 0.15$, (b) $W/h = 0.3$, (c) $W/h = 0.45$, and (d) $W/h = 0.6$.

The statistics of the fluctuations in the volume hologram can be tailored so that the resulting scattering pattern can take a variety of shapes. The diffusers are subject to two constraints upon the controlled scattering:

1. Light is scattered relative to its incoming direction, NOT relative to the surface of the diffuser. This means that maximum pattern control requires collimated light (i.e., parallel rays as from a searchlight). Deviations from this ideal input blur the output pattern. Noncollimated light has a spread that is the superposition (technically, the convolution) of the input light and the controlled scattering.
2. The scattering pattern is arbitrarily specifiable, but has the form of an autocorrelation function, as explained below. This means that the maximum brightness of the scattering pattern is in the same direction as that of the input light ray, and that the brightness of the scattered light falls off smoothly to zero with increased scattering angle away from the original direction of the incoming light, with no discontinuities in the illumination pattern. Unlike conventional diffuse scattering, the pattern need not be rotationally symmetric, so that elliptical patterns are easily generated from round apertures.

The overall illumination pattern from a light with a holographic diffuser is the resultant of the diffuser's scattering pattern and the total number of directions encompassed by the light striking the diffuser.

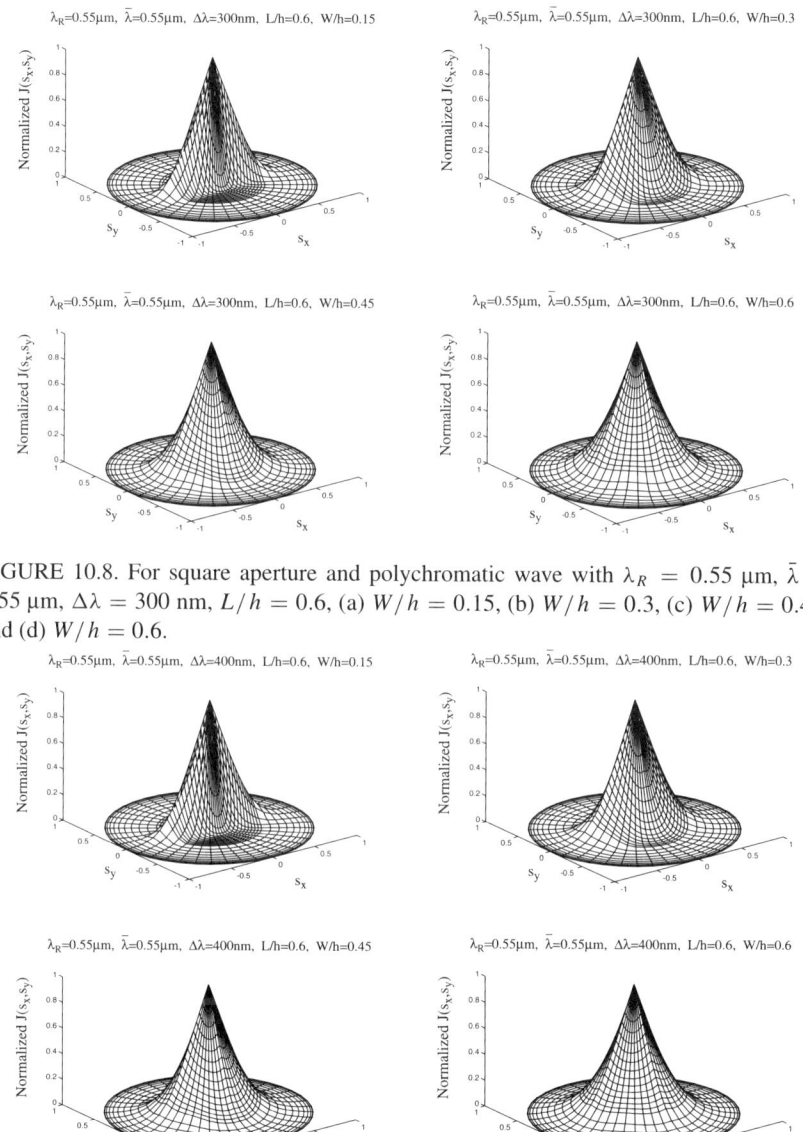

FIGURE 10.8. For square aperture and polychromatic wave with $\lambda_R = 0.55$ μm, $\bar{\lambda} = 0.55$ μm, $\Delta\lambda = 300$ nm, $L/h = 0.6$, (a) $W/h = 0.15$, (b) $W/h = 0.3$, (c) $W/h = 0.45$, and (d) $W/h = 0.6$.

FIGURE 10.9. For square aperture and polychromatic wave with $\lambda_R = 0.55$ μm, $\bar{\lambda} = 0.55$ μm, $\Delta\lambda = 400$ nm, $L/h = 0.6$, (a) $W/h = 0.15$, (b) $W/h = 0.3$, (c) $W/h = 0.45$, and (d) $W/h = 0.6$.

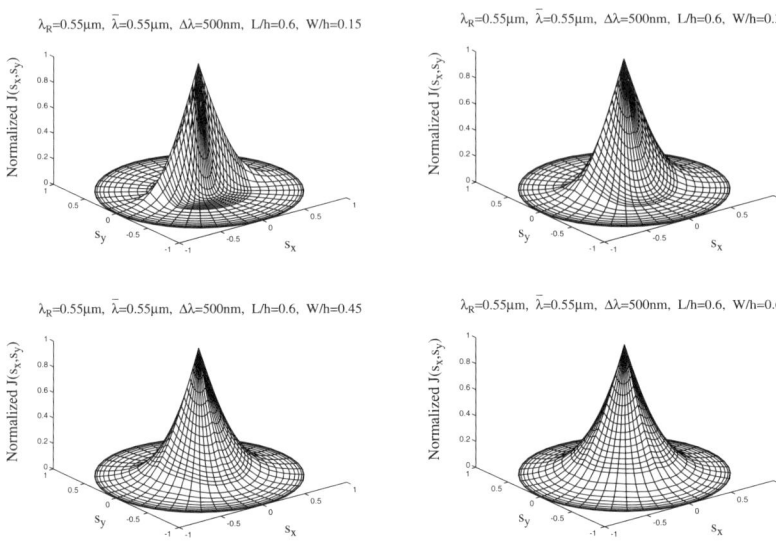

FIGURE 10.10. For square aperture and polychromatic wave with $\lambda_R = 0.55$ μm, $\bar{\lambda} = 0.55$ μm, $\Delta\lambda = 500$ nm, $L/h = 0.6$, (a) $W/h = 0.15$, (b) $W/h = 0.3$, (c) $W/h = 0.45$, and (d) $W/h = 0.6$.

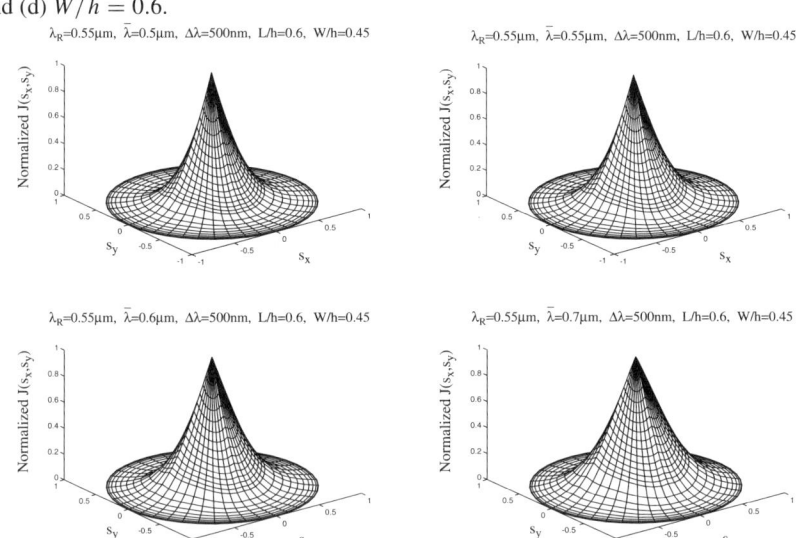

FIGURE 10.11. For square aperture and polychromatic wave with $\lambda_R = 0.55$ μm, $L/h = 0.6$, $W/h = 0.45$, $\Delta\lambda = 500$ nm, (a) $\bar{\lambda} = 0.5$ μm, (b) $\bar{\lambda} = 0.55$ μm, (c) $\bar{\lambda} = 0.6$ μm, and (d) $\bar{\lambda} = 0.7$ μm.

10.5.1   Fabrication of Light Shaping Diffusers

Although many years have passed since the discovery of holography in 1948 by Denis Gabor,[13] its potential for light transformation has never been fully exploited. Holography has found many applications ranging from 3D displays, lidar, and information storage to advanced information processing. Unfortunately, most of these have remained laboratory models or prototypes. Two of the main reasons for this have been material problems and prohibitive production costs. To date, surface-relief holography is the only application that has achieved mass production with reasonable manufacturing cost and market size, but today, embossed holograms are found on almost every credit card and even on many magazine covers. Thus far, no other known holographic technique has offered similar mass-production advantages.

10.5.2   Mass Production of Light Shaping Diffusers

Mass replication of light shaping diffusers requires tight control of the processes of uniform coating, holographic exposure to generate the diffuser pattern in the coating, and wet processing.

The master, a surface-relief hologram, is embossed into a deformable plastic such as polycarbonate, polyester, mylar, acrylic, or TPX. Embossing requires a metal submaster in which the surface holographic topography has been reproduced. POC's mastering technique is illustrated in Fig. 10.12.

Hard Embossing

Using a metal shim, POC's hard embossing machine (see Fig. 10.13) has the capability to produce 22-in. × 22-in. sheets of LSD in less than 20 minutes.

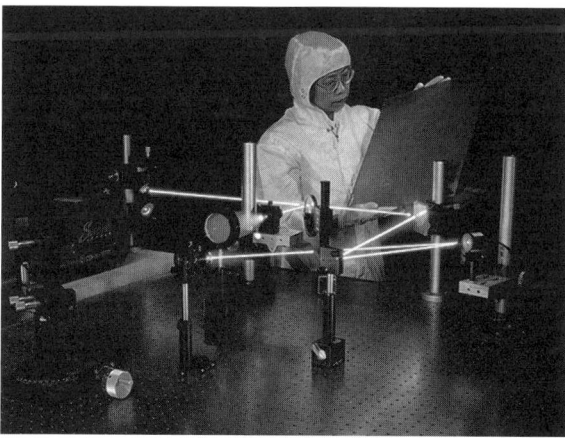

FIGURE 10.12. POC's clean room facility, where LSD masters are produced.

FIGURE 10.13. POC's embossing machine installed in 1996.

Soft Embossing on Hard Substrates

LSD topography, i.e., surface-relief structures, can be replicated in ultraviolet (UV) curable epoxy, on hard and thick substrates such as glass, 30/60/125-mil acrylic/polycarbonate, and other substrates, either on POC's custom-made replication machine or manually using a metal shim and UV curing.

Soft Embossing on Flexible Substrates

Light shaping diffusers can also be embossed into a continuous roll of flexible substrate. POC has developed a soft embossing machine that consists of an un-winding station, coating station, UV curing module, and winding station, which also has slitting and preliminary inspection capability (see Fig. 10.14). This machine can produce diffuser at a rate of 40 to 53 feet/minute in widths of 9 in. to

FIGURE 10.14. POC's highly automated LSD replication machine, with an annual replication capability of about 75 million linear feet.

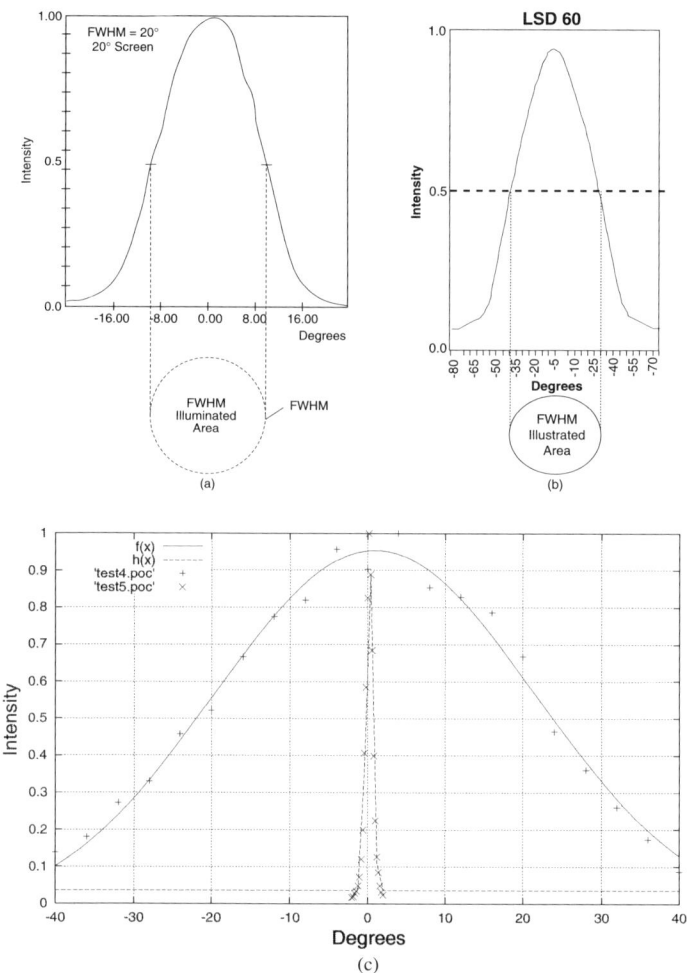

FIGURE 10.15. Scatter plots for three diffuser angles. (a) Angular spread (scatter) for 10° diffuser manufactured at POC and measured using POC's custom fabricated angle measurement instrument. (b) Angular scatter for 60° diffuser. Scatter plots for three diffuser angles. (c) Elliptical scattering by 48.6° × 1.02° diffuser.

40 in. As of late 1999, we produce 22-in. to 26-in. width LSD. The single machine shown in Fig. 10.14 can produce roughly 75 million linear feet of LSD per year.

10.5.3   Characteristics of LSD

For the past 10 years, POC has advanced through product development, pilot plant production, and now mass production of a truly unique product. POC is now able to produce diffusers with the following characteristics:

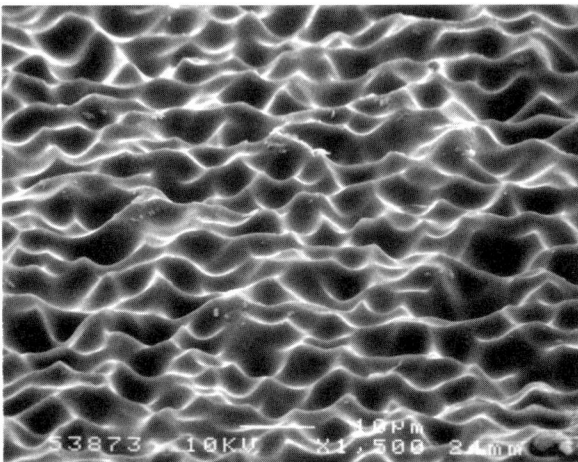

FIGURE 10.16. Circular diffuser angle = 60°.

FIGURE 10.17. Elliptical diffuser angle 40° × 0.2°.

- *Regulated Scattered Beam Divergence:* POC now produces diffusers with ±0.1° to ±60° angular range, i.e., circular 0.2° to 90°; elliptical: minor 0.2° to 65° and major: 10° to 95°. Typical angular plots for 10° circular, 60° circular, and 48.6° × 1.03° are shown in Fig. 10.15.
- *Anisotropic Angular Spectrum of Scattered Beam:* Circular and extreme elliptical profiles are routinely achievable. Figures 10.16 through 10.18 show scanning electron micrographs of circular, elliptical, and high aspect ratio diffuser.
- *Supreme Scattering Efficiency:* POC's LSD offers extraordinarily high scattering efficiency. As a result, an observer cannot see any specular beam, i.e., no zero order at all. Such high efficiency without loss in transmission is, to our knowledge, only achievable with POC's diffusers. As an illustration, Fig. 10.19

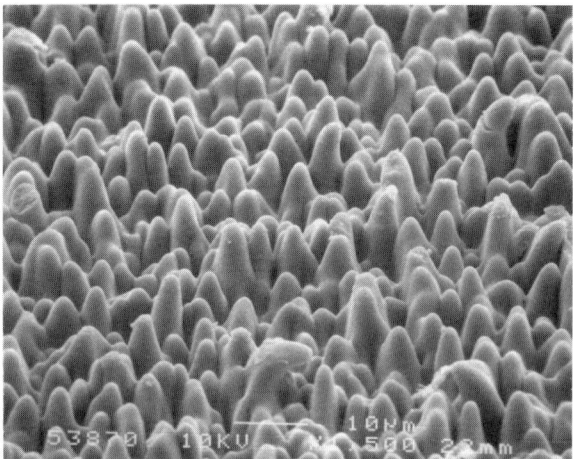

FIGURE 10.18. High aspect ratio diffuser.

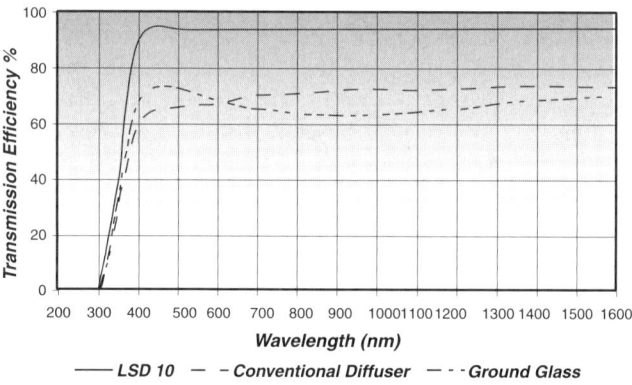

FIGURE 10.19. Comparison of POC's LSD transmission with ground glass and plastic diffuser. POC diffuser offers significantly better transmission without loss in diffusion.

compares the transmission of 5° LSD with that of conventional plastic and ground glass diffusers. These characteristics are made possible by LSD's high aspect ratio and GRIN sculpted surface topography.

- *Significantly Reduced Fresnel Reflection:* In specially designed samples, LSD can function as antireflection or antiglare structures, approaching 100% scattering efficiency. Again, this is a direct result of the GRIN structure.
- *High-Resolution Diffuser Screen:* Because its scattering angle can exceed 160°, the resolving spatial elements of the diffuser structure are within wavelength range, exceeding 1000-l/mm spatial resolution.
- *Large Screen (5 ft. × 4 ft.):* Unlike competitive screens, increasing the diffuser screen size does not degrade resolution or aspect ratio.

- *Environmental Survivability:* POC's surface-relief topography remains unaffected by temperature, vacuum, or humidity. Survivability depends on the material used to produce LSDs. For example, plastics-based LSDs survive $-30°C$ to $100°C$ over several hundred hour time cycles, and LSDs fabricated on metal, graphite, or thermally stable substrates can survive temperatures of $500°C$ to $1000°C$.

To the best of our knowledge, none of these characteristics has been duplicated anywhere, placing our diffuser technology in an exceptional position worldwide. A number of display manufactures have expressed interest in implementing LSD products in their applications.

10.5.4   LSD Testing and Applications

Ground or milk glass, bleached silver-halide plates, and surface photoresist plates diffuse light primarily at their surface. This reduces the path length of the light affected by the diffuser, thereby reducing the number of applications in which they are useful and increasing losses to reflection in unwanted directions.

A diffuser that increases the optical path interaction length of the light in the diffuser makes it possible to design diffusers for many applications that are currently not well served by state-of-the-art diffusers, and it increases the efficiency of light transmission through the diffuser. A diffuser that can be rotated would deliver regulated spatial coherence useful in a number of applications.

Currently, POC produces light shaping diffuser material in a wide variety of angles, many of which are available off-the-shelf and some produced to order (see Table 10.1).

References

[1] M.J. Lahart and A.S. Marathay, "Image speckle patterns of weak diffusers," *J. Opt. Soc. Am.* **65**, 769 (1975).

[2] J. Palmer, "Measurement of transmission, absorption, emission, and reflection," *Handbook of Optics*, Vol. II, (McGraw Hill, New York, 1995).

[3] J.C. Dainty, *Optica Acta*, **17**, 761 (1970).

[4] L.I. Goldfischer, "Autocorrelation function and power spectral density of laser-produced speckle patterns, *J. Opt. Soc. Am.* **55**, 247 (1965).

[5] S. Lowenthal and H. Arsenault, "Image formation for coherent diffuse objects: statistical properties," *J. Opt. Soc. Am.* **60**, 1478 (1970).

[6] W.H. Carter and E. Wolf, "Coherence and radiometry with quasi homogeneous planar sources," *J. Opt. Soc. Am.* **67**, 785 (1977).

[7] M.G. Miller, A.M. Schneiderman, and P.F. Kellen, "Second order statistics of laser-speckle patterns," *J. Opt. Soc. Am.* **65**, 779 (1975).

[8] M. Kowalczyk, "Spectral and imaging properties of uniform diffusers," *J. Opt. Soc. Am.* A**1**, 192 (1984).

 [9] L.B. Lesem, "The kinoform: a new wavefront reconstruction device," *IBM J. Res. Develop.* (March 1969).
[10] H.J. Caulfield, "Kinoform Diffusers," *Proc. SPIE* **25**, 111 (1971).
[11] R.J. Collier, C.B. Burckhardt, and L.H. Lin, *Optical Holography*, (Academic Press, New York, 1971).
[12] M. Born and E. Wolf, *Principles of Optics*, 6th ed. (Pergamon Press, Oxford, 1980).
[13] D. Gabor, "A new microscopic principle," *Nature*, **161**, 777 1948.
[14] J.W. Goodman, *Statistical Optics*. (Wiley, New York, 1985).

TABLE 10.1. Substrates and Diffusion Angles of Available POC Light Shaping Diffusers

Delivery: 4 Weeks
Web Replicated (Thin Film)

Angle	PE	PC	PCS
1°	0.003	0.005	0.010
5°	0.005	0.010	0.020
10°	0.007		
20°	0.010		
30°			
60°			
80°			
95° × 25°			
95° × 35°			
40° × 0.2°			
41° × 2.5°			

1-3 Weeks
Hand or Web Replicated (Thin and Rigid Substrate)

Angle	PE	PC	Clear	Acrylic (% Tint) 30%	50%	70%	90%	Glass
0.5°	0.003	0.005	0.030	0.060	0.060	0.060	0.060	BK-7
1°	0.005	0.010	0.060	0.080	0.080	0.080	0.080	B270
2°	0.007	0.030	0.080	0.125	0.125	0.125	0.125	&
5°	0.010	0.037	0.125	0.250	0.250	0.250	0.250	Customer
6°		0.040	0.250					Supplied
7°		0.060						
10°		0.125						
12°		0.250						
13°								
(Cont'd)								
15°	40°	0.2° × 20°						
20°	46°	0.2° × 40°	2° × 33°					
25°	60°	0.5° × 10°	5° × 20°					
28°	80°	2° × 30°	10° × 75°					
30°	0.2° × 10°	3° × 30°	35° × 95°					

Available Sizes

• All Web and Hand Replicated Products Available up to 40"x40"
• Manufacturing capability to 70" (nonproduction)

1-5 Days Kits		3 Weeks Embossed						8-12 Weeks Injection Molding	
Angle	PC		Angle	PE	PC	PS	AC	AC	PC
0.5°	0.030		1°	0.010	0.010	0.030	0.030	(Standard angles and size up to 10"×10")	
1°			5°		0.030	0.080	0.060		
5°			10°		0.037		0.080		
10°			20°		0.040		0.125		
15°	(Cont'd)		30°		0.060		0.250		
20°	30° × 5°		60°		0.125				
25°	40° × 0.2°		95° × 25°		0.250				
30°	60° × 2°		95° × 35°						
40°	60° × 10°		40° × 0.2°						
60°	80° × 20°		41° × 2.5°						
80°	80° × 40°		20° Prismatic						
10° × 0.2°	70° × 35°								
10° × 2°	95° × 25°								
20° × 10°	95° × 35°								
25 mm			4"×4"–Intermediate sizes cut as needed						
50 mm			6"×6"						
			8"×8"						
			10"×10"						

11

Holographic Nonspatial Filtering for Laser Beams

Jacques E. Ludman and Juanita R. Riccobono

11.1 Introduction

The laser owes much of its usefulness to its inherent high degree of spectral purity and spatial uniformity. This spectral purity, however, also produces a large coherence length, which makes laser light very sensitive to the spatial imperfections (such as dirt and scratches) present in most optical components, including the laser optics themselves.

Spatial nonuniformities can seriously degrade the signal-to-noise ratio (SNR) of laser-based sensing systems. For example, a high-resolution image can lose information if the input laser beam has an intensity minimum at a critical feature location. It is also possible for the laser light to introduce erroneous information due to the inability to distinguish whether a particular intensity variation is produced by the image or by a laser beam imperfection. Aside from SNR degradation, spatial imperfections in laser beams degrade the accuracy of power delivery in high-power applications and the effective pointing stability in low-power systems.

Clearly, for practically any application, the input laser beam should be as clean as possible. Currently, this is accomplished by spatial filtering.[1] There are two types of spatial filters in widespread use. The most common is the so-called "pinhole" filter, which consists of a lens with a pinhole in the focal plane. For more demanding applications, the pinhole is replaced by a single-mode optical fiber. The problem with both types of spatial filters is the requirement for matching spatially and angularly a tightly focused laser beam to a small target. This problem is aggravated by the fact that positioning is 3D, because tightly focused laser beams have short Rayleigh lengths. Experimentally, the initial alignment (or realignment) of spatial filters is difficult and tedious and requires a manual search through 3D position space in an attempt to maximize the throughput intensity. Furthermore, maintaining proper alignment is often a problem.

Moreover, the focusing requirements of both pinhole and fiber spatial filters are problematic with high-power lasers. A high-power laser can damage a pinhole or even create a new one. They are usually aligned with the power turned down. If the pinhole position changes or if there is a slight misalignment, a high-power laser can damage the pinhole or damage the edges when the power is increased. Fibers have many of these problems and in addition have other unique ones because the focused energy can produce nonlinear optical interactions (e.g., in Brillouin and Raman scattering) and index changes, or even permanent damage due to heating. Because these devices are in the focal plane of the converging lens, there are problems due to reflections that are, by definition, reflected back into the laser, occasionally with disastrous effects.

To solve the numerous problems inherent in spatial filters, we have developed nonspatial filtering.[2-4] The basic idea of nonspatial filtering is similar to spatial filtering. In spatial filters, because a laser beam with arbitrary spatial variations in intensity can be expressed as linear combinations of plane waves, a Fourier transform lens is used to convert the propagation angle into position. The pinhole then selects the spatial on-axis component and blocks the others. In contrast, nonspatial filtering is done directly on the laser beam propagation angle. A holographic filter element is inserted in the laser beam path to selectively diffract light propagating at a particular on-axis angle, and transmit without diffraction the unwanted light (to be discarded, but not reflected back into the laser or absorbed by the filter). In particular, we use a very thick hologram as the filter element. The Bragg selectivity of such a hologram guarantees that only waves propagating at a particular angle (namely, the Bragg angle) are diffracted and deflected. Two-dimensional filtering is achieved by using two such holograms in series. Because the nonspatial filter operates with an unfocused beam and relies on Bragg selectivity to remove excess or arbitrary spatial variations in intensity, the majority of the problems due to focusing are alleviated.

11.2   Theory

The theory of wave propagation in thick hologram is well known.[5, 6] For simplicity, the special case of no absorption with only phase modulation due to the presence of index modulation is considered here. For a monochromatic hologram with a grating periodicity of Λ the angle of incidence, θ_b, for a perfectly phase-matched Bragg diffraction is given by:

$$\cos(\theta_b) = \frac{\lambda}{2n\Lambda}.$$

Here, the angle of incidence θ_b is the angle between the incident ray and the holographic fringes within the media. Also, λ is the wavelength of incident light, and n is the average index of refraction of the hologram. Here, we assume that the

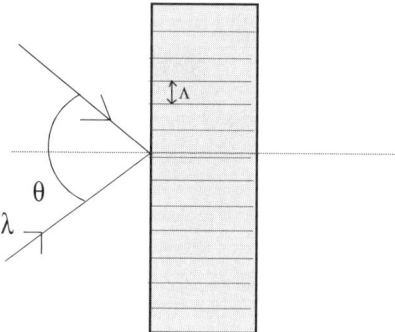

FIGURE 11.1. Setup scheme for hologram recording with symmetric incident laser beams. The optimum playback is at 90° diffraction ($\theta_0 = 45°$); so the fabrication θ_0 must be chosen to take account of any wavelength difference between recording and playback.

hologram fringes are perpendicular to the surface of the hologram. In this case, θ_b is related to q in Fig. 11.1 by Snell's law ($\sin \theta/2 = n^* \sin \theta_b$). The diffraction efficiency, η, is then given by:

$$\eta = \sin^2 \left(\frac{\pi n_1 d}{\lambda \cos \theta_b} \right).$$

Here, n_1 is the amplitude of the sinusoidal index modulation, and d is the thickness of the hologram. The smallest value of n_1 for which maximum diffraction can be obtained varies inversely with the thickness, and it is very small for very thick holograms. The full-width at half-maximum (FWHM) of the angular selectivity is given by:[5]

$$\delta\nu_{\text{FWHM}} \cong \frac{\lambda}{d}.$$

For the typical angular deviation used in the nonspatial filter ($\theta \approx 45°$), the beam deviation is approximately 90°, and for such substantial deviations, the relative angular selectivity and wavelength selectivity are similar.[6] The wavelength sensitivity of a 1-mrad beam cleanup filter would also be 10-3. Laser linewidths are several orders of magnitude narrower than this and are not differentiated by the filter. For high wavelength sensitivity, if desired, the deviation must be close to 180° (reflective geometry). The highest angular and lowest wavelength selectivity occur for small angle deviations, but practical considerations lead to the choice of a larger angle.

11.3   Practical Issues

Theoretical analyses[3, 9] show that a holographic grating to be used as a nonspatial filter with an angular half-width selectivity of 0.5 to 2 mrad, should have a thickness of 0.5 to 2 mm and a spatial frequency approaching 3500 mm^{-1}. For high-efficiency holograms, this medium must have a refractive index modulation of about 1 to 4×10^{-4}. Such 3D gratings can be recorded in thick recording media using impregnated porous glasses[7, 13] and also on photopolymers with diffusion amplification (PDA).[8, 14–16] By comparison, a 40 × objective with a 5-micron pinhole yields an angular selectivity of approximately 1 mrad.

Thick holograms have been made using porous glass impregnated with a photosensitive material.[2–4] The porous glass contains boron, sodium, and about 70% SiO_2 (before being rendered porous). The light-sensitive material is the photopolymer PDA. The average pore diameter is about 30 nm, much smaller than the optical wavelength. Holograms can be recorded at a wavelength of 488 nm, which results in about 2000 lines/mm. Typical hologram thickness is about 0.7 mm. These holograms are most transparent (96%) at 1500 nm, a wavelength of particular interest in fiber-optic communications. The transparency drops on either side of this wavelength. At 1900 nm, the transparency is about 50%, whereas at 540 nm, it is about 70%.

There is an optimal combination of recording parameters (specimen thickness, spatial frequency, and exposure) that produce highly selective and highly efficient nonspatial filter holograms.[3, 4, 9] Values for refractive index modulation are of the order of $\sim 10^{-3}$, which is sufficient for holograms (0.5 to 2 mm thick) with diffraction efficiency approaching 100%. Holograms with a thickness of 2 mm yield filters with an angular selectivity contour of 0.5 mrad. These values are in agreement with the theoretical calculations. However, the experimental angular selectivity is affected by nonuniform index modulation.[3, 4] Nonuniformity increases the angular selectivity. There are at least two sources of nonuniform index modulation. They are variable exposure, either due to beam inhomogeneaties or beam absorption by the hologram during exposure, and nonuniformities in the absorption by the photosensitive media during exposure and processing.

The theoretical[5, 6] angular selectivity contour is compared with the experimental angular selectivity of a hologram recorded in porous glass impregnated with PDA (Fig. 11.2). The half-width of the experimental contours of angular selectivity of the 3D grating are somewhat larger than theory predicts. A part of this difference is due to the angular divergence of the test laser (3×10^{-4} rad). In addition, the shape of the angular selectivity contour can be affected by changing the preparation of the porous photosensitive material. A grating with a maximum index modulation in the filter center has a smooth contour of angular selectivity without lateral maxima, as shown here.[3, 4]

These nonspatial holograms fabricated in porous glass impregnated with PDA have achieved a bandwidth of angular selectivity contour of 10^{-3} rad and diffraction efficiencies of 60% to 80%.[3, 4]

11.4   Advanced Two-Dimensional Filtering

A single nonspatial filter cleans a laser beam in one dimension only.[2–4, 9–11] To filter a laser beam symmetrically, as with a conventional spatial filter, two nonspa-

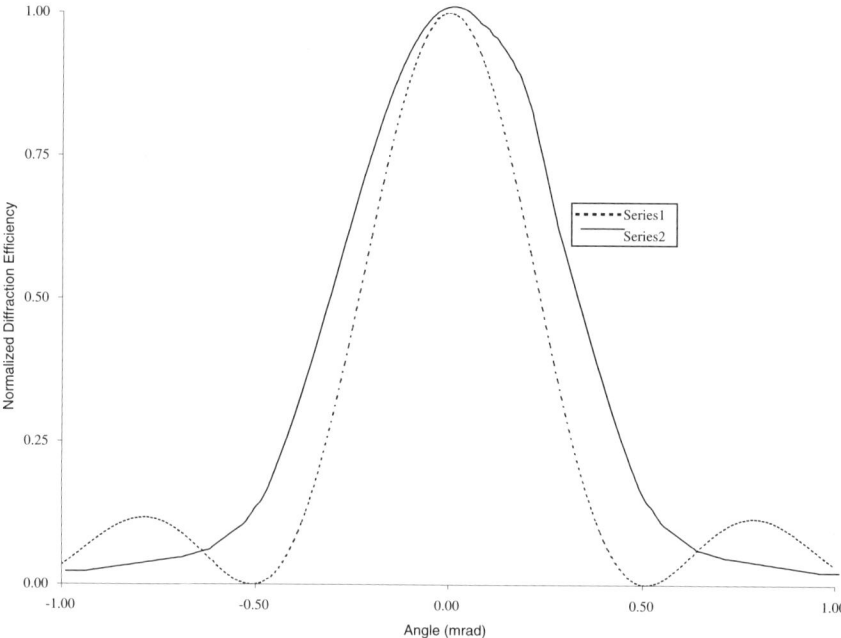

FIGURE 11.2. Calculated (1) and experimentally measured (2) contours of angular selectivity.

tial filters are combined such that their grating vectors are orthogonal.[10, 11] This can be done in such a way that this two-filter nonspatial device is wavelength independent.[12]

The basic structure of this wavelength-independent design is shown in Fig. 11.3. The two nonspatial filters that make up the device are parallel to each other and are identical except that the second filter is rotated 180° around the x-axis. Figure 11.4 shows the device with the two filters separated so that the wave vectors of the individual 1D filters are visible. The two filters have identical structures and some unusual features. The input beam is incident in the x-z plane, but it diffracts to an output in the y-z plane. To achieve this unusual relationship between the input and output beams, the grating fringes run diagonally across the holographic material (Figs. 11.3 and 11.4). The grating vector, however, is confined to the x-y plane and is tilted 45° to the x- and y-axes. The unusual orientation of the wave vectors and their associated grating vector allow for the simple design: using two identical filters, rotating the second filter 180° around the x-axis, and sandwiching the two filters together.

This simplistic design yields two important results.[12] The first is that the grating vectors of the two filters are perpendicular to each other, a requirement for 2D filtering. In addition, the output from the first grating is automatically a Bragg-matched input for the second grating. This is due to the inherent stability of the PDA recording material, before, during, and after recording.

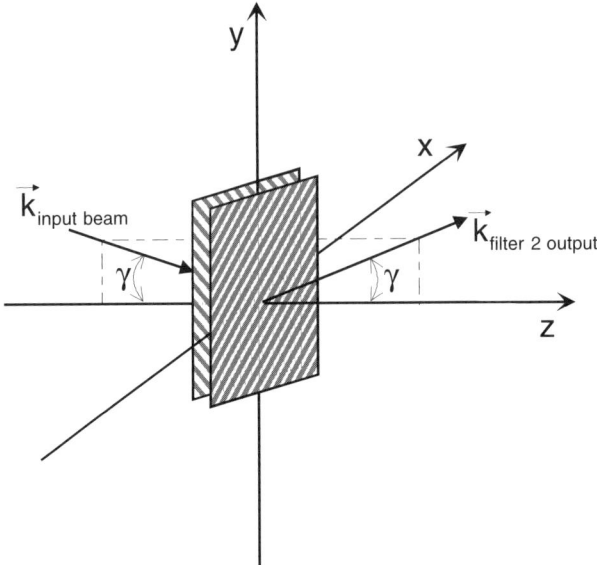

FIGURE 11.3. Two-dimensional nonspatial filtering device.

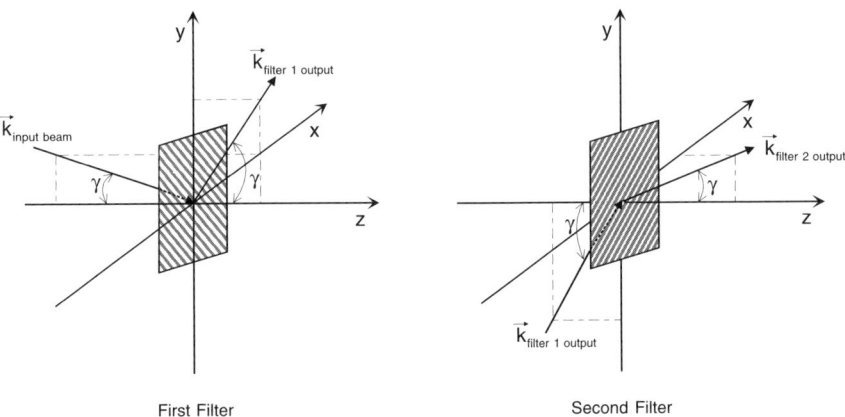

First Filter   Second Filter

FIGURE 11.4. The device from Fig. 11.3 with separation between the filter elements showing the direction of beam propagation.

Previously, the filters were tuned for optimum performance using two independent angular adjustments.[2-4] Each angular adjustment rotated the associated grating around an axis in the plane of the holographic material defined by the direction of the fringes. This adjustment allowed the user to Bragg-match the input beam angle to the grating. An orthogonal rotation in the same plane (i.e., around the grating vector) has no affect on the grating's Bragg condition and, therefore, its diffraction. Here, the grating vectors of the two filters are always perpendicular;

so the adjustment of one filter does not affect the other and vice versa. This means that the two filters can be glued together to make a single 2D filtering device.

Another method of adjustment is also possible because the output of the first filter is an automatic Bragg- match for the second filter when the input to the first filter is the same as its x-z plane recording beam. It is simply necessary to find the Bragg-match for the x-z input beam by rotating the filter pair about the y-axis. As soon as the input is matched, the output of the first grating diffracts at the second grating. Although this arrangement appears to eliminate one adjustment step from the two-angular adjustment process, the filter must be preadjusted to lie in the x-y plane. It is important to note that the rotation is adjusting both gratings simultaneously, because the y-axis is 45° to the direction of both filters. For high-fidelity gratings, this is fairly trivial despite the unusual geometry. This unusual geometry, however, makes the result for an input at a different wavelength somewhat less intuitive.

11.5   Conclusion

The holographic nonspatial filter described here is currently a 2D, viable device. It is clear that in many ways this device will function better than will the spatial filters it replaces. Besides being simpler and easier to operate than other systems, it avoids many of the substantive, fundamental problems of spatial filters in high-power or high-energy laser beam cleanup. For example, the nonspatial filter is relatively insensitive to lateral and axial shifts, but the spatial filter is very sensitive to spatial misalignments (Table 11.1).[12] Both spatial and nonspatial systems suffer losses (Table 11.2);[12] however, nonspatial systems do not incur the large power loss from expanding the beam and cropping it as do spatial systems.

Two nonspatial filters, properly oriented, are capable of cleaning the beam in both the horizontal and vertical directions. The intensity profile from one of the filters shows the nonspatial filtering effect (Fig. 11.5) in the vertical direction. This can be compared with the profile of the transmitted or input beam (Fig. 11.6). For each filter, only one adjustment is necessary, as each hologram is insensitive to the angle orthogonal to the one to which it is highly sensitive. Untrained workers learned within minutes how to establish alignment within seconds. Hologram

TABLE 11.1. Comparison of the alignment sensitivities of nonspatial filters and spatial filters.

Sensitivity or Result	Nonspatial Filter	Spatial Filter
Alignment	2 independent angular adjustments	three coupled translation adjustments
Sensitivity to Lateral Shifts	centimeters	microns or less
Sensitivity to Axial Shifts	millimeters to centimeters	microns or less
Spatial Misalignment	no effect	loss of power
Axial Misalignment	loss of power	loss of power

TABLE 11.2. Comparison of losses for nonspatial and spatial filtering technologies.

Loss	Nonspatial Filter	Spatial-pinhole	Spatial-fiber
Reflection	two PDA-air surfaces two PDA-glue surfaces	two Lens-air surfaces	two Lens-air surfaces two Fiber-air surfaces
Absorption	PDA Glue	Lens	Lens Fiber
Power (lens-aperture cropping)	N/A	Lens f/# Pinhole aperture	Lens f/# Fiber aperture
Diffraction	two Gratings	N/A	N/A

FIGURE 11.5. Profile of the diffracted beam, shows nonspatial filtering in the vertical direction.

FIGURE 11.6. Profile of the transmitted beam, identical to the same of the input beam.

alignment is much simpler than is pinhole alignment. The output beam is deviated from the input beam, but it remains in the same plane. With the advanced 2D filter, the two filters may be joined into one unit and oriented as a unit for any wavelength.

In addition to the advantages of this new technology, additional features are unique to the holographic filter. The primary one is the fact that the amplitude of the index modulation may be varied as a function of the depth of the filter. Such variations lead to modifications of the angular selectivity profile. The output laser beam can have a contour as well as an angular divergence as required. Additional benefits of this technology are that the output may be tailored in shape as well. The output beam may be designed to be parallel, diverging, or converging. Other potential applications include use as an etalon, a Raman filter, or for spectroscopic analysis.

Acknowledgments

Many colleagues have contributed to this work. We would like to acknowledge the intellectual and experimental contributions of John Caulfield, Jean-Marc Fournier, Phil Hemmer, Michele Henrion, Yuri Korzinin, Nadya Reinhand, Irina Semenova, Selim Shahriar, Gennadi Sobolev, and Svetlana Soboleva.

References

[1] E. Hecht and A. Zajac, *Optics* Addision-Wesley, Reading, MA, (1975).

[2] J.E. Ludman, J.R. Riccobono, H.J. Caulfield, J.M. Fournier, I.V. Semenova, N.O. Reinhand, P.R. Hemmer, and S.M. Shahriar. "Porous-matrix holography for nonspatial filtering of lasers," *IS&T/SPIE Symposium on Electronic Imaging: Science and Technology*, February 5–10, 1995, San Jose, CA, *Proc. SPIE*, **2406**, 76–85 (1995).

[3] J.E. Ludman, J.R. Riccobono, N.O. Reinhand, Y.L. Korzinin, I.V. Semenova, and S.M. Shahriar, "Nonspatial filter for laser beams," *Quantum Electron.* **26**, 1093–1096 (1996).

[4] J.E. Ludman, J.R. Riccobono, N.O. Reinhand, I.V. Semenova, Y.L. Korzinin, and S.M. Shahriar, "Holographic nonspatial filter," *Application and Theory of Periodic Structures*, July 10–12, 1995, San Diego, CA, *Proc. SPIE*, **2532**, 481–490 (1995).

[5] H. Kogelnik, "Coupled waves theory for thick hologram gratings," *Bell System Tech. J.* **48**, 2909–2947 (1969).

[6] J.E. Ludman, "Approximate bandwidth and diffraction efficiency in thick holograms," *Am. J. Phys.* **50**, 246 (1982).

[7] S.A. Kuchinskii, V.I. Sukhanov, and M.V. Khazova, "Principles of hologram formation in capillary composites," *Opt. Spectrosc.* **72**, 716–730 (1992).

[8] A.V. Veniaminov, V.F. Goncharov, and A.P. Popov, "Hologram amplification due to diffusion destruction of opposite phase structures," *Opt. Spectrosc.* **70**, 864–869 (1991).

[9] J.E. Ludman, J.R. Riccobono, N.O. Reinhand, I.V. Semenova, Y.L. Korzinin, S.M. Shahriar, H.J. Caulfield, J.M. Fournier, and P. Hemmer, "Very thick holographic nonspatial filtering of laser beams," *Opt. Eng.* **36**, 1700–1705 (1997).

[10] Y. Korzinin, V. Alekseev, A. Kursakova, E. Gavrilyuk, N. Reingand, I. Semenova, J. Ludman, and J. Riccobono, "Holographic spatial-frequency filter for laser radiation", *J. Opt. Technol.* **64**, 341–345 (1997).

[11] Y. Korzinin, N. Reinhand, I. Semenova, J. Ludman, J. Riccobono, S. Shahriar, and J. Caulfield, "Two dimensional holographic nonspatial filtering," *Proc. SPIE* **2688**, 109–122 (1996).

[12] M. Henrion, J. Ludman, G. Sobolev, S. Shahriar, S. Soboleva, and P. Hemmer, "Two-dimensional holographic nonspatial filtering for laser beams," *Proc. SPIE* **3417**, 195–206 (1998).

[13] S.A. Kuchinskii, V.I. Sukhanov, and M.V. Khazova, "Principles of the formation of holograms in capillary composites," *Opt. Spektrosk.* **72**, 716 (1992).

[14] A.P. Popov, V.F. Goncharov, A.V. Veniaminov, and V.A. Lyubimtsev, "High-efficiency narrowband spectral selectors," *Opt. Spectrosc.* **66**, 3–4 (1989).

[15] A.P. Popov, V.F. Goncharov, A.V. Veniaminov, and V.A. Lyubimtsev, "Hologram amplification due to diffusion destruction of opposite phase structures," *Opt. Spectrosc.* **70**, 864–869 (1991).

[16] A.P. Popov, A.V. Veniaminov, and V.F. Goncharov, "Photophysical mechanisms of effective phase holograms recording in polymers with chemically attached photoactive centers," *Proc. SPIE* **2215**, 113–124 (1994).

12

Particle Holograms

Chandra S. Vikram

12.1   Introduction

The history of particle holograms goes back to evolution of modern holography. By
particle, here one means small objects, such as particles, bubbles, droplets, boundary layers, and so on, generally in a dynamic volume. Gabor's historical[1−3] attempts
to increase resolution in electron microscopy involved passing monochromatic
beam through opaque lines on a clear transparency. The subsequent complex pattern, or the interference between the directly transmitted beam and one diffracted by
the lines was stored. The processed transparency, upon illumination by a monochromatic light reconstructed images—although noisy because two diffracted orders
(famous twin-image problem) and the directly transmitted light were all along one
line. Nevertheless, this historical work opened the door for modern holography
in general but particle field holography explicitly. Later, Thompson[4, 5] discovered
that if the objects are small so that they are in the far-field from the recording
plane, then the practical role of the bothersome twin images becomes negligible.
Basically, what that means is just passing a suitably pulsed laser light through the
volume containing the particles to store the hologram. Now, upon reconstruction,
the images can be studied at leisure. As we know, a conventional imaging system
such as a well-corrected lens can yield aperture-limited resolution. Nevertheless,
using such optimum resolution leads to a narrow depth of field. In a dynamic volume, such as a spray, in which one wants to study the droplets, the longitudinal
position is one of the parameters to be studied. Thus, a tool with good resolution
throughout the volume and not only the fixed object plane is required. Incidentally,
an incoherent imaging system resolving a diameter d yields the depth of field d^2/λ,

where λ is the wavelength of light used. For a nominal object diameter of 4 μm and wavelength 0.6328 μm, the depth of field is only about 25 μm! However, in holography, as stated above, the entire scene image can be frozen and one can study in detail anywhere in the volume. Practically allowable depth of the scene volume is not infinite but still useful in a large number of situations. In this chapter, we cover some recent aspects from a futuristic point of view. These cover in-line as well as off-axis approaches. We start with a brief description of the in-line approach, which is unique and historically important for particle fields.

12.2   In-Line Fraunhofer or Far-Field Holography

The simplest form of in-line Fraunhofer holography of particle fields is shown in Fig. 12.1. A collimated (collimation not necessary but convenient to analyze the reconstruction), pulsed (duration suitable to freeze a dynamic field) laser light is just passed through the test volume and the hologram stored. Upon reconstruction, we obtain real images, virtual images, and directly transmitted light. In fact, a screen can be used to observe the real images as such. Nevertheless, the common procedure is to use a closed circuit TV system to observe the images. The hologram (or the TV camera) on a x-y-z translation stage can cover the image volume. The system helps in the image magnification, pinpointing the particle location in the image volume, and further data storage and processing tasks. The in-line approach, although very convenient and common, posed certain limitations. For example, to have dominant reference beam, the particle field should not be dense. Generally, at most, 20% of the recording beam cross section may encounter particles. The off-axis[6] (or independent reference beam) approach becomes useful. Besides routine engineering applications, innovations, new applications, and system developments have been performed on a continuous basis.[7-10]

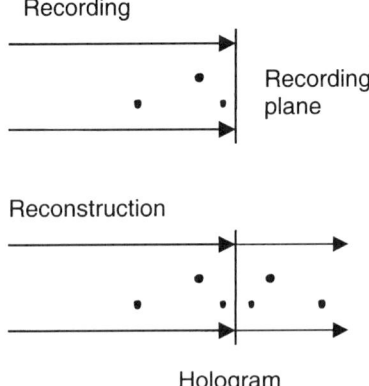

FIGURE 12.1. Schematic diagram of in-line Fraunhofer holography of particle fields.

12.3   System Enhancement

System developments for enhanced capabilities and new applications remain one of the popular subjects of study. Schaller and Stojanoff[11] applied the technique to study a diesel jet injected into a high-pressure test chamber. They used two reference angles for two recordings for velocity measurements. This way two images can be separated upon reconstruction. The effective hologram aperture for an individual particle in in-line hologram is generally small compared with the entire hologram (covering the cross section of the test volume). Nevertheless, there can be further restrictions for the particles in the cross section near the outer boundaries of the hologram. Effects of such asymmetries and aperture limits have been discussed in detail by Fang and Hobson.[12] They obtained an approximate expression for the loss of resolution of objects away from the optical axis. The ideal resolution is divided by $1 + (d/H)^2$, where d is the distance of the micro-object in the cross section from the optical axis and H is the hologram half-width. Obviously, objects close to the optical axis ($d = 0$) are ideal to attain maximum resolution. Lai and Lin[13] used phase steps for image subtraction of double-exposure holograms. The background noise is rejected this way for improved signal-to-noise ratio of particle images. Royer[14] described the role of holography in particle image velocimetry. Trolinger et al.[15, 16] applied holography for microgravity applications. More than 1000 holograms aboard Space Shuttle Discovery were recorded to study several microgravity particle dynamics and crystal growth-related parameters. An example of their holographic particle image velocimetry is presented in Fig. 12.2. Polystyrene spheres in fluid act as tracers. Their positions at different times yield several useful parameters. Some particle/pairs are seen in-focus. Thus, their behavior in the volume is revealed. In connection to measure behind-armor debris, Anderson et al.[17] summarize the progress on cylindrical holography. As shown in Fig. 12.3, the laser beam enters a cylindrical section near the target. The section is composed (half each) of a diffusively reflecting surface and the holographic film. A particle field generated by the impact by the projectile is stored using the reference beam (light directly reaching the film). In this unique off-axis arrangement, the image of the particle field can be viewed over the entire half-cylinder of the film. Lai and Lin[18] emphasized the role of divergent beam to significantly improve recordable farthest far-field distance in in-line Fraunhofer holography. Thus, rather than nominal maximum of 50 far fields, the micro-object can be far away. The approach allows significant enhancement of the allowable test-section depth. Amara and Ozkul[19] used a coherence-coded approach and photorefractive BSO crystal for velocimetry applications. Again, in connection to holographic particle image velocimetry of turbulent flows, Scherer and Bernal[20] use two (in different directions) in-line holograms of the flow field simultaneously. Three-component velocity measurements are thus obtained. For high Reynolds number pipe flow studies, Chan and Li[21] devised a side-scattering system. Polystyrene spherical particles of about 40 μm in diameter were seeded in distilled water with flow Reynolds number as 31,000. As seen in Fig. 12.4, scattered light from the pipe is stored using an off-axis reference beam. Velocities within 5% error have

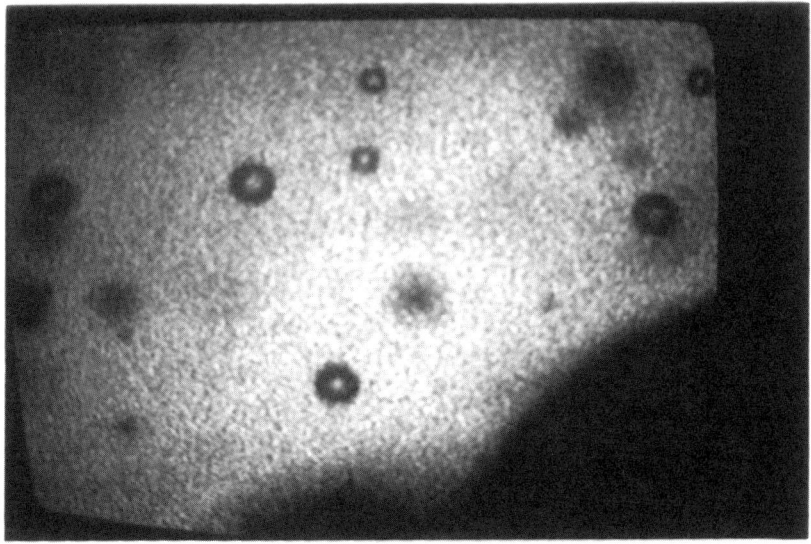

FIGURE 12.2. Reconstructed, double-exposure hologram of polystyrene spheres (600 μm and 400 μm in diameter) made in crystal growth chamber in space. The crystal front is seen in the bottom right region of the picture (figure courtesy of Dr. James D. Trolinger, MetroLaser, Irvine, California).

been obtained. Chan et al.[22] earlier described several design aspects of this mode. Liu and Hussain[23] describe in detail the role of high pass filtering in off-axis holographic particle image velocimetry. Significant improvements in signal-to-noise ratio are possible in this forward scattering approach. Wallace[24] reports the role of a spinning disk for high-speed holographic movies of ballistic events. A tremendous amount of information is possible.

Lebrun et al.[25] compared the Fraunhofer diffraction pattern and simulation based on Lorenz–Mie theory. The aim was to measure wire diameters accurately from the diffraction pattern.

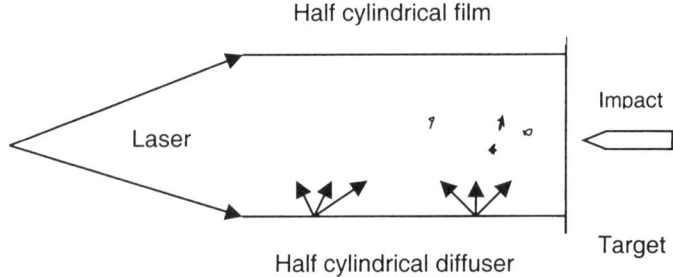

FIGURE 12.3. Arrangement represents cylindrical holography of behind-armor bebris.

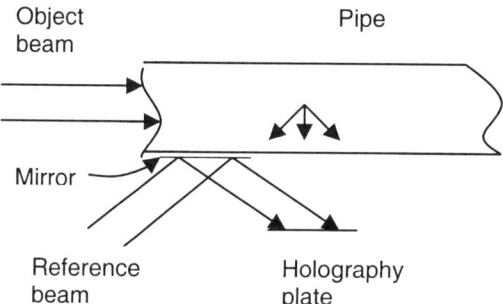

FIGURE 12.4. Schematic diagram of a side-scattering holographic particle imaging technique.

12.4 Aberrations and Control

Ultimate resolution depends on the aberrations as it becomes a dominant factor when the hologram aperture size is increased. Although significant developments on this aspect have been reported earlier,[7, 10] some recent works are described here. Coupland and Halliwell[26] described the role of optical autocorrelation of the wavefront amplitude in a real image region to achieve immunity to aberrations. Hobson et al.,[27] in connection to the study of chemical cycles in aquatic systems, performed holography of aquatic particles in water. However, for high resolution and precision necessary in such studies, recording and reconstruction wavelengths must be different. The image aberrations can be minimized this way. The replay wavelength should be the construction wavelength divided by the refractive index (of water). Complete correction demands an entire holographic system in water. However, for a more practical holographic plate in air, the aberration balancing is achieved by suitable hologram-window separation compared with the window thickness. By these techniques, high-quality imaging in water using in-line as well as off-axis approaches has been achieved.[27] Fang and Hobson[28] recently provided spherical aberration-related aspects on the in-line approach and further established validity of the approach. More critical studies with the view of designing of underwater holographic camera have been presented by Hobson and Watson.[29] Thus, their replay wavelength change approach in conjunction with geometrical considerations has been able to obtain high-fidelity imaging of underfluid objects. These studies help in critical studies of biological microparticles, which is becoming increasingly important.[27–31]

Holographic optical elements can also be used for aberration correction of large-aperture conventional optical systems.[32] Such elements can be used to correct telescopes, microscopes, and so on. The elements provide the desired phase-conjugation to neutralize aberrations often associated with large apertures. Blood cells and microchips have been clearly imaged this way. Anderson and Knize[32] corrected a telescope with large aberrations (over 2000 waves) to diffraction-limited operation for high-bandwith optical communications.

12.5   Innovative Approaches

Innovative approaches continue to appear in literature. Imaging through scattering media is an important subject with potential applications in biology, meteorology, and so on. Different gating techniques have been successfully applied to the problem and refinements, for example, by Hyde et al.[33] are available on a continuous basis. In their approach, object and reference beams are a train of ultrashort low coherent optical pulses. A hologram is recorded on a photorefractive crystal when both beams arrive at the same time. By varying the relative path lengths between reference and object beams, light from a particular section of the object is stored. Whole-field 2D image sections with known depths can thus be acquired in a few seconds. Arons and Dilworth[34] combined short coherence length and Fourier-synthesis holography, eliminating unwanted sampling effects.

Measuring total power contained within the reconstructed image for the size analysis has also been proposed.[35] Using an aperture-limited hologram and a circular-cross-section opaque object as an example, such a possibility has been highlighted. Detuned interference filters can be used to modify diffraction patterns. For in-line holography, they can thus be used to increase the effective aperture and hence resolution. The image formation and analysis of such an approach has been described in detail.[36]

Baldwin et al.[37] proposed and experimentally demonstrated a novel spatial filtering procedure to enhance the signal-to-noise ratio of the image of particle field. As shown in Fig. 12.5, a Fourier transform absorption hologram of the particle field id stored. As we know, the low-frequency region in the recording plane contributes to unwanted background and noise in the reconstructed image. However, by setting the exposure conditions, the low-frequency region can be overexposed so that diffraction efficiency in this region becomes very low. Upon reconstruction, the hologram will automatically act as a high-pass filter.

Nishihara et al.[38] determine particle diameter by fast Fourier transform of the hologram. The Fourier transform pattern of the hologram preserves the properties of the original fringe pattern. However, background and noise of the hologram

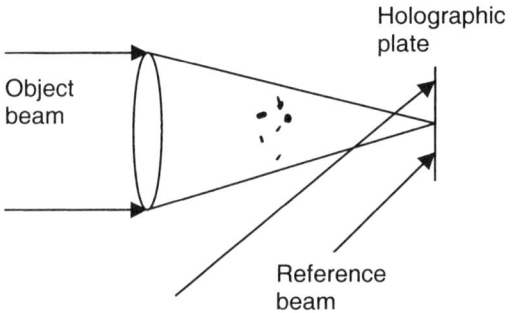

FIGURE 12.5. Recording configuration for exposure-induced high-pass filtering.

contribute to the low-frequency region in the Fourier transform. Thus, the Fourier transform provides a convenient way to perform the analysis. It has been found that the Fourier transform has a white ring zone. The diameter of this ring provides a powerful and reliable tool to obtain the particle diameter. Optical scanning microscopy[39] has been successfully applied in twin-image elimination.[39−41] The technique has also been applied to fluorescence microscopy by Schilling et al.[42] with significant system improvements incorporated by Indebetouw et al.[43] Besides fluorescence applications, this type of holography is continuously applied to new situations such as in optical image recognition of 3D objects.[44] Lebrun et al.[45] demonstrated the use of optical wavelet transform for the reconstruction. The holographic plate is directly placed on the input plane of a wavelet transform optical system. The system consists of a Vander Lugt correlator with an electrically addressed spatial light modulator. A noiseless, selective, and fast reconstruction tool is obtained. Demonstrated for glass fibers, the approach opens a door for a new tool and further research.

12.6   X-ray and Electron Holography

Because resolution is dependent on the wavelength, there have always been attempts for optimum use of x- rays and electron beams in particle field holography. Xiao et al.[46] compared Gabor and lensless Fourier transform x-ray holograms. Kodama et al.[47] presented an algorithm to eliminate artifacts such as a twin-image in in-line x-ray holography. Xu[48] proposed using the amorphous sample on a crystal to resolve its electron density. The arrangement (Fig. 12.6) employs a x-ray diffractometry type of setup with collimated monochromatic radiation from a normal synchrotron as a source. Bragg diffraction from the crystal in the majority of the cross section provides the reference beam. The light from the sample (micrometer size) acts as in-line object beam. Tanji et al.[49] described differential microscopy in off-axis electron holography using a biprism. More recently, Lai et al.[50] used off-axis reconstruction of in-line holograms for the twin-image elimination in soft x-ray holography. Watanabe et al.[51] devised a cooled backilluminated charge-coupled device (CCD) camera to store and numerically reconstruct a soft-x-ray Gabor hologram. Meyer and Heindl[52] described the role of a neural network to use the off-axis electron hologram for the image reconstruction. The process is less sensitive to noise present in the conventional reconstruction. Xiao et al.[53] introduced a two-hologram way to minimize the twin-image effect in in-line x-ray holography. Their computer simulations show that the image contrast can be significantly improved by the technique. Yang et al.[54] presented a digital reconstruction process for the twin image problem. The algorithm is practical for flat objects (for which the object-hologram distance is known) but requires less time than does the phase-retrieval method. Duan et al.[55] emphasized the role of Moiré fringes for accurate measurement of phase shifts in electron holography.

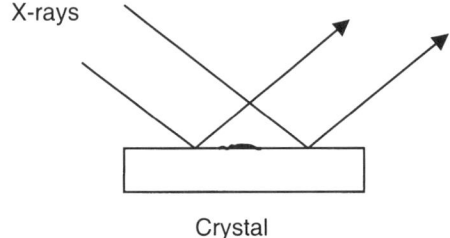

X-rays

Crystal

FIGURE 12.6. Schematic diagram (not to scale) of in-line x-ray holography of an amorphous sample placed on a crystal.

On the applications side, Frost et al.[56] used electron holography to study the leakage field of magnetic force microscopy thin-film tips. Rau et al.[57] characterized stacked gate oxides by electron holography. Lin and Dravid[58] used transmission electron holography to map the potential of graphite nanotubes. Wang et al.[59] obtained the phase image of silicon nanospheres using a field-emission transmission electron microscope. Johnson and Dravid[60] provided the first direct real space evidence for breakdown of an internal grain boundary barrier using electron holography. Tegze et al.[61] used x-ray holography at atomic resolution. They obtained the position of a cobalt atom in cobalt oxide with an accuracy of 0.5 Å.

The above examples show continued interest in x-ray and electron holography. Not only new storage and processing techniques but applications as well are frequently appearing. Two recent books on electron holography are also available.[62, 63]

12.7   Digital Holography

Numerical reconstruction of holograms has shown significant progress in the recent past.[64-76] In such holography, it is possible to store the hologram in a CCD array and then numerically reconstruct the image or other desired parameters. The ultimate resolution is limited by low (compared with that with holography materials) spatial resolution of CCDs. However, the technique has already been established as very useful in many situations. Not only the storage is fast and convenient, but the optical reconstruction is avoided as well with immediate results. For in-line Fraunhofer holography applications, Owen and Zozulya[77] successfully introduced digital holography. They investigated individual particles over a deep depth (25 cm) with a resolution of 5 μm. They presented hardware and software for complete (size and position) analysis of particles. They have also successfully operated the system in field tests on a remotely operated vehicle. By comparing different hologram reconstructions, they also derive velocity vectors for marine mass transport.

12.8   Related Developments

There have been developments that can be related to particle field holography. In the past, the Fraunhofer diffraction pattern has been used for the analysis. More recently, Barnes et al.[78] successfully demonstrated a fast droplet size and refractive index tool from the diffraction data. Obtaining the correct image location has always been important for automated analysis systems in particle holography. The concepts of defocus corrections of standard imaging, say, one by Raveh et al.,[79] can be attempted to reconstructed images as well. Kim and Poon[80] used power fringe-adjusted filtering and Wigner analysis for correlation to obtain the position. Although particle fields are generally used for display applications, spectacular views are possible and shown, for example, in cylindrical holography.[17] In that connection, an application on imaging in dense scattering media as a projection medium is interesting.[81] As shown in Fig. 12.7, the real images can be projected in a dense fog, smoke, water, dilute milk, and so on. The possible applications are entertainment, advertisement, medical diagnostics, image representations for surgical procedures, and education.[80]

12.9   Conclusions

Studying particle fields is one of the earliest applications of holography. The enthusiasm still persists. The reason is that it works fairly easily with the ability to solve equally complex problems. Aperture-limited resolution in a volume (as against in a plane of conventional photography) is readily achieved. On the other hand, complex problems with the need of applying such a tool are continuously appearing. New system developments are thus being reported on a continuous basis. Rapidly changing digital storage or processing technology not only provide new research tools, but innovative applications as well. Critical accuracy needs in certain applications are directing researchers to achieve true "aperture-limited" resolution, i.e., without aberrations. With nonoptical wavelengths, besides pio-

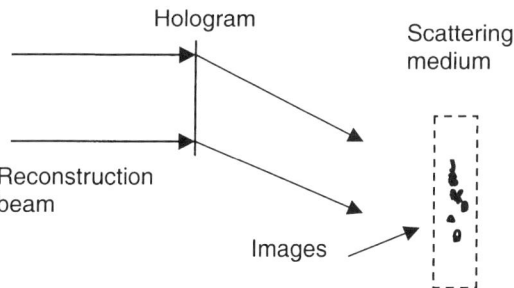

FIGURE 12.7. Diagram represents projection of real holographic images in a scattering volume.

neering applications, new resolution landmarks are being achieved. Because of the very nature of the subject and universal applicability, particle field holography is likely to remain an important tool in optical metrology for a long time.

References

[1] D. Gabor: A new microscopic principle, *Nature* **161**, 777–778 (1948).

[2] D. Gabor: Microscopy by reconstructed wavefronts, *Proc. Royal. Soc. A* **197**, 454–487 (1949).

[3] D. Gabor: Microscopy by reconstructed wavefronts: II, *Proc. Phys. Soc. (London) B* **64**, 449–469 (1951).

[4] B.J. Thompson: Fraunhofer diffraction patterns of opaque objects with coherent background, *J. Opt. Soc. Am.* **53**, 1350 (1963).

[5] B.J. Thompson: Diffraction by opaque and transparent particles, *J. Soc. Photo-optical Instrumentation Engineers* **2**, 43–46 (1964).

[6] R.E. Brooks, L.O. Heflinger, R.F. Wuerker, and R.A. Briones: Holograpic photography of high- speed phenomena with conventional and Q-switched ruby lasers, *Appl. Phys. Lett.***7**, 92–94 (1965).

[7] C.S. Vikram: *Particle Field Holography* (Cambridge Univ. Press, Cambridge 1992).

[8] P.J. Santangelo and P.E. Sojka: Holographic particle diagnostics, *Progress in Energy and Combustion Sci.* **19**, 587–603 (1993).

[9] B.J. Thompson: Holographic methods, in *Liquid- and Surface-Borne Particle Measurement Handbook*, J.Z. Knapp, T.A. Barber, A. Lieberman, ed., (Marcel Dekker, New York 1996) 197–233.

[10] C.S. Vikram: Particle holograms, in *Optical Measurement Techniques and Applications*, P.K. Rastogi, ed., (Artech, Boston 1997) 277–304.

[11] J.K. Schaller and C.G. Stojanoff: Holographic investigations of a diesel jet injected into a high- pressure test chamber, *Part. Part. Syst. Charact.* **13**, 196–204 (1996).

[12] X. Fang and P.R. Hobson: Effect of transverse object position on the reconstructed image of an aperture-limited in-line hologram, *Opt. Laser Technol.* **28**, 573–578 (1996).

[13] T. Lai and W. Lin: Effective implementation of subtraction holography for noise suppression in particle fields, *Appl. Opt.* **35**, 4334–4336 (1996).

[14] H. Royer: Holography and particle image velocimetry, *Measurement Sci.Technol.* **8**, 1562– 1572 (1997).

[15] J.D. Trolinger, R.B. Lal, D. McIntosh, and W.K. Witherow: Holographic particle-image velocimetry in the First International Microgravity Laboratory aboard the Space Shuttle Discovery, *Appl. Opt.* **35**, 681–689 (1996).

[16] J. Trolinger, M. Rottenkolber, and F. Elandaloussi: Development and application of holographic particle image velocimetry techniques for microgravity applications, *Measurement Sci. Technol.* **8**, 1573–1584 (1997).

[17] C. Anderson, J. Gordon, D. Watts, and J. Marsh: Measurement of behind-armor debris using cylindrical holograms, *Opt. Eng.* **36**, 40–46 (1997).

[18] T. Lai and W. Lin: Recordable depth of view and allowable farthest far-field distance of in-line far-field holography for micro-object analysis, *Appl. Opt.* **36**, 4419–4424 (1997).

[19] M.K. Amara and C. Ozkul: Holographic particle image velocimetry: simultaneous recording of several light sheets by coherence encoding in a photorefractive crystal, *J. Opt. (UK)* **28**, 173– 180 (1997).

[20] J.O. Scherer and L.P. Bernal: In-line holographic particle image velocimetry for turbulent flows, *Appl. Opt.* **36**, 9307–9318 (1997).

[21] K.T. Chan and Y.J. Li: Pipe flow measurement by using a side-scattering holographic particle imaging technique, *Opt. Laser Technol.* **30**, 7–14 (1998).

[22] K.T. Chan, T.P. Leung and Y.J. Li: Holographic imaging of side-scattering particles, *Opt. Laser Technol.* **28**, 565–571 (1996).

[23] F. Liu and F. Hussain: Holograpic particle velocimeter using forward scattering with filtering, *Opt. Lett.* **23**, 132–134 (1998).

[24] J. Wallace: Spinning disk records holographic movies, *Laser Focus World* **35–8**, 16–18 (1999).

[25] D. Lebrun, S. Belaid, C. Ozkul, K.F. Ren, and G. Grehan: Enhancement of wire diameter measurements: comparison between Fraunhofer diffraction and Lorenz-Mie theory, *Opt. Eng.* **35**, 946– 950 (1996).

[26] J.M. Coupland and N.A. Halliwell: Holograpic displacement measurements in fluid and solid mechanics: immunity to aberrations by optical correlation processing, *Proc. R. Soc. Lond. A* **453**, 1053– 1066 (1997)

[27] P.R. Hobson, E.P. Krantz, R.S. Lampitt, A. Rogerson, and J. Watson: A preliminary study of the distribution of plankton using hologrammetry, *Opt. Laser Technol.* **29**, 25–33 (1977).

[28] X. Fang and P.R. Hobson: Effect of spherical aberration on real-image fidelity from replayed in-line holograms of underwater objects, *Appl. Opt.* **37**, 3206–3214 (1998).

[29] P.R. Hobson and J. Watson: Accurate three-dimensional metrology of underwater objects using replayed real images from in-line and off-axis holograms, *Measurement Sci. Technol.* **10**, 1153–1161 (1999).

[30] V.V. Dyomin and V.V. Sokolov: Holographic diagnostics of biological microparticles, *Proc. SPIE* **2678**, 534–542 (1996).

[31] V.V. Dyomin: Development of methods for optical diagnostics of microstructure parameters of water suspensions, *Proc. SPIE* **2678**, 543–547 (1996).

[32] G. Andersen and R.J. Knize: Holograpically corrected telescope for high-bandwidth optical communications, *Appl. Opt.* **38**, 6833–6835 (1999).

[33] S.C.W. Hyde, N.P. Barry, R. Jones, J.C. Dainty, and P.M.W. French: High resolution depth resolved imaging through scattering media using time resolved holography, *Opt. Commun.* **122**, 111– 116 (1996).

[34] E. Arons and D. Dilworth: Improved imagery through scattering materials by quasi-Fourier-synthesis holography, *Appl. Opt.* **35**, 3104–3108 (1996).

[35] C.S. Vikram: Image power for size analysis in in-line Fraunhofer holography, *Opt. Lett.* **21**, 1073–1074 (1996).

[36] C.S. Vikram: Image formation in detuned interference-filter-aided in-line Fraunhofer holography, *Appl. Opt.* **35**, 6299–6303 (1996).

[37] K.C. Baldwin, M.J. Ehrlich, and J.W. Wagner: Exposure-induced high-pass filtering of a volume by means of an absorption hologram recording technique, *Appl. Opt.* **35**, 227–321 (1996).

[38] K. Nishihara, S. Hatano, and K. Nagayama: New method of obtaining particle diameter by the fast Fourier transform pattern of the in-line hologram, *Opt. Eng.* **36**, 2429–2439 (1997).

[39] T.-C. Poon, M.H. Wu, K. Shinoda, and Y. Suzuki: Optical scanning holography, *Proc. IEEE* **84**, 753– 764 (1996).

[40] K.B. Doh, T.-C. Poon, and G. Indebetouw: Twin-image noise in optical scanning holography, *Opt. Eng.* **35**, 1550–1555 (1966).

[41] K. Doh, T.-C. Poon, M.H. Wu, K. Shinoda, and Y. Suzuki: Twin-image elimination in optical scanning holography, *Opt. Laser Technol.* **28**, 135–141 (1996).

[42] B.W. Schilling, T.-C. Poon, G. Indebetouw, B. Storrie, K. Shinoda, Y. Suzuki, and M.H. Wu: Three- dimensional holographic fluorescence microscopy, *Opt. Lett.* **22**, 1506–1508 (1997).

[43] G. Indebetouw, T. Kim, T.-C. Poon, and B. Schilling: Three-dimensional location of fluorescent inhomogeneities in turbid media by scanning heterodyne holography, *Opt. Lett.* **23**, 135–137 (1998).

[44] T.-C. Poon and T. Kim: Optical image recognition of three-dimensional objects, *Appl. Opt.* **38**, 370– 381 (1999).

[45] D. Lebrun, S. Belaid, and C. Ozkul: Hologram reconstruction by use of optical wavelet transform, *Appl. Opt.* **38**, 3730–3734 (1999).

[46] T. Xiao, J. Chen, Z. Xu, and P. Zhu: Gabor and lensless Fourier transform X-ray holograms: a comparison, *J. Mod. Opt.* **43**, 607–615 (1996).

[47] I. Kodama, M. Yamaguchi, N. Ohyama, T. Honda, K. Shinohara, A. Ito, T. Matsumura, K. Kinoshita, and K. Yada: Image reconstruction from an in-line X-ray hologram with intensity distribution constraint, *Opt. Commun.* **125**, 36–42 (1996).

[48] G. Xu: Atomic resolution x-ray hologram, *Appl. Phys. Lett.* **68**, 1901–1903 (1996).

[49] T. Tanji, Q. Ru, and A. Tonomura: Differential microscopy by conventional electron off-axis holograpy, *Appl. Pys. Lett.* **69**, 2623–2625 (1996).

[50] S. Lai, B. Kemper and G. von Bally: Off-axis reconstruction of in-line holograms for twin-image elimination, *Opt. Commun.* **169**, 37–43 (1999).

[51] N. Watanabe, K. Sakurai, A. Takeuchi, and S. Aoki: Soft-x-ray Gabor holography by use of a backilluminated CCD camera, *Appl. Opt.* **36**, 7433–7436 (1997).

[52] R.R. Meyer and E. Heindl: Reconstruction of off-axis electron holograms using a neural net, *J. Microsc.* **191**, 52–59 (1998).

[53] T. Xiao, H. Xu, Y. Zhang, J. Chen, and Z. Xu: Digital image decoding for in-line X-ray holography using two holograms, *J. Mod. Opt.* **45**, 343–353 (1998).

[54] S. Yang, X. Xie, Y. Zhao, and C. Jia: Reconstruction of near-field in-line hologram, *Opt. Commun.* **159**, 29–31 (1999).

[55] X.F. Duan, M. Gao, and L.-M. Peng: Accurate measurement of phase shift in electron holography, *Appl. Phys. Lett.* **72**, 771–773 (1998).

[56] B.G. Frost, N.F. van Hulst, E. Lunedei, G. Matteucci, and E. Rikkers: Study of the leakage field of magnetic force microscopy thin-film tips using electron holography, *Appl. Phys. Lett.* **68**, 1865–1867 (1996).

[57] W.-D. Rau, F.H. Baumann, J.A. Rentschler, P.K. Roy, and A. Ourmazd: Characterization of stacked gate oxides by electron holography, *Appl. Phys. Lett.* **68**, 3410–3412 (1996).

[58] X. Lin and V.P. Dravid: Mapping the potential of graphite nanotubes with electron holography, *Appl. Phys. Lett.* **69**, 1014–1016 (1996).

[59] Y.C. Wang, T.M. Chou, M. Libera, and T.F. Kelly: Transmission electron holography of silicon nanospheres with surface oxide layers, *Appl. Phys. Lett.* **70**, 1296–1298 (1997).

[60] K.D. Johnson and V.P. Dravid: Grain boundary barrier breakdown in niobium donor doped strontium titanate using in situ electron holography, *Appl. Phys. Lett.* **74**, 621–623 (1999).

[61] M. Tegze, G. Faigel, S. Marchesini, M. Belakhovsky, and A.I. Chumakov: Three dimensional imaging of atoms with isotropic 0.5 ? resolution, *Phys. Rev. Lett.* **82**, 4847–4850 (1999).

[62] E. Volkl, L.F. Allard, and D.C. Joy, eds., *Introduction to Electron Holography* (Kluwer Academic/Plenum, New York 1999).

[63] A. Tonomura: *Electron Holography, 2nd ed.* (Springer-Verlag, New York 1999).

[64] J. Pomarico, U. Schnars, H.-J. Hartmann, and W. Juptner: Digital recording and numerical reconstruction of holograms: a new method for displaying light in flight, *Appl. Opt.* **34**, 8095–8099 (1995).

[65] U. Schnars, T.M. Kreis, and W.P.O. Juptner: Digital recording and numerical reconstruction of holograms: reduction of the spatial frequency spectrum, *Opt. Eng.* **35**, 977–982 (1996).

[66] I. Yamaguchi and T. Zhang: Phase-shifting digital holography, *Opt. Lett.* **22**, 1268–1270 (1997).

[67] E. Marquardt and J. Richter: Digital image holography, *Opt. Eng.* **37**, 1514–1519 (1998).

[68] B. Nilsson and T.E. Carisson,: Direct three-dimensional shape measurement by digital light-in-flight holography, *Appl. Opt.* **37**, 7954–7959 (1998).

[69] T. Zhang and I. Yamaguchi: Three-dimensional microscopy with phase-sifting digital holography, *Opt. Lett.* **23**, 1221–1223 (1998).

[70] R.M. Powell, *Digital Holography*, Master of Engineering Thesis, Electrical and Computer Engineering, University of Florida, Gainsville (1998).

[71] G. Pedrini, P. Froning, H.J. Tiziani, and F.M. Santoyo: Shape measurement of microscopic structures using digital holograms, *Opt. Commun.* **164**, 257–268 (1999).

[72] E. Cuche, F., Bevilacqua, and C. Depeursinge: Digital holography for quantitative phase-contrast imaging, *Opt. Lett.* **24**, 291–293 (1999).

[73] Y. Takaki and H. Ohzu: Fast numerical reconstruction technique for high-resolution hybrid holographic microscopy, *Appl. Opt.* **38**, 2204–2211 (1999).

[74] E. Cuche, P. Marquet, and C. Depeursinge: Simultaneous amplitude-contrast and quantitative phase- contrast microscopy by numerical reconstruction of Fresnel off-axis holograms, *Appl. Opt.* **38**, 6994– 7001 (1999).

[75] S. Schedin, G. Pedrini, H.J. Tiziani, and F.M. Santoyo: Simultaneous three-dimensional dynamic deformation measurements with pulsed digital holography, *Appl. Opt.* **34**, 7056–7062 (1999).

[76] P. Froning, G. Pedrini, H.J. Tiziani, and F.M. Santoyo: Vibration mode separation of transient phenomena using multi-pulse digital holography, *Opt. Eng.* **38**, 2062–2068 (1999).

[77] R.B. Owen and A.A. Zozulya: A new digital holographic sensor for marine particulates, *Opt. Eng.* **39**, 2187–2197 (2000)

[78] M.D. Barnes, N. Lermer, W.B. Whitten, and J.M. Ramsey: A CCD based approach to high-precision size and refractive index determination of levitated microdroplets using Fraunhofer diffraction, *Rev. Sci. Instrum.* **68**, 2287–2291 (1997).

[79] I. Raveh, D. Mendlovic, Z. Zalevsky, and A.W. Lohmann: Digital method for defocus corrections: experimental results, *Opt. Eng.* **38**, 1620–1626 (1999).

[80] T. Kim and T.-C. Poon: Extraction of 3–D location of matched 3–D object using power fringe-adjusted filtering and Wigner analysis, *Opt. Eng.* **38**, 2176–2183 (1999).

[81] H.-K. Liu and N. Marzwell: Holographic imaging in dense artificial fog, *NASA Tech. Briefs* **20**(6), 64 (1996).

13

Holographic Antireflection Coatings

Jacques E. Ludman, Timothy D. Upton,
H. John Caulfield, and David W. Watt

13.1 Introduction

13.1.1 Why Holographic Antireflection Coatings?

In all applications of optics, it is vital to control the reflectivity of components. Mirror reflections should be high; lens reflections should be low; and beamsplitter reflections should be intermediate. From such needs has grown a large optical coating industry.

To first order, we can analyze the spectral transmission and reflection (they are complements in a lossless system) by simple Fourier analysis. Similar analyses apply to Fabry–Perot filters, DFB (distributed feedback) mirrors and lasers, and so on. One can view it as a consequence of the duality of time and frequency. The multilayer structure produces multiple time delays in transmission and reflection to an incident temporal delta function. The Fourier transform of that pulse train gives the transmission or reflection spectrum. As the Fourier transform of a perfect sine wave is a pair of delta functions, narrow band filters aim at achieving a highly periodic structure.

A hologram formed by two plane waves is highly periodic. Is it possible that holography may be a good way to produce antireflection coatings? If so, how may it be more desirable than the well-established manufacturing methods? These were among the questions we sought to answer.

FIGURE 13.1. Multilayer film refractive index modulation

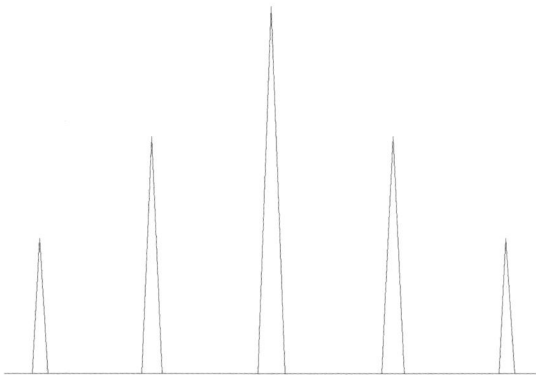

FIGURE 13.2. Fourier transform of the multilayer film.

13.1.2   Antireflection Coatings

There are several types of antireflection coatings[1, 2] and it is important to mention the main types of coatings. The most common, because it is the easiest (easiest but still difficult) to manufacture, is the multilayer film. The index of refraction pattern of a multilayer film is shown in Fig. 13.1.

This pattern is a comb function convolved with a rect function; so its Fourier transform is a comb function multiplied by a sinc function (Fig. 13.2). One problem such films have is that they are not altogether stable. They can flake apart at the boundary between layers.

To get a more stable film and a different spectral response curve, manufacturers sometimes seek to allow a continuous variation in index of refraction by controlled, differential evaporation of two materials at once (Fig. 13.3). This can give an almost delta function spectral response. Such films are called rugate, but rugate films are much harder to make well.

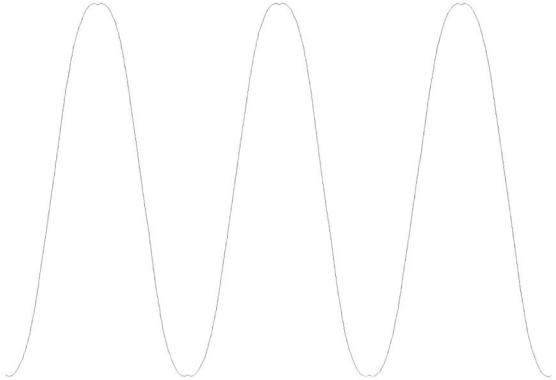

FIGURE 13.3. Continuously variable index modulation.

Both a rugate film and a two-plane-wave hologram have a sinusoidal index of refraction pattern. This suggests that holographic patterns with fringes parallel to the surface have the potential of constituting excellent rugate antireflection coatings.

There are three reasons that holographic rugate may be superior to evaporated rugate.

(1) *It can be made on curved surfaces.* Making a rugate film on a planar substrate is very hard. Making it on a sphere presents almost impossible problems, but it is easy to make fringes parallel to a spherical surface by holography. One beam of the same curvature as the spherical surface must be incident from outside the sphere. The other of the opposite curvature must be incident from the other side. That is, we need converging and diverging beams of the same curvature as the surface incident from both sides. One converges to the center of the surface. The other diverges from the same center. The simplest geometry is to use a converging beam and a mirror as in Fig. 13.4.

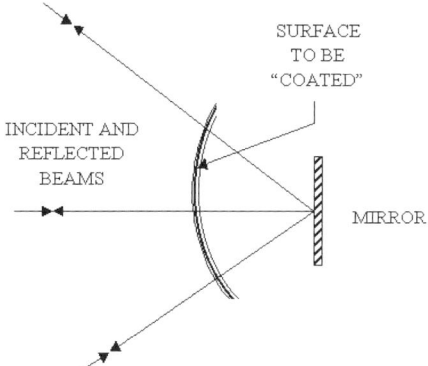

FIGURE 13.4. Holographically coating curved surfaces.

(2) *Holography is faster, simpler, and cheaper.* Holography is a single-step photographic process, whereas evaporation requires a long programmed run using
multiple sources. Holograms are manufactured in air. The process of vapor
deposition requires a good, clean vacuum and excellent, expensive control.
Holography does not. Even if we make a huge 10% error in the holographic
coating, the reflection goes up from 0% to only about 0.4%. This is because
we are making a very inefficient (roughly 4% typically) hologram to remove
the surface-reflected light interferometrically.

(3) Holography finds large surfaces much easier to handle than does evaporation.
It is easier and cheaper to make a large beam of light than to make a large
vacuum system to deposit films.

13.2   Theory

The principle of holographic antireflection coatings is to record a hologram that,
when illuminated, creates a wave that is of the correct amplitude and phase to
exactly cancel the Fresnel surface reflection. For an incident wave normal to the
interface between two dielectrics, the reflected amplitude E_r in terms of the incident
amplitude E_i is given by[3, 4]

$$\frac{E_r}{E_i} = \frac{m_1 - n_2}{n_1 + n_2},\tag{1}$$

where n_1 is the refractive index of the first material and n_2 is the refractive index of the second material. Equation (1) is an expression of the Fresnel equation
for normally incident radiation between two dielectric materials. The ratio of the
reflected energy to the incident energy is the reflectivity ρ, which is given by[3, 4]

$$\rho = \left(\frac{E_r}{E_i}\right)^2 = \left(\frac{n_1 - n_2}{n_1 + n_2}\right).\tag{2}$$

Equation (1) gives the phase of the reflected wave relative to the incident wave. At
the interface between two dielectrics for $n_1 > n_2$, the phase of the reflected wave
is the same as the incident wave. For $n_1 < n_2$, the phase of the reflected wave
is shifted by π. Equation (2) indirectly gives the required hologram efficiency
as a function of refractive index. For an air-glass interface with $n_1 = 1.0$ and
$n_2 = 1.5$, the reflected energy is 4.0% and the transmitted energy is 96.0%. To
create a reconstructed wave of energy equal to the surface reflection, the required
hologram efficiency is 4.2%. For attenuating materials, the simple refractive index
n is replaced by the complex index $\rfloor = n - i\kappa$, where κ is the extinction coefficient,
and (1) and (2) are used in the same way.

Figure 13.5 shows an antireflection hologram. The amplitude of the reconstructed wave is proportional to the modulation and thickness of the hologram, and
the phase is determined by the position of the modulation variation with respect
to the surface of the hologram. The shaded regions are of a higher than average
refractive index, and the unshaded regions are of a lower than average refractive

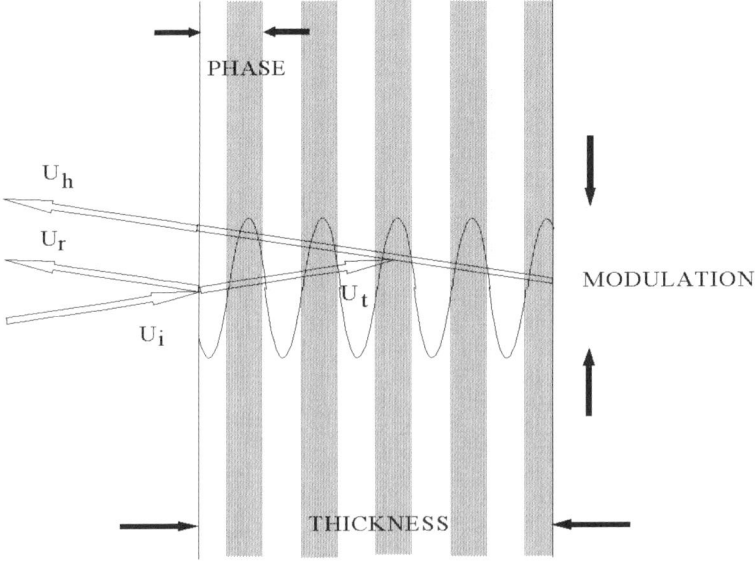

FIGURE 13.5. Antireflection hologram.

index. The locations of the maximum gradient of the modulation may be thought of as interfaces between regions of high and low refractive index. The incident wave U_i is normally separated into a transmitted wave U_t and a reflected wave U_r at the interface between the two different materials. In the antireflection hologram, the transmitted wave additionally creates a holographically reconstructed wave U_h. With proper design, the reconstructed wave is of equal amplitude and shifted in phase by π relative to the surface reflection.

The waves may be described simply using a complex wave notation by[3]

$$U_{i,r,t,h} = E_{i,r,t,h}e^{i\varphi_{i,r,t,h}},$$

where φ is the phase of the wave and the subscripts refer to the incident, transmitted, reflected, and holographically reconstructed waves, respectively. In the case of $n_1 < n_2$, $\varphi + r = \varphi_i + \pi$, and $\varphi_t = \varphi_i$. For the holographically reconstructed phase, $\varphi_h = \varphi_t + \Delta$, where Δ is the additional phase shift as the wave travels to and from the first location of maximum negative index gradient (Fig. 13.5). For complete cancellation $\varphi_h = \varphi_r + \pi$, and for the case of $n_1 < n_2$, $\Delta = 2\pi$. This requires that the maximum negative index gradient be located $\lambda/2$ below the surface of the hologram.

13.3 Fabrication

The antireflection hologram is recorded using two collimated waves traveling in opposite directions, as shown in Fig. 13.6. The center wavelength of the hologram

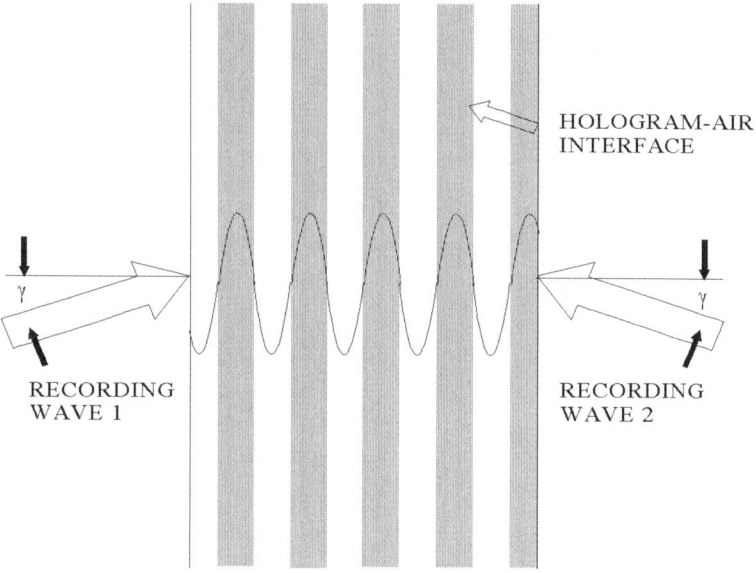

FIGURE 13.6. Antireflection hologram recording geometry.

may be varied by changing the angle of incidence γ. The phase of the reconstructed wave is determined by the position of the modulation variation with respect to the surface of the hologram, and the amplitude of the wave is proportional to the modulation and thickness of the hologram.

The position of the modulation variation is set by the phase difference between the two recording waves. A simple way to record an antireflection hologram is to use a single beam and the back surface hologram—air interface to form the second wave (Fig. 13.6). In addition to the simplicity of the setup, this method has the significant advantage that any variation in the surface of the emulsion is matched by variation in the hologram. However, with this recording, geometry, the modulation is a maximum at the surface of the emulsion, and $\Delta = \pi/2$ instead of the required 2π. It is possible to control the position of the modulation using a device similar to that shown in Fig. 13.7. This device allows the phase difference between the recording beams to be adjusted by changing the optical path length of the second wave. However, a device such as this requires a wave-shaping optical element to match the surface variation of the emulsion to produce a uniform antireflection hologram. In other words, any variation in the surface of the hologram must be matched by the holographically created wave to give proper cancellation over the entire surface. For this reason, using the hologram—air interface to create the second wave is the method of choice. To adjust the hologram so that $\Delta = 2\pi$ using this method, a postrecording SiO_2 coating step is used; instead of locating the modulation variation correctly with respect to the surface, the surface is effectively moved so that the phase is correct.

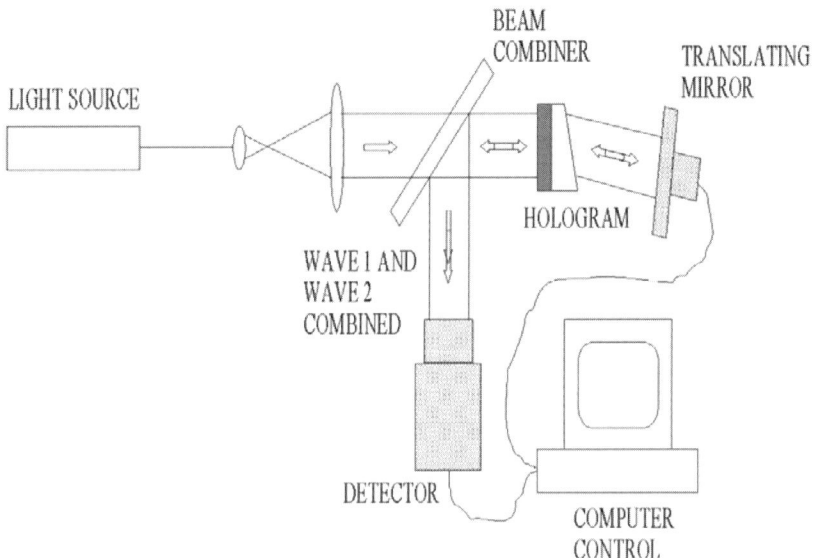

FIGURE 13.7. Modulation position control device.

The modulation of the hologram is controlled by the chemistry of the emulsion, the exposure energy, and any postexposure processing. A major consideration is the holographic material used. Antireflection holograms have been recorded in Northeast Photosciences (Hollis, NH) dichromated gelatin, Agfa silver halide, and DuPont (Wilmington, DE) photopolymer. Dichromated gelatin is typically high efficiency and extremely sensitive to changes in temperature and humidity. The chemistry of the dichromated gelatin was adjusted to reduce the efficiency to the required 4% with limited success. A major drawback of dichromated gelatin is the variability in efficiency with environment. Silver halide has lower maximum efficiency and is less sensitive to environmental changes, which makes it more suited to holographic antireflection. Of these, the most environmentally stable and consistent in terms of modulation and efficiency are the photopolymer emulsions.

Figure 13.8 shows a typical efficiency curve for DuPont 10-μm blue/green sensitized reflection film as a function of wavelength. The transmission of the hologram was measured using a Varian Cary 5 spectrophotometer. Because the maximum efficiency of this film is around 50%, it is difficult to produce repeatable low-efficiency holograms. However, the photopolymer emulsion is available in liquid form for spin coating directly on glass. A study is planned to determine the maximum efficiency as a function of emulsion thickness using liquid photopolymer spin coated to a range of thicknesses. This will indicate the correct thickness for producing repeatable 4% efficient holograms. This should be possible with an emulsion of around 1 μm thick.

Narrowband holographic antireflection coatings will have applications primarily in coherent optical systems. In reflection holography and many transmission

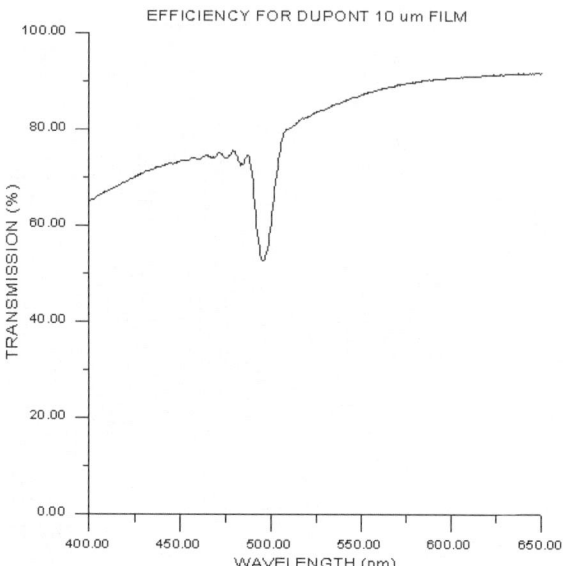

FIGURE 13.8. Typical efficiency curve.

applications, the surface quality of the hologram is not important. However, in coherent applications, surface flatness is important. The prelaminated DuPont films
have shown unacceptable surface quality; an initially clean and uniform coherent
beam becomes noisy and distorted on transmission. Laminating the photopolymer
film between two glass plates with Epoxy Technology 301 two-part epoxy prior
to recording has reduced the distortion. Surface problems should be eliminated by
spin coating the photopolymer directly on glass.

13.4   Simulations

The previous analysis has focused on narrowband antireflection holograms. The
theory of broadband holographic antireflection coatings is inherently more difficult. It is interesting to contrast the nature of multilayer antireflection coatings with
that of holographically generated antireflection coatings. Multilayer coatings have
refractive index modulation of order 1 and small thicknesses; the layers vary in
thickness from about 10 to 100 nm, and the entire structure is generally less than
1 μm thick. Holograms are thicker, ranging from about 2 to 20 μm in thickness,
and they have a much lower refractive index modulation of order 0.01. In neither
case is there an obvious method to optimize construction of the refractive index
profile, and the potential of holograms as broadband antireflection devices is by
no means clear. To investigate the optimization of the refractive index profile, a
genetic algorithm was used to generate a hypothetical broadband coating. Figure 13.9 shows the theoretical optimum design found using the algorithm with a

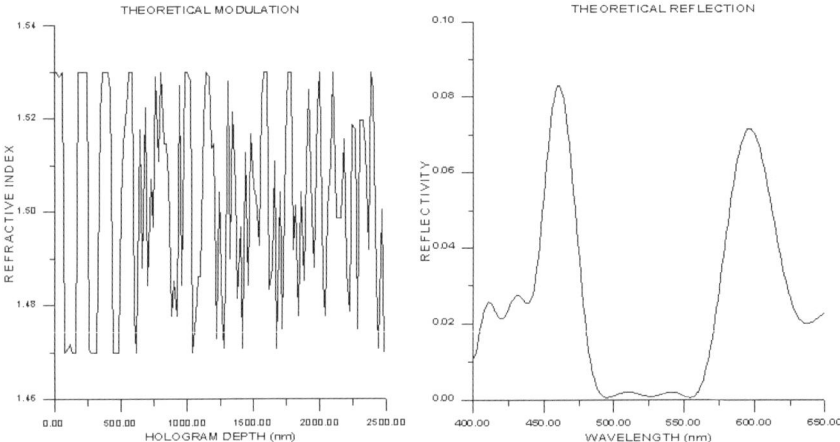

FIGURE 13.9. Theoretical optimum hologram design.

refractive index modulation of 0.03 in an emulsion that is 2.5 μm thick. A 96-nm thick layer of MgF$_2$ has been applied to the surface of the emulsion. The algorithm used is incorporated into the PGA genetic algorithm package that is available in the public domain via the Internet from Argonne National Laboratory (Argonne, IL.). In the example shown here, the coating is optimized to minimize reflectivity in a 75-nm band. The refractive index takes on a complex structure, which requires a continuous spectrum of recording wavelengths.

Figure 10 shows the refractive index modulation and hologram reflection possible with a single exposure using a 50-nm band recording source. In this case, a 120-nm-thick layer of MgF$_2$ has been applied to the surface of the emulsion.

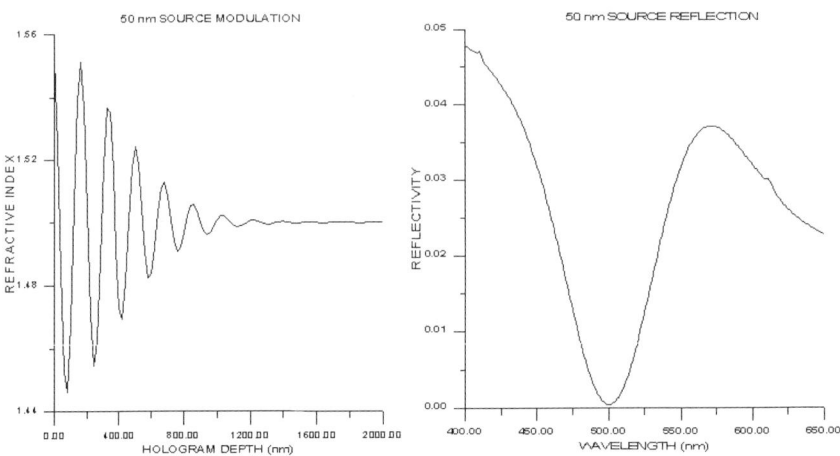

FIGURE 13.10. Hologram recorded with a 50-nm source

Using a postrecording coating procedure with materials such as SiO_2 or MgF_2 has several benefits. First, as already mentioned, it allows the phase to be adjusted to the required value. Second, it creates a durable protective layer for the antireflective coating. Third, using a coating of MgF_2 for broadband holographic antireflection coatings dramatically reduces the demands on the hologram to perform all the required cancellations; the MgF_2 layer works together with the hologram to cancel the surface reflection.

13.5   Conclusions

There is widespread interest in highly effective narrowband antireflection coatings for optical elements in low-energy laser applications. Designers and researchers using these types of systems require optical elements with losses of 0.1% or less at a single wavelength. For holographic antireflection coatings to meet this requirement, the hologram must not only be fabricated correctly, but the holographic material must also have very high transmission. The material also must be of optical quality and be designed to have consistent maximum efficiency of 4%. DuPont liquid photopolymer spin coated on glass meets these requirements. Narrowband holographic antireflection coatings are unique because arbitrarily small reflectivities may be achieved at a single wavelength.

The recording geometry has been designed to self-correct for any surface variation; a postrecording coating procedure with $SiO2$ for narrowband antireflection holograms and MgF_2 for broadband antireflection holograms is used to adjust the phase of the reconstructed wave. Additionally, this coating increases the durability of the hologram and, if MgF_2 is used, increases the effectiveness of the hologram in broadband antireflection. Multiple exposures and more significantly broadband sources are used to determine the region of antireflection.

Fundamental questions concerning the recording of broadband antireflection holograms remain. Theoretically, an antireflection hologram can be recorded to reduce the reflection to less than 0.1% over a wavelength range of 75 nm. This coating, however, requires a continuous range of recording wavelengths. In practice, using two or fewer broadband recording sources is the most feasible option. The short coherence length of broadband sources requires that the emulsion–air interface be used to form the second recording wave, and the emulsion must be uncoated during recording. More simulations and experiments using broadband sources are required to determine the most elegant and effective design for broadband antireflection holograms.

Acknowledgments

This work was supported by the U.S. Dept. of Defense under Contract DASG60-96-C-0160.

References

[1] R.E. Hummel and K.H. Guenther, eds., *Handbook of Optical Properties: Thin Films for Optical Coatings* (CRC Press, New York, 1995).

[2] M.J. Weber, ed., *CRC Handbook of Laser Science and Technology: Optical Materials Part 3, Applications, Coatings, and Fabrication*. (CRC Press, New York, 1987).

[3] E. Hecht, *Optics*. (Addison-Wesley, Reading, MA, 1990).

[4] R. Siegel and J.R. Howell, *Thermal Radiation Heat Transfer* (Hemisphere, New York, 1981).

[5] C.M. Vest, *Holographic Interferometry* (Wiley, New York, 1979).

Part V

Physics and Holography

14

Holography and Relativity

Nils Abramson

14.1   Holographic Uses of the Holodiagram

Both holography and the special theory of relativity are based on interferometry, which in turn depends on pathlengths of light. Let such a pathlength be represented by a string. Fix the ends of the string with two nails on a blackboard, and while keeping the string stretched, draw a curve with a chalk. The result will be an ellipse with its two focal points at the nails (Fig. 14.1). Thus, we understand that if a light source is at one focalpoint, (A) the pathlength for light to the the other focalpont (B) via any point on the ellipse will be constant. If we could draw a curve with the chalk in three dimensions, the result would be a rotational symmetric ellipsoid still with the focalpoints (A) and (B). Now shorten the string by the coherence length of a He-Ne-laser and produce another ellipse, and finally, instead lengthen the string by another coherence length. If we place a He-Ne-laser at (A), a hologram plate at (B) and a reference mirror at the middle ellipse, we can record any object within the two outermost ellipses because it will be within the coherence length. If the coherence length, or the pulse length, of the laser is very short, we will at reconstruction of the hologram see the object intersected by a thin ellipsoidal shell. This method named Light-in-flight recording by holography can be used either to observe the 3D shape of a wavefront (or pulse front) or of a real object.

If we instead produce a set of ellipses by many times lengthening the string by the wavelength of light (λ), one interference fringe would form for every ellipsoid intersected by each point on an object as it has moved between two exposures in holographic interferometry. Based on this idea, we produce the holodiagram.[1] The separation of the ellipses at a point (C) compared with their separation at the x-axis

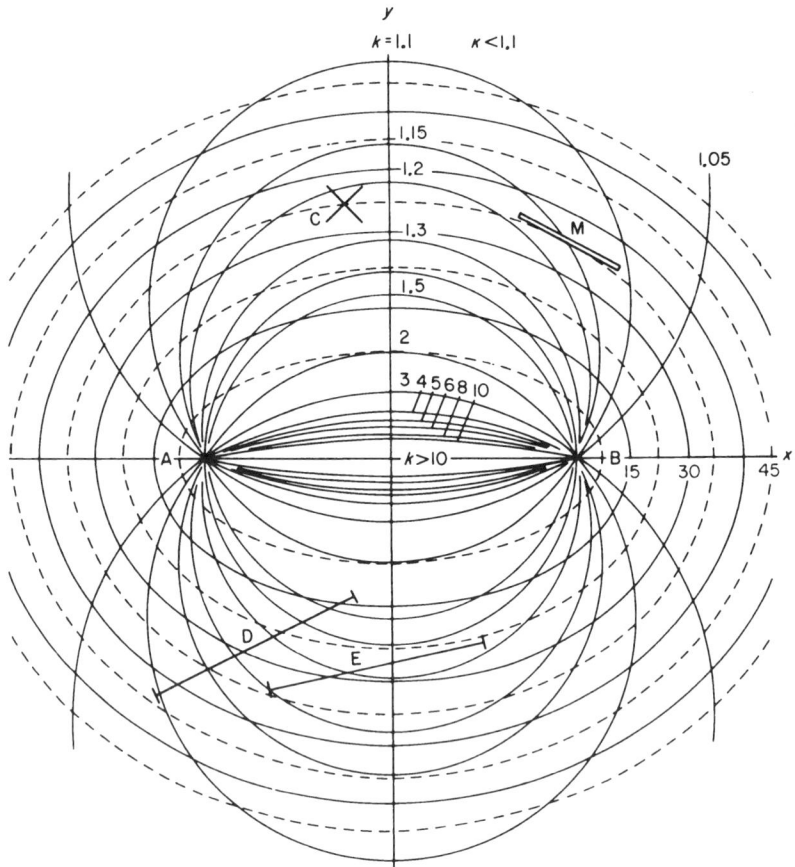

FIGURE 14.1. A string is fixed with one end at (A) and the other at (B). Keeping the string stretched, a set of ellipses are drawn; for each adjacent ellipse, the string length is increased by a certain constant value (ΔL). The separation of the ellipses varies with a factor (k), which is constant along arcs of circles. In this holodiagram, let (A) be the spatial filter in a holographic setup and (B) be the center of the hologram plate. If (ΔL) is the coherence length of light, the diagram can be used to optimize the use of a limited coherence in recording large objects. If (ΔL) instead represents the wavelength of light, the ellipses can be used to evaluate interference fringes in holographic interferometry or conventionl interferometry with oblique illumination and observation. Finally, if (ΔL) represents a very short coherence length, or pulse length, the ellipses visualize the spherical wavefront from (A) as seen from (B), deformed by the limited speed of the light used for the observation.

is named the k-value, which depends on the angle ACB. As the peripherical angle on a circle is constant, the k-value will be constant along arcs of circles passing through (A) and (B), as seen in Fig. 14.1. The moiré effect of two sets of ellipsoids visualize interference patterns in holographic interferometry.[2]

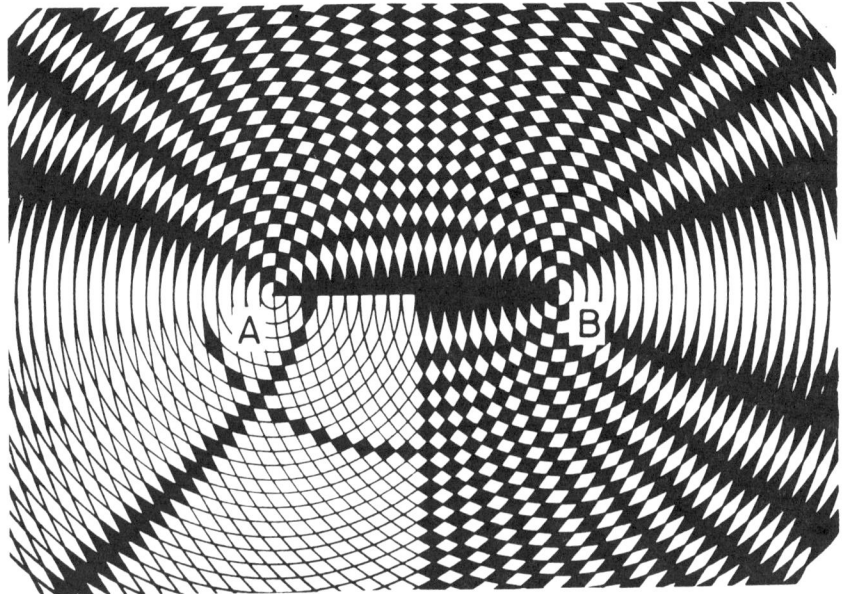

FIGURE 14.2. An alternative way to make the holodiagram is by drawing two sets of concentric circles, where for each adjacent circle, the radius is increased by a certain constant value. One set of ellipsoids and one set of hyperboloids are formed. If (A) and (B) are two sources of coherent light, or two points of observation, the ellipsoids move outward with a velocity *greater than the speed of light*, whereas the hyperboloids will be stationary. These hyperboloids represent the diffraction-limited resolution of a lens and are used for in-plane measurements in holography, moiré and speckle techniques. If (A) is a source of coherent light and (B) is a coherent point of observation, the hyperboloids move with a speed *greater than light* to the right, whereas the ellipsoids will be stationary. These ellipsoids represent the interferometric-limited resolution and are used for out-of-plane and 3D measurements.

Another approach to this holodiagram is to draw one set of equidistant concentric circles centered at (A) and another set at (B), as seen in Fig. 14.2. A number of rhombs, or diamonds, are formed where the circles intersect. By painting every second rhomb black, one set of ellipses and one set of hyperbolas are produced.[3] The two diagonals of a rhomb are perpendicular, and thus, the ellipses and the hyperbolas intersect at right angles. One diagonal of the rhombs represents the separation of the ellipses and the other the separation of the hyperbolas. The radii from (A) and (B) are named R_A and R_B, respectively. The set of ellipses is represented by $R_A + R_B = $ constant and the set of hyperbolas by $R_A - R_B = $ constant. If the separation of the concentric circles is $0,5\lambda$, the separation of the ellipses (one diagonal of the rhomb) at point C is

$$D_{\text{ell}} = \frac{0,5\lambda}{\cos \alpha} = k \times 0,5\lambda, \tag{1}$$

while the separation of the hyperbolas (the other diagonal) is

$$D_{hyp} = \frac{0,5\lambda}{\sin \alpha}, \qquad (2)$$

where α is half the angle ACB of Fig. 14.1. The factor (k) of (1) is identical to the (k) of Fig. 14.1 and its value is

$$k = \frac{1}{\cos \alpha}. \qquad (3)$$

Interferometric measured displacement (d) is calculated $d = nk\lambda$, where n is the number of interference fringes between the displaced point and a fixed point on the object. In holographic, or speckle, interferometry, (A) is a light source from which sperical waves radiate outward, whereas (B) is a point of observation or a light sink toward which spherical waves move inward. In that case, the hyperbolas will move with a *velocity greater than light* while the ellipses are stationary and (1) is used to find the displacement normal to the ellipsoids (out-of-plane displacement). If both (A) and (B) are light sources, the ellipsoids will move with a *velocity greater than light*, whereas the hyperbolas (Youngs fringes) are stationary and are used to evaluate the displacement normal to the hyperboloids (in-plane displacement). The situation will be similar if (A) and (B) both are points of observation while the light source could be anywhere. Finally, if A-B represents the diameter of a lens, the minimum separation of the hypebolas will represent the diffraction-limited resolution of that lens[4] (except for a constant of 1,22).

14.1.1   Light-in-Flight Recording by Holography

A hologram is recorded only if object light and reference light simultaneously illuminate the hologram plate. Thus, if the reference pulse (t) is one picosecond (10^{-12} sec $= 0,3$mm) long, the reconstructed image of the object will be seen only where it is intersected by this light slice, the thickness of which depends on the length of the pulse. If the reference pulse illuminates the hologram plate at an angle, e.g., from the left, it will work like a *light shutter* that with a *velocity greater than light* sweeps across the plate. Thus, what happens first to the object will be recorded furthest to the left on the plate, whereas what happens later will be further to the right. If the hologram plate is studied from left to right, the reconstructed image functions like a movie that with picosecond resolution in slow motion shows the motion of the light pulse during, e.g., one nanosecond (300 mm). The described method is named "Light-in-flight recording by holography" and results in a frameless motion picture of the light as it is scattered by particles or any rough surface. It can be used to study the coherence function of pulses,[5, 6] the 3D shape of wavefronts (Fig. 14.3), or of physical object[7] (Fig. 14.4).

   If (A) and (B) are close together and if there are scattering particles, e.g., smoke in the air, those who observe the reconstructed hologram would find themselves in the center of a spherical shell of light with a radius $R = 0,5ct$, where t is time interval between emitance of light (at A) and recording (at B). If, however,

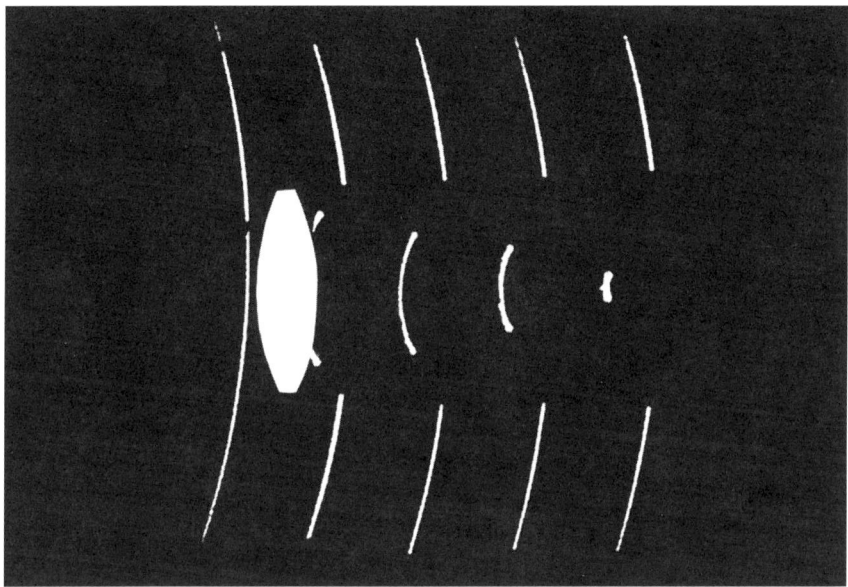

FIGURE 14.3. A Light-in-flight recording of light focused by a lens. One single picosecond spherical pulse from a modelocked laser at (A) of Fig. 14.1 illuminated a white screen at an oblique angle. The screen was placed so that its normal passed through the hologram plate at (B). Part of the pulse was after appropriate delay used as a reference beam at (B). A cylindric lens was fixed to the screen, and by multiple exposures of the reconstructed image, the focusing effect of the lens was recorded.

(A) and (B) are separated, they would find themselves in the focalpoint (B) of an ellipsoid, where (A) is the other focalpoint. The string length (refering to Fig. 1) is $R_A + R_B = ct$. The thickness of the ellipsoidal shell would be $k \times 0, 5c\Delta t$, where (Δt) is pulse length (or coherence length) and k is the usual k-value. If the distance A-B were infinite, observers at (B) would find themselves in the focalpoint of a paraboloid. Thus, we have shown that a sphere of observation appears distorted into an ellipsoid and a flat surface of observation into a paraboloid, and the sole reason for these distortions is caused by the separation A-B.

In Fig. 14.5, we see at the top the ordinary holodiagram in which the eccentricity of the ellipse is caused by the static separation of (A) and (B). However, it is unimportant what observers are doing during the time between emittance and detection of light pulses. Thus, the observers could just as well be running from (A) to (B) so that the eccentricity of their observation ellipsoid is caused solely by the distance they have covered until they make the observation. If their running speed is close to the velocity of light, this new dynamic holodiagram is identical to the ordinary, static holodiagram. This fact has inspired the new graphic approach to special relativity that will be explained in Section 3, but let us first study the development of Einstein's Special Relativity Theory.

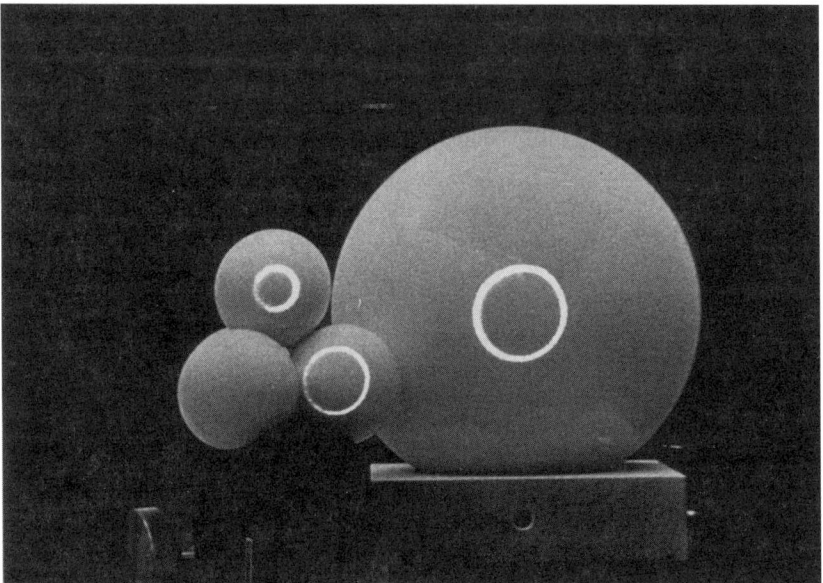

FIGURE 14.4. Light-in-flight recording of a set of spheres illuminated by a three-picosecond-long laser pulse. The light source (A) and the point of observation at the hologram plate (B) were close together and far from the object. Thus, the intersecting ellipsoidal light slice can be approximated into a spheroidal light slice with a thickness of 0.5 mm and a large radius.

14.2   Introduction to Einsteins Special Relativity Theory

Special relativity is based on Einstein's two postulates of 1905:

(1) The same laws of electrodynamics and optics will be valid for all frames of reference for which the equations of mechanics hold good.
(2) Light is always propagated in empty space with a definite velocity c, which is independent of the state of motion of the emitting body.

In this paper, I would prefer to express these two postulates using the following words:

(1) If we are in a room that is totally isolated from the outside world, there is no experiment that can reveal a constant velocity of that room.
(2) When we measure the speed of light (c) in vacuum, we always get the same result independent of any velocity of the observer or of the source.

To make such strange effects possible, it is assumed that the velocity results in that time moves slower (so that seconds are longer), which is named relativistic time dilation, and it was assumed (by Lorentz) that lengths (rulers) are shortened in the direction of travel (Lorentz's contraction). The contraction effect was presented by Lorentz[8] (1895) to explain the experiments by Michelson-Morley (1881), and

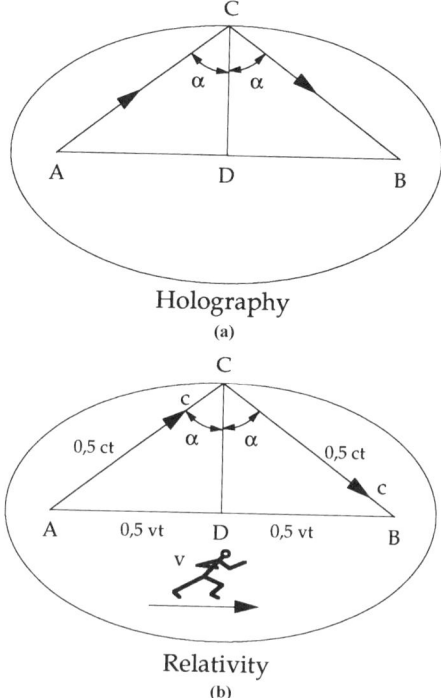

Holography

(a)

Relativity

(b)

FIGURE 14.5. (a) The ordinary static holodiagram, in which (A) is the light source, and (B) is the point of observation, e.g., the center of the hologram plate, whereas (C) is an object for which the k-value is $1/\cos\alpha$. (b) The dynamic holodiagram in which an experimenter emits a picosecond pulse when he is at (A), whereafter he runs with a velocity close to the speed of light and makes a picosecond observation at (B). The k-value is as before

$$= \frac{1}{\cos\alpha} = \frac{ct}{\sqrt{c^2t^2 - v^2t^2}} = \frac{1}{\sqrt{1-\frac{v^2}{c^2}}}.$$

it was later adapted by Einstein (1905) in his famous Special Theory of Relativity.[9] Time dilation is more "real" than is the Lorentz's contraction because it produces a permanent result, a lasting difference in the reading of a stationary and a traveling clock, whereas the Lorentz's contraction is much more "apparent" because it produces no permanent result; there is a difference in length only as long as there is a difference in velocity of two rulers.

14.2.1 The Michelson-Morley Experiment

One way to understand special relativity is to see how time dilation and Lorentz's contraction of objects parallel to motion can be used to explain the null results of the Michelson-Morley experiment, which was performed to measure the velocity of earth in relation to an assumed ether. The result was that the expected influence of such an ether on the velocity of light was not found. Let us in the following, study this double pass example, in which one arm of a Michelson interferometer was perpendicular to the velocity of the earth surface, whereas the other was parallel. When the interferometer is at rest, the lengths of the two arms are identical.

Now, let me, the author, and you, the reader, be stationary in the stationary space and study from there the moving interferometer. The velocity of the interferometer from left to right is (v), whereas the speed of light is (c). The time for light to travel along the perpendicular arm when the interferometer is stationary is (t_0). When it is moving at the velocity (v), the pathlength becomes longer and therefore the travel time (t_V) becomes longer. However, this increase in traveling time of the light must not be observable by the traveling observer because that would be against the postulate (1) of special relativity. But why is this increase in traveling time not observed by the observer who is traveling with the interferometer? The accepted solution is that this increase in traveling time is made invisible because all clocks moving with the interferometer are delayed by a certain value so that each second becomes longer. A great number of experiments have supported this statement, and the slowing of time is named time dilation. If the time dilation is 4/3, a traveling clock will show 3 seconds when a stationary clock shows 4 seconds.

The delay along the parallel arm of the interferometer has to be exactly the same as that of the perpendicular arm; otherwise, the difference in arriving time could be observed interferometrically and interpreted as a change in the speed of light, which would be against Einstein's postulates (1) and (2). When the light is moving in the same direction as the interferometer, the travel time will be longer than when it is moving in the opposite direction. The total time will be longer than that of the perpendicular arm. Thus, judging from the time of flight, the parallel arm appears elongated in comparison to the perpendicular arm. In order to fulfill postulates (1) and (2), it was decided by Lorentz that this, to the traveling observer, apparent elongation of the parallel arm was compensated for by introducing a corresponding assumed contraction namely the Lorentz contraction.

14.2.2 Discussion

The time dilation has been proved by many experiments, and as it produces a permanent delay, there is no reason to doubt its reality. It is independent of the sign of the velocity of the interferometer, and it produces the correct result for a single-pass measurement as well as for the demonstrated double pass. The slowing of time results in a longer wavelength of the light from the source traveling with the interferometer, consequently, the number of waves in the perpendicular arm, and thus, the phase after a single or a double pass through that arm is independent of the velocity (v) of the interferometer. In this paper, we will just accept the time dilation as derived from the effect of the arm perpendicular to velocity and, in the following, study solely the arm parallel to velocity.

The Lorentz' contraction of the parallel arm is more complicated and cannot be measured directly as it is not permanent but disappears when the velocity (v) disappears. Thus, I find the discussion of assumed contractions or elongations of a moving object meaningless as they are by definition invisible. I look at them as only theoretical tools, and we will in the following solely study how the stationary world appears deformed when studied and measured by an observer traveling with a velocity close to that of light.

14.3   Introducing Spheres of Observation Transformed into Ellipsoids of Observation

Let us compare static and dynamic separation of illumination and observation. A man is making experiments based on gated viewing, which means that a short pulse of light (picosecond pulse) is emitted and, after a short time (e.g., 20 nsc), he makes a high-speed recording with a picosecond exposure time. If the illumination point source (A) and the observation point (B) are close together, the experimenter will find himself surrounded by a luminous spherical shell with a radius of 3 m. This spherical shell can be seen only if something scatters the tight, e.g., if the experiment is performed in a smoke-filled or dusty space. If there are large objects in the space, he will see these objects illuminated only in those places where they are intersected by the sphere. The experiment described can be used to map the space around the experimenter. It is identical to well-known radar methods. By changing the delay between emission and recording, intersections of differently sized spheres can be studied. In this way, the outside world]is mapped in polar coordinates.

   If the illumination point (A) and the observation point (B) are separated, the situation will be different. As the luminous sphere around (A) grows, the observer will see *nothing* until the true sphere reaches (B). Then, he will find himself *inside an ellipsoidal luminous shell*. One focal point of the ellipsoid will be (A), and the other (B). By changing the delay between emission and recording, intersections of ellipsoids with different sizes, but identical focal points, can be studied. In this way, the space around the experimenter can be mapped in bipolar coordinates. The experimenter should know the separation of (A) and (B) so that his mapping will be correct. If he erroneously believes the separation to be zero, he will misjudge the ellipsoids as spheres and make errors, especially in the measurement of lengths parallel to the line AB. He will also make angular errors because of the angular differences between points on the spherical and the ellipsoidal shells.

   We shall take a closer look at the possibilities of using the concept of the ellipsoids to visualize special relativity more generally. Our goal is to find a simple graphical way to predict the apparent distortions of objects that move at velocities close to that of light and to restore the true shape of an object from its relativistically distorted ultrahigh-speed recording.

14.3.1   Dynamic Transformation

We have already described that if the illumination point (A) is separated from the observation point (B), the gated viewing system produces recordings of intersections of ellipsoids having (A) and (B) as focal points. Now, let me, the author, and you, the reader, be stationary in a stationary space and study what a traveling experimenter (the traveler) will see of our stationary world when he travels past at relativistic velocity using picosecond illumination and observation.

Instead of an assumed contraction of fast moving objects, I introduce the idea that the travelers spheres of observation by the velocity are transformed into ellipsoids of observation. One advantage is that this new concept is easier to visualize and that it makes possible a simple graphic derivation of distortions of time and space caused by relativistic velocities. Another advantage is that it is mentally easier to accept a deformation of spheres of observation than a real deformation of rigid bodies, depending on the velocity of the observer.

Our calculations refer to how a stationary observer (the rester) judges how a traveling observer (the traveler) judges the stationary world. The reason why we have restricted ourselves to this situation exclusively is that it is convenient to visualize ourselves, you the reader and I the author, as stationary. When we are stationary, we find it to be a simple task to measure the true shape of a stationary object. We use optical instruments, measuring rods, or any other conventional measuring principle. We believe that we make no fundamental mistakes and, thus, accept our measurements as representing the true shape. No doubt the traveler has a much more difficult task. Thus, we do not trust his results but refer to them as apparent shapes.

In Fig. 14.6a, we see a stationary car in the form of a cube. An experimenter emits an ultrashort light pulse from the center of the cube (A) and some nanoseconds later makes an ultrashort observation (B) from that same place. The true sphere of light emitted from (A) will reach all sides of the cube at the same time. Thus, as he makes the observation from (B), he will simultaneously on all sides see bright points growing into circular rings of light. However, if the car is moving at velocity (v) close to (c), the true sphere, as seen by a stationary observer, will not move with the car but remain stationary. The result will be that the sphere reaches the point (E) on the side of the car earlier than, e.g., point (D), as seen in Fig. 14.6b. However, referring to Einsteins postulates (1) and (2), this fact must in some way be hidden to the traveling observer.

Our explanation is that the traveler's sphere of observation is transformed into an ellipsoid of observation, its focal points being the point of illumination (A) and the point of observation (B) separated by the velocity (v), as seen in Figs. 14.5 and 14.7. The minor diameter of the ellipsoid is, however, unchanged and identical to the diameter of the sphere. The ellipsoid reaches (E) earlier than, e.g., (D), and therefore, the different sides of the cube are observed at different points of time. However, to the traveller, all sides appear to be illuminated simultaneously, as by a sphere of light that touches all sides at the same time. The reason why the ellipsoid to the traveler appears spherical is that (A) and (B) are focal points, and thus, ACB = ADB = AEB = AFB.

The reason why the cubic car is shown elongated is that time varies linearly along the car. The point (E) appears further to the left because it was illuminated and thus recorded early when the whole moving car was further to the left, whereas

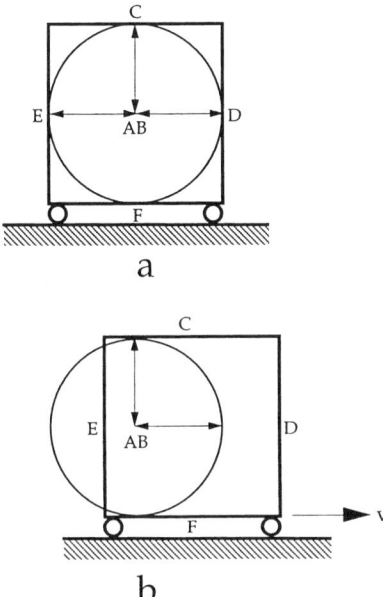

FIGURE 14.6. (a) (A) is the point source of light, and (B) is the point of observation. To the traveling observer, the sphere of light appears centered to the cubic car, independent of its velocity. Thus, he observes that all sides are touched by this sphere simultaneously. (b) To the stationary observer, the sphere of light appears fixed to his stationary world. Thus, the sphere will reach (E) earlier than (D). This fact must be hidden from the traveling observer; otherwise, he could measure the constant velocity of the car, which would be against Einstein's Special Relativity Theory.

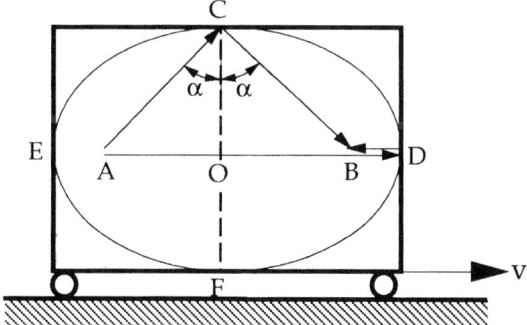

FIGURE 14.7. Our new explanation is that to the traveling observer, his sphere of observation is by the velocity transformed into an ellipsoid of observation. However, to the traveler, this ellipsoid appears spherical because (A) and (B) are focal points of the ellipsoid and thus ACB = ADB = AEB. Thus, to him, the car appears cubic because he observes that all sides of the car are touched by this sphere simultaneously just as when the car was stationary. To the stationary observer, this simultaneity is however not true.

the point (D) was illuminated and recorded later when the car had reached further to the right. As both the car and the ellipsoid of observation are elongated in the direction of travel, there is no need for an assumed contraction to make the velocity invisible to the traveler.

As time passes, the ellipsoid will grow and produce an ellipse on the ceiling of the car. Even this phenomenon will be invisible to the traveler because it has been mathematically proven that the intersection of an ellipsoid by a plane appears circular when observed from any focal point of the ellipsoid.[10] Further on, if the car had not been cubic but instead had consisted of a reflecting sphere with the light source in its center, the distortion into an ellipsoid would still be invisible. The traveler would find no difference in the reflected rays.

14.3.2   Time Dilation and Apparent Lorentz's Contraction

Let us now see if this new idea about the observation ellipsoid produces the same results as derived from the Michelson interferometry experiment. We who are stationary, the resters, understand that the traveler's observation ellipsoid has its focal points at (A) and (B) and that light with the speed of (c) travels ACB of Fig. 14.7. The traveler, on the other hand, believes his observation sphere to be centered at (B) and that the light simply has traveled with the speed of (c) in the path ACB of Fig. 14.6a. Thus, the time dilation is the time t_v it takes for light to pass ACB divided by the time it takes to pass OCO (Fig. 14.7), where $OC = ct_0$, $CB = ct_V$, and $OB = vt_V$. Applying the Pyhtagorian theorem on the triangle, OCB results in the accepted value of the time dilation:[11]

$$\text{Time Dilation}: \frac{t_v}{t_0} = \frac{CB}{CO} = -\frac{1}{\sqrt{1 - \frac{v^2}{c^2}}} = \frac{1}{\cos \alpha} = k, \qquad (4)$$

where k is the usual k-value of the holodiagram. Stationary objects as measured by the observation ellipsoid, or by any measuring rod carried in the car, will appear to the traveler contracted by the inverted value of the major diameter ED to the minor diameter CF of his observation ellipsoid:

$$\text{Lorentz's Contraction}: \frac{L_v}{L_0} = \frac{CO}{CB} = \sqrt{1 - \frac{v^2}{c^2}} = \cos \alpha = \frac{1}{k}. \qquad (5)$$

This result is identical to the accepted value of the Lorentz contraction,[12] but our graphic derivation shows that this is true only for objects that just pass by. Objects in front appear elongated by OC/BD, whereas objects behind appear contracted by OC/BE.

When the traveler emits a laser beam in his direction OC, it will in relation to the stationary world have the direction AC and its wavelength will be changed by the factor AC/OC. When looking in the direction OC, his line of sight will be changed to BC and the wavelength of light from the stationary world will be changed by the factor BC/OC (the same change will happen to his laser beam). This factor is well known as the "relativistic transversal Doppler shift." Finally, using our approach, we find that (v) can never exceed (c) because $v = c \sin \alpha$.

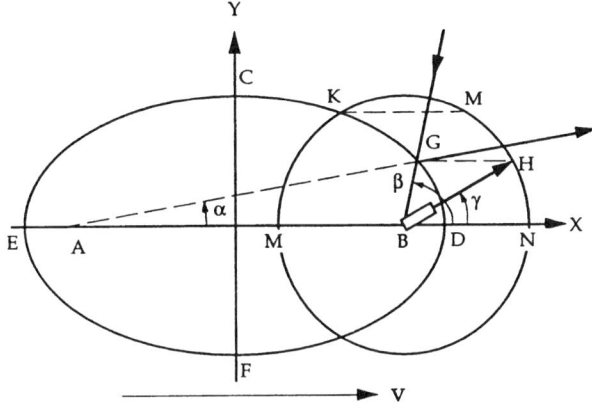

FIGURE 14.8. To the traveler, an arbitrary point (G) of the stationary world appears to exist at (H), which is found by drawing a line of constant Y-value from (G) to the sphere. As the traveler directs his telescope in his direction BH, his line of sight in the rester's universe will be GB. As the traveler directs his laser in his direction BH, the direction of the laser beam will to the rester appear to be AG. The Lorentz contraction is MN/ED.

14.3.3 Graphic Calculations

Let us now examine the emitted light rays in more detail and assume that the traveler who is moving from left to right directs a laser in the direction BH of Fig. 14.8. The direction of the beam in relation to the stationary world will then be AG. Point (G) is found by drawing a line parallel to the line of travel (the x-axis) from the point (H) on the sphere of observation to the corresponding point (G) on the ellipsoid of observation. As the point of observation is identical in space and time in the two systems, the center of the sphere should coincide with the focal point of the ellipsoid of observation.[13] From Fig. 14.8, the angle of outgoing (emitted) light (α) and incoming (lines of sight) light (β) are calculated.

$$\tan \alpha = \frac{\sin \gamma \sqrt{1 - \frac{v^2}{c^2}}}{\frac{v}{c} + \cos \gamma}, \tag{6}$$

$$\tan \beta = \frac{\sin \gamma \sqrt{1 - \frac{v^2}{c^2}}}{-\frac{v}{c} + \cos \gamma}. \tag{7}$$

These two equations solely derived from Fig. 14.8 are identical to accepted relativistic equations (See, e.g., Rindler.[14]) From Fig. 14.9a, it is easy to see that the emitted rays are concentrated forward as if there was a focusing effect by a positive lens. This phenomenon, which was pointed out by Einstein, results in the light energy from a moving source appearing to be concentrated forward for two reasons. The light frequency is increased by the Doppler effect, and the light rays are aberrated forward. This explains why the electron synchrotron radiation appears sharply peaked in intensity in the forward tangential direction of motion

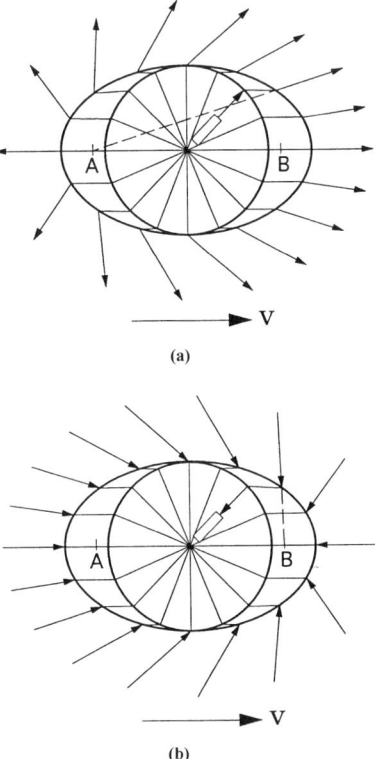

(a)

(b)

FIGURE 14.9. (a) The graphic method to find the direction of light rays in Fig. 14.8. is independent of where along the x-axis the circle is situated. Thus, for simplification, it can just as well be placed in the center of the ellipse. Thus, we see how the outgoing laser beams are concentrated forward as by a positive lens. (b) Again, using the method of Fig. 14.8, we find that the lines of sight are diverged backward so that the view forward appears demagnified as by a negative lens.

of the electrons. In the backward direction, we have the opposite effect. The light is defocused as if by a negative lens.

In Fig. 14.9b, the line of sight is seen aberrated backward along a line through (G) from focal point (B) of Fig. 14.8. The traveler is still moving to the right and his direction of observation, the telescope axis, is in the direction BH. Thus, the stationary world around the traveler appears concentrated in the forward direction as if demagnified by a negative lens. In the backward direction, we have the opposite effect. The stationary world appears magnified by a positive lens.

Let us again study Fig. 14.8 and calculate the Doppler ratio, which is the Doppler-shifted wavelength divided by the original wavelength. The traveler observes the wavelength (λ_v) as he measures the true wavelength (λ_0) from the stationary world.

Using some trigonometry, we get the following expression for the Doppler ratio:

$$\frac{\lambda_y}{\lambda_0} = \frac{BG}{BH} = \frac{1 - \frac{v}{c}\cos\gamma}{\sqrt{1 - \frac{v^2}{c^2}}} = k\left(1 - \frac{v}{c}\cos\gamma\right), \tag{8}$$

where (λ_v) is the wavelength as seen by the observer who travels in relation to the light source and (λ_0) represents the wavelength as seen by the observer who is at rest in relation to the light source. This equation, solely derived from Fig. 14.8, is identical to accepted relativistic equations (e.g., Rindler.[15])

Now, let us study Fig. 14.8 again and seek the direction in which the traveler should experience zero Doppler shift. He should not look backward because in that direction there is a red shift. Nor should he look directly sideways because even then there is a red shift, the relativistic transverse red shift. As forward as the blue shift is he should look slightly forward. The way to find the zero Doppler shift for incoming light (a) is as follows: (K) is the point of intersection for the traveler's sphere of observation as seen by the traveler, and the ellipsoid of observation as seen by the rester. Draw a line parallel to the x-axis from (K) to the corresponding point on the sphere (M). Thus, BM is the direction the traveler should look to see zero Doppler shift. The line BM is directed slightly forward, and from that fact, we understand that red shift has to be predominant in the universe even if there was no expansion from a Big Bang but only a random velocity increasing with distance.

14.3.4   Transformation of an Orthogonal Coordinate System

We shall now demonstrate how the diagram can be used for practical evaluation of the true shape of rigid bodies, whose images are relativistically distorted. The true shape is defined as the shape seen by an observer at rest in relation to the studied object. Again, let a traveling experimenter at high velocity pass through

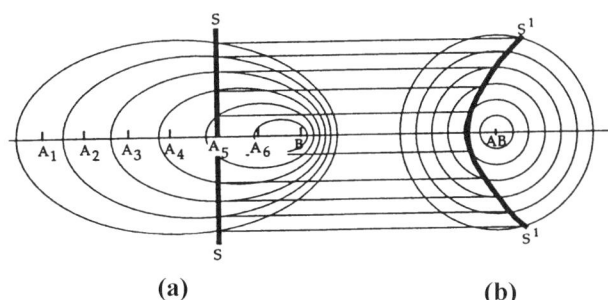

(a)   (b)

FIGURE 14.10. The traveler of Fig. 14.8 emits ultrashort light pulses at A_1, A_2, A_3, A_4, A_5, and A_6. Finally, he makes one ultrashort observation at (B). The vertical straight line S - S in the stationary world appears to the traveler to be distorted into the hyperbolic line $S^1 - S^1$.

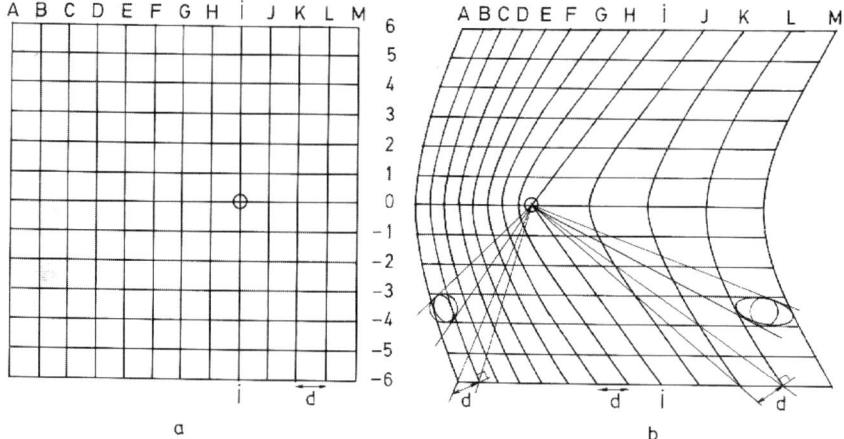

FIGURE 14.11. Orthogonal coordinate system of the stationary world (a) appears distorted into that of (b) to the traveler who exists at the small circle $(i, 0)$ and is moving to the right at a speed of $0,6\ c$. From the diagram, we find that flat surfaces parallel to motion are not changed but those perpendicular to motion are transformed into hyperboloids, whereas the plane $(i - i)$ through the observer is transformed into a cone. The back side can be seen on all objects that have past this cone. The separation of advancing hyperboloids is increased, whereas that of those moving away is decreased.

the stationary space (Fig. 14.10). He emits six-picosecond light pulses with a constant time separation of (t). After another time delay of (t), he makes one single picosecond observation at (B). Figure 14.10 shows how his spheres are transformed into ellipsoids. Let us look at one stationary straight line that is perpendicular to the direction of travel and see how it appears to the traveler. From every point at which the stationary line (S - S) of Fig. 14.10a is intersected by an ellipsoid, a horizontal line is drawn to Fig. 14.10b until it intersects the corresponding sphere. The curve connecting these intersections in Fig. 14.10b then represents the straight line of Fig. 14.10a as it appears distorted to the traveler.

In Fig. 14.11a, a stationary orthogonal coordinate system is shown, and in Fig. 14.11b, we see the corresponding distorted image as observed by the traveler, who is passing with the constant speed of 0.6 c from left to right and is represented by the small circle (i, o). The identical transformation would of course occur if the observer was stationary, and instead, the orthogonal coordinate system passed him from right to left. The results of our Fig. 14.11 (from Abramson[13]) was later confirmed by Fig. 4.13b of Mork and Vargish,[16] which, however, represents slightly different conditions.

From Fig. 14.11b, we find the following: The shape and separation of flat surfaces parallel to motion are unchanged. However, flat surfaces normal to the direction of motion appear transformed into hyperboloids, whereas the plane $(i - i)$ through the observer is transformed into a cone. The back side can be seen on all objects that have passed this cone. The separation of advancing hyperboloids is increased, whereas that of those moving away is decreased.

14.4   Intersecting Minkowski Light Cones

In discussions with my colleague Torgny Carlsson, it became clear that the ellipsoids of observation represent intersections of one cone by another.[17] These cones are identical to the Minkowski light-cones. Therefore, it is very interesting to revive the Minkowski diagram,[18] which was invented in 1908 to visualize relativistic relations between time and space.

Let us study the Minkowski diagram of Fig. 14.12. The x- and y-axes represent two dimensions of our ordinary world, whereas the z-axis represents time (t), multiplied by the speed of light (c), just to make the scales of time and space of the same magnitude. Thus, in the x-ct coordinate system, the velocity of light is represented by a straight line at 45° to the ct-axis. As all other possible velocities are lower than that of light, they are represented by straight lines inclined at an angle of less than 45° to the ct-axis.

We will first concentrate on the illumination cone. A spherical wave front is emitted in all directions from a point source (A) and expands with the speed of light. In our chosen coordinate system, which is limited to only two space coordinates and one time coordinate, this phenomenon is represented by a cone with its apex

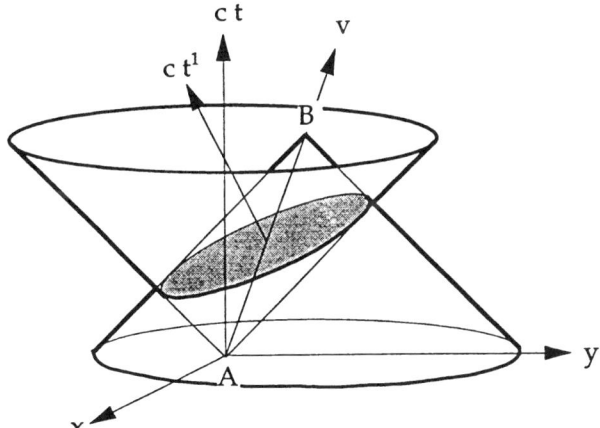

FIGURE 14.12. An ultrashort light pulse is emitted at (A) that is the apex of a Minkowski light cone. In our coordinate system, x-y represent two axes of our ordinary world, whereas the third axis ct represents time. The widening of the cone upward represents the radius of the sphere of light as it increases with time. An ultrashort observation is later made at (B), the apex of an inverted cone. The only way for light to be transmitted from (A) to (B) is by scattering objects placed where the two cones intersect. If his velocity (v) is high, this intersection will be an ellipse that is inclined in relation to our stationary world.

at (A), expanding in the direction of the positive time axis. The passing of time is represented by cross sections of the cone by planes parallel to the x-y plane at increasing ct-values. Thes intersections will, when projected down to the x-y plane, produce circles of increasing radius that in our 3D world represent the expanding spherical wavefront from the point source at (A).

Let us now introduce an observation cone that intersects the illumination cone. If a point of illumination represents a point source of light, a point of observation represents a point sink of light, a point toward which spherical waves are shrinking. In the Minkowski diagram, it is represented by a cone that is inverted in relation to the light cone, referred to as the observation cone (B in Fig. 14.12) , which like the light cone has a cone angle of 90° and the observer at the apex (B). Thus, an observer at (B) can see nothing outside this cone because of the limited speed of light (c). The only general way for light to pass from the illumination cone to the observation cone is by deflection, e.g., by scattering from matter that exists at the intersection of the two cones. The only exception is when the two cones just touch each other, which is the only case when light may pass directly from (A) to (B) of Fig. 14.12. The intersection of the two cones produces an ellipse that represents an ellipsoid of observation. In other words, it represents the travelers surface of simultaneity, which to him, who is situated at the apex (B) of the cone, of course appears spherical. The traveler's time axis (ct^1) is at a different direction than is the time axis of the resters x-y plane (ct). This result is in full accordance with Fig. 14.7, where different sides of the car are observed as they are at different points of time because time varies in a linear fashion along the line of travel.

To simplify the diagram of Fig. 14.12, we have drawn only two space dimensions and time. If all three space dimensions had been included, the intersection of the two cones would represent the ellipsoid of observation, the apex (A) and (B) of the two cones being the focal points. The situation will be the same whether the separation of (A) and (B) are static or caused by an ultrahigh velocity. Thus, our concept of ellipsoids of observation applies just as well to the evaluation of apparent distortions of relativistic velocities as to radar, gated viewing, holographic interferometry, and holography with picosecond pulses.[19] As soon as a separation exists between (A) and (B), the spheres of observation are transformed into ellipsoids of observation.

14.5   Conclusion

Inspired by the holographic uses of the holodiagram, we have introduced the new concept of spheres of observation that by velocity are transformed into ellipsoids of observation. In this way, time dilations and apparent length contractions are explained as results of the eccentricity of ellipsoids. Our approach explains how a sphere of light can appear stationary in two frames of reference that move in relation to each other. It visualizes and simplifies in a graphic way the apparent distortions of time and space that are already generally accepted using the Lorentz transformations. When the ellipsoids are used to explain the null result of the Michelson-Morley experiment, there is no need to assume a real Lorentz contraction of rigid bodies, caused by a velocity of the observer. We have used our graphic concept to calculate the apparent Lorentz contraction, the transversal relativistic red shift, the relativistic aberration of light rays, and the apparent general distortion of objects. In all cases, our results agree with those found in other publications. However using our approach, we look at those relativistic phenomena as solely caused by the influence of velocity on the measurement performed.

Appendix

Until now, we have used the concept of ellipsoids of observation to explain apparent distortions of fast-moving objects as measured by light. However, these ellipsoids of observation apply just as well for observations using other fields, e.g., electric or gravitational fields. Thus, the precession of the perihelion of Mercury[20] can only be explained by an asymmetry in the forces acting on the planet during its orbit. Such an asymmetry is visualized in Fig. 14.13, where because of the limited velocity of gravitation, the gravitational force is different as the planet advances toward the sun and as it retreats from the sun. Using similar approaches, it is pos-

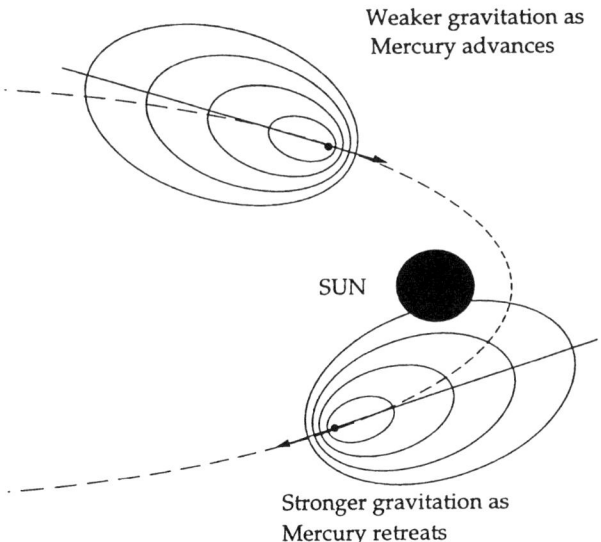

FIGURE 14.13. It is here assumed that the concept of the ellipsoids of observation apply to all fields moving with the velocity of light, such as electric or gravitational fields. Thus, precession of the perihelion of Mercury (the rotation of its elliptic orbit) can be explained by the asymmetry of the gravitational forces as the planet advances toward, and as it moves away from the sun.

sible to develop our methods to visualize kinetic energy and magnetic forces as functions of the eccentricity of ellipsoids.

References

[1] N. Abramson, "The Holo-diagram: a practical device for the making and evaluation of holograms," *Proc. Applications of Holography*, Strathclyde, Glasgow, 45–55 (1968).

[2] N. Abramson, "Moiré patterns and hologram interferometry," *Nature* **231**, 65–67 (1971).

[3] N. Abramson, "The holo-diagram. IV: a practical device for simulating fringe patterns in hologram interferometry," *Appl. Opt.* **10**, 2155–2161 (1971).

[4] N. Abramson, "The holo-diagram VI: practical device in coherent optics," *Appl. Opt.* **11**, 2562–2571 (1972).

[5] D.I. Staselko, Y.N. Denisyuk, and A.G. Smirnow," Holographic recording of the time-coherence pattern of a wave train from a pulsed laser source" *Opt. Spectrosc.* **26**, 413 (1969).

[6] Y.N. Denisyuk, D.I. Staselko and R.R. Herke, "On the effect of the time and spatial coherence of radiation source on the image produced by a hologram," *Proc. Applications of Holography*, Section 2, Besancon, 1–8 (1970).

[7] N. Abramson, "Light-in-flight recording: high-speed holographic motion pictures of ultrafast phenomena," Appl. Opt. 22, 215–232 (1983).

[8] H.A. Lorentz, "Michelsons' Interference Experiment, 1885," in *The Principle of Relativity* (Dover New York, 1952), 7.

[9] A. Einstein, "On the electrodynamics of moving bodies, 1905," in *The Principle of Relativity*, (Dover New York, 1952), 37–65.

[10] D. Coxeter, Prof. Emeritus, Math. Dept., University of Toronto, Toronto, Ontario, Canada M5S 3G3. Private communication.

[11] P.A. Tipler, College Physics (Worth, New York, 1987), 719.

[12] P.A. Tipler, College Physics (Worth, New York, 1987), 720.

[13] N. Abramson, "Light-in-flight recording 4: visualizing optical relativistic phenomena," *Appl. Opt.* **24**, 3323–3329 (1985).

[14] W. Rindler, *Special Relativity* (Wiley, New York, 1960), 49.

[15] W. Rindler, Special Relativity (Wiley, New York, 1960), 47.

[16] D.E. Mork and T. Vargish, Inside Relativity, (Princeton University Press, Princeton, NJ, 1987).

[17] N. Abramson, "Minkowski diagram in relativity and holography," *Appl. Opt.* **27**, 1825–1832 (1988).

[18] H. Minkowski, "Space and Time, 1908" in *The Principle of Relativity* (Dover New York, 1952), 75.

[19] N.H. Abramson, *Light in Flight or the Holodiagram-The Columbi Egg of Optics*, Vol. PM27 (SPIE Press, 1996).

[20] A. Einstein, "The Foundations of the General Theory of Relativity, 1916," in *The Principle of Relativity* (Dover, New York, 1952), 111–164.

15

Quantum Holograms

George Chapline and Alex Granik

15.1   Introduction

In this chapter, we offer some remarks on the possibilities of using quantum states for electromagnetic fields to encode geometrical information. What makes such remarks timely are the impressive achievements of the past few years relating to the time-domain quantum state holography. It is not our purpose here to review in detail those marvelous results. Instead, we will focus in this essay on the question of to what extent it may be possible to use quantum holographic techniques to permit reconstruction of the wavefunction for a quantum mechanical surface.

The simplest context in which quantum effects are important in holography is conventional holography, in which photon counting statistics are important, for example, in x-ray holography.[1] The most immediate practical effect of photon statistics in these circumstances is that the resolution and information content of the hologram are limited. In the x-ray holography of biological structures[1] or the unit cells of crystals,[2] the limitations on the resolution due to photon statistics are particularly important because of the desire to use holography to reveal the details of mesoscopic structures inaccessible to optical or even electron microscopy.

The problem of photon counting statistics in holography is accompanied by the conceptual problem of quantizing holographic representations of objects. This problem is a special case of the usual problem of quantizing the electromagnetic field and is perhaps most interesting in the case of a smooth surface with nontrivial topology, because holographic representations for such surfaces require multiple directions of illumination. In Section 15.2, we show that in a paraxial ray-type approximation for the electromagnetic field, the usual mode variables for the electromagnetic field can be replaced by slowly varying field variables $\hat{E}$ and $\hat{E}^{\dagger}$ whose variations represent the structure of the hologram. These field variables depend on both the position on the hologram and the orientations of the illuminating and reference laser beams. In the following, our primary focus will be on the fact that quantum fluctuations of these variables give rise to "fuzziness" in the reconstructed surface. In our view, the fuzziness in the reconstructed surface can be interpreted by saying that quantization of classic holography amounts to replacing a classic surface with a quantum wavefunction $\Psi(\tau_1, \tau_2, \ldots, \tau_n)$, where the τ_i are parameters specifying the shape of the surface. Thus, from the point of view of pure mathematics, quantum holography may be a pedestrian way of introducing the concept of quantum geometry.

The production of a classic hologram using ordinary light is, of course, very reminiscent of the superposition principle for quantum states that is one of the fundamental building blocks of quantum mechanics. Indeed, an ordinary hologram can be thought of as representing the result of many measurements of the spatial wavefunction for a single photon.[3] This analogy is the inspiration for the recently burgeoning field of "quantum state holography" (cf. Leichtle et al.[4] and references therein), where one makes use of time-domain correlations between a reference wavefunction and an object wavefunction to reconstruct the object wavefunction. Because there are cases in which the reconstructed wavefunction carries a lot of spatial information (see, e.g., Weinacht et al.[5]), time-domain quantum holography has already approached very closely our concept of quantum of quantum holography as the reconstruction of a wavefunction for a quantum mechanical surface. In addition to our focus on the use of a wavefunction to describe a geometrical object, though, we wish to also emphasize the potential importance of using spatial domain correlations to reconstruct the object wavefunction.

Following a brief review of time-domain quantum state holography in Section 15.3, we describe in Sections 15.4 and 15.5 our concept of quantum holography as the reconstruction of the "object" wavefunction for a quantum mechanical surface using the spatial pattern of interference between the object wavefunction and a coherent reference wavefunction.

15.2   Quantization of Holographic Representations

As a quick reminder, all information concerning an arbitrarily curved surface in three dimensions can be enclosed onto a flat plane or the surface of a sphere by recording photographically or otherwise the interference of light scattered off the surface with a reference beam. Of course, if the surface is reentrant, then the interference pattern must be recorded for various orientations of the illuminating beam in order to capture the entire surface. In order to physically describe a hologram, we must introduce a slowly varying "envelope" electric field whose magnitude corresponds to the intensity of the interference pattern. To this end, we write the vector potential on the recording surface as a function of position x on the hologram in the form

$$\vec{A}(x) = -(i/k)\vec{E}(x, t)\exp[i(k_1 z - \omega_1 t)] + h.c., \tag{1}$$

where the factor $\exp[i(k_1 z - \omega_1 t)]$ represents the rapidly varying phase of the reference and scattered beams. It is straightforward to quantize these fields by substituting expression (1) into the standard radiation gauge Lagrangian for the electromagnetic field. One finds the following commutation relations for the x (or y) component electric field operators on the recording surface:

$$[\hat{E}(x, t), \hat{E}^{\dagger}(x', t)] = C\delta(x - x'). \tag{2}$$

In classic holography, the electric field receives contributions from various points on the surface. In a similar way, in quantum holography, the electric field operator $\hat{E}(x, t)$ associated with photons scattered from points r on a surface can be written in the form:

$$\hat{E}(x, t) = i\frac{1}{\sqrt{V}}(\frac{\hbar\omega}{2\epsilon_0})^{1/2}\sum_{r_s}\hat{a}_{\vec{k}}e^{i\vec{k}\bullet(\vec{x}-\vec{r}_s)}, \tag{3}$$

where $\hat{a}_{\vec{k}}$ is the photon creation operator for mode $\vec{k}$. The quantum properties of multimode electromagnetic fields have been discussed by many authors. We prefer the treatment of Ou et al.[6] They introduced two slowly varying Hermitian quadrature operators $\hat{E}_1(x, t)$ and $\hat{E}_2(x, t)$:

$$\hat{E}_1(x, t) = \hat{E}(x, t)e^{-i\beta} + \hat{E}^{\dagger}(x, t)e^{i\beta},$$
$$\hat{E}_2(x, t) = -i\hat{E}(x, t)e^{-i\beta} + i\hat{E}^{\dagger}(x, t)e^{i\beta}. \tag{4}$$

When the total electric field at x is the sum of a coherent field $\vec{E}_0\exp[i(k_1 z - \omega_1 t - \phi)]$ and a scattered field of the form (4), then it can be shown[6] that the current–current correlation in a photodetector is a linear function of $\sin[2(\beta - \phi)]$ and $\cos[2(\beta - \phi)]$. The linear coefficients can be expressed in terms of correlation functions for the quadrature operators $\hat{E}_1(x, t)$ and $\hat{E}_2(x, t)$. As a corollary, when the photoelectric current $\vec{J}(t)$ can be expressed as a sum over pulses due to arrival of photons at times t_j, i.e. $\vec{J}(t) = \sum_{j=1}\vec{k}(t - t_j)$, then the mean square current can be expressed in the form:

$$\langle(\Delta J)^2\rangle = \eta|\vec{E}_0|^2\int_0^{\infty}k^2(t')dt' + \eta^2|\vec{E}_0|^2q^2\langle(\Delta E_1)^2\rangle, \tag{5}$$

where η is the quantum efficiency and q is the total charge resulting from one photon. The first term on the right-hand side of (5) is just the usual shot noise, and the only contribution to the noise when coherent light is being used to make the hologram. The presence of the second term on the right-hand side of (5) is a signal that nonclassic quantum states are being used to make the hologram. The most famous of the nonclassic states of light are the "squeezed states" for which the contours of the Wigner function for the quadrature operators for a single mode are ellipses.[7] If $< (\Delta E_1)^2 >$ is negative, then the fluctuations in current density will be sub-Poissonian. Thus, production of holograms with nonclassic states of light

nonlinear crystal

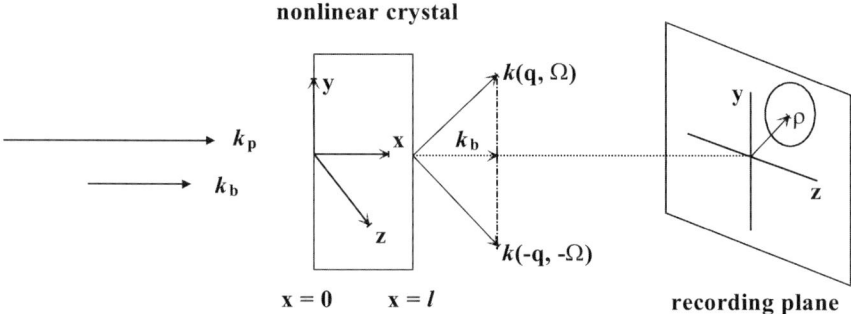

FIGURE 15.1. A schematic representation of the experiment with the multimode squeezed light. A classic light wave (pump) of frequency ω_p passes through a nonlinear crystal and decays into pairs of conjugate photons of frequencies $\omega_b + \Omega$ and $\omega_b - \Omega$ with $\omega_b = \omega_p/2$. The recording plane $\rho = \rho(x, y)$ is made of a set of highly efficient photon detectors.

is of interest from the point of view of mitigating the effects of photon counting statistics.

To our knowledge, the first proposal to construct spatial domain holograms with nonclassic quantum states was that of Kolobov and Sokolov.[8] Their primary interest was in the possibility of producing spatial images with sub-Poissonian levels of noise. However, they did mention the possibility of using squeezed states of light to make holograms, and what they actually calculated was the squeezed light analog of the classic hologram from a point scatterer. In particular, for the geometry shown in Fig. 15.1, they studied the interference of parametric downconverted photons with the coherent pump wave that was used to create these photons. The squeezing parameter for the downconverted photons as a function of angle is shown in Fig. 15.2.

The result of greatest interest is their calculation for how fluctuations in pho-todetector current after homodyning with a plane pump wave would depend on angle. In particular, for the noise in the detector photocurrent after homodyning with a coherent pump wave, they obtained the following result:

$$\langle J^2 \rangle = \langle J \rangle \{1 - \eta[\cos^2(\psi)e^{2r} + \sin^2(\psi)e^{-2r} - 1]\}, \qquad (6)$$

where $\psi = Re\beta$ is the angle of rotation of the squeezing ellipse and $\exp(\pm r) = |e^{-i\beta}| \pm |e^{i\beta}|$. The noise spectrum as a function of phase mismatching and angle is shown in Fig. 15.3.

Although the results of Kolobov and Sokolov clearly show that by restricting the ranges of angles and of phase mismatching it would be possible to emphasize either amplitude or phase squeezing, they were primarily interested in the possibilities of using amplitude squeezing. Indeed, to date, almost all of the interest in practical uses of nonclassic states of light has been connected with the possibility of choosing the parameter β in (4) so that the photon counting statistics are sub-Poissonian. In the context of holography, one may say that the practical goal of such efforts is to simply approach the classic ideal of using a Maxwellian electromagnetic field to

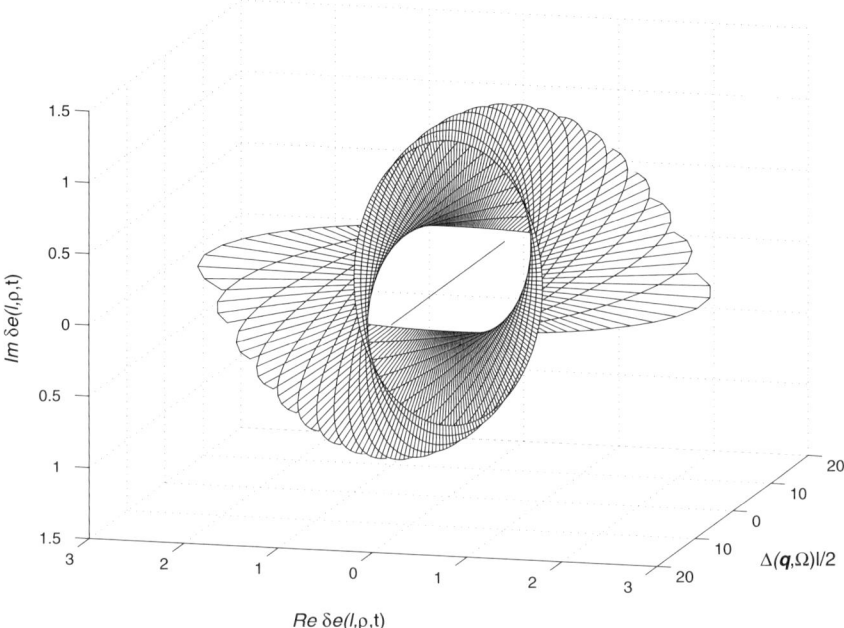

FIGURE 15.2. Squeezing ellipses for various values of the wave mismatch $\Delta(\vec{q}, \Omega)l/2$; $\Delta(\vec{q}, \Omega) = k_x(\vec{q}, \Omega) + k_x(-\vec{q}, -\Omega) - k_p$.

exactly encode information about the shape of a fixed surface. In the following, however, we would like to emphasize that the use for holography of nonclassic states of light that are squeezed with respect to phase rather than intensity is also of considerable intellectual and practical interest. In particular, in Sections 15.4, 15.5, and 15.6, we will argue that phase-squeezed states of light can be used to construct a quantum description for a curved 2D surface, and that this provides a new approach to parallel information processing. First, though, we will briefly review the techniques that have been developed over the past decade for reconstructing quantum wavefunctions.

15.3   Quantum State Holography

Classic holography was born out of need to restore lost phase information that encodes a complete description of relative positions of different parts of a recording surface, its distances from a light source, and so on. Such a restoration is achieved by a parallel encoding via interference of both the path and phase differences of two coherent light waves. As we have already indicated in Section 15.2, one may also say that classic holography provides a complete description of geometry of the

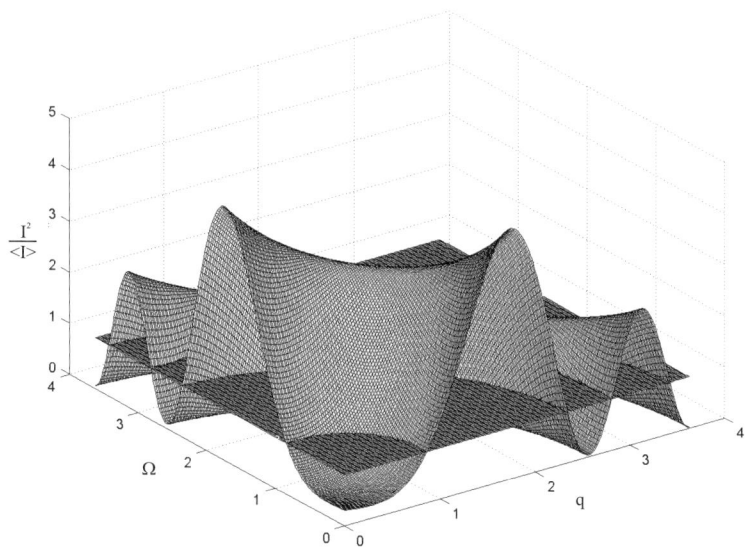

FIGURE 15.3. Spectrum (6) of current density fluctuations.

recorded object on a grand scale. On the other hand, in terms of information transfer, holography can be viewed as a reconstruction lost in the process of measurement (recording).

The latter provides a direct connection to processes in quantum mechanics, in which by definition, any measurement destroys part of a complete information about a quantum system that is contained in a wavefunction. Hence, to be able to "measure" the wavefunction would be equivalent to extracting all possible information about a quantum system. The problem of such measurement, that is, of finding the amplitude and phase of the wavefunction, has attracted a lot of attention in the last few years. Some time ago it was proposed[9] to interfere the object wavefunction (the one to be "measured") and a known reference wavefunction.

Still, there is a substantial difference between classic and quantum holography. In the former, the restoration of the lost information is easily achieved by simply illuminating the recorded hologram by the reference light beam. But how is it possible to "illuminate" the result of interference of two wave functions? One such methods was proposed by Leichtle et al.,[4] who called it *quantum state holography*. This quantum state holography uses two time-delayed laser pulses. One of them generates an object wavefunction, and the other one serves as a reference wave function. If the path lengths of the pulses are different, then there exists a time delay τ between them. The resulting incoherent fluorescence (time and space integrated) of a generated excited state is to be recorded as a function of the time delay τ. The resulting record serves as a quantum state hologram. More precisely, we should have called it a time-domain quantum state hologram. To date, the demonstrations of reconstruction of the wavefunction based on the above interference have been restricted to a few comparatively simple systems, e.g., a single mode of the electromagnetic field,[10-14] vibrational states of a molecule,[15] a single ion in a Paul trap,[16] and an atomic beam.[17]

Another technique, the tomographic method,[18] reconstructs the Wigner distribution from the marginal distributions for all angles $0 \leq \phi \leq \pi$ that enter the expressions for the observable function $q(\phi) = p \cos(\phi) - x \sin(\phi)$. The name of the method is derived from the fact that its technique is analogous to the one used in medical tomography.

Here, we will not attempt to give a complete account of all the work in this area, but instead concentrate on what represents quantum state holography *per se*. By analogy with conventional holography, the former can be defined as a process of reconstruction of a wavefunction on the basis of parallel encoding (via interference) of the square moduli and phase differences between the object and the reference wavefunctions.

In detail, the concept works as follows. The reference wavefunction $|\Psi_r\rangle = e^{i\phi(\tau)} \sum_n b_n |n\rangle$ is coherently added to the object wavefunction $|\Psi_0\rangle = e^{-i\omega_n(t)} \sum_n a_n |n\rangle$. As a result, the population of the nth state becomes

$$P_n = |a_n|^2 + |b_n|^2 + 2Re(a_n b_n^* e^{-i[\omega_n t + \phi(\tau)]}). \tag{7}$$

Because the emitted time-integrated energy (fluorescence) is

$$F_{\text{tot}} \propto \sum_{mn} P_n | < m|n > |^2 (E_n - E_m)^4, \tag{8}$$

relations (7) and (8) lead to the following expression:

$$F_{\text{tot}} = \sum_n (|k_n a_n|^2 + |k_n b_n|^2) + F_{\text{int}}(\tau). \tag{9}$$

Here, E_n is the energy of the nth state and $k_n = \{\sum_m | < m|n > |^2 (E_n - E_m)^4\}^{1/2}$. The overlap between the two wave packets results in the interference term E_{int}, which depends on the time delay τ and is crucial for the time-domain quantum state holography:

$$F_{\text{int}}(\tau) = 2Re(a_n b_n^*) e^{-i[\omega_n t + \phi(\tau)]}). \tag{10}$$

The first two parts of the total energy (9) can be measured directly by simply measuring the fluorescence of the object and reference wave packets. The interference part $E_{\text{int}}(\tau)$ of the total energy (10) contains both the known quantities b_n (the reference wave packet $|\Psi_r\rangle$ is known) and the unknown coefficients a_n representing the complex-valued amplitudes of the object wavefunction $|\Psi_0\rangle$ to be recovered.

In contradistinction to classic holography, in which the recovery of the object wave is done by physical means (that is, by illuminating the hologram by the reference beam), here, one can recover the object wavefunction (that is, to find the unknown coefficients a_n) only numerically. The coherent interaction of the time-delayed coherent pulses transfers the relative phase between two pulses $\phi(\tau) =$

$\phi_0 + \omega_L \tau$ (here, ω_L is the source frequency) onto the wave packets. The interactive part of fluorescence $E_{int}(\tau)$ then can be measured at a fixed delay time τ for two different phases ϕ_0, say, $\phi_0 = 0$ and $\phi_0 = -\pi/2$. The linear combination of the respective fluorescences defines the recordable signal

$$S(\tau) = \frac{1}{2}[F_{int}(\tau, \phi_0 = 0) + i\, F_{int}(\tau, \phi_0 = -i\pi/2)]$$
$$= \sum_n k_n^2 a_n b_n^* e^{-i(\omega_n + \omega_L)\tau}. \tag{11}$$

If there are N measurements corresponding to different time delays τ_m ($m = 1, 2, \cdots N$), then (11) yields N linear equations with respect to N unknown amplitudes a_i. These equations can be written in the matrix form as follows:

$$\hat{S} = \hat{B}\hat{A}, \tag{12}$$

where $S = S(\tau_m) = S_m$, and $\hat{B}$ is a $m \times n$ matrix $B_{mn} = k_n^2 B_n^* e^{-i(\omega_n + \omega_L)\tau_m}$. Solving the system (12), one obtains the amplitudes a_i

$$\hat{A} = \hat{B}^{-1}\hat{S}. \tag{13}$$

This restoration algorithm was used in the remarkable experiments carried out by Weinacht et al.[5, 19] In one of their experiments, they reconstructed the quantum state of a bound electron in a caesium atom. The "object wave" in the case of a highly excited atom (a Rydberg atom) was a quantum state of the bound electron and the "reference wave" was a known state.

In the experiment (shown schematically in Fig. 15.4), the atom was excited by two laser pulses emitted by the same source and separated by a time delay τ. One of the pulses passed through a pulse shaper controlling its form, amplitude, and frequency. The other pulse went directly to the atom. In this setup, the first pulse served as an "object wave," and the second one as a "reference wave." Measurements of the population of the individual energy eigenstates were performed by applying a varying electric field across the capacitor. In the final phase of the experiment, the complete object wavefunction, that is, its amplitude and phase, has been recovered with the help of the algorithm developed in Leichtle et al.[4]

In the second experiment, Weinacht et al.[19] extended their technique to the problem of preparation of a desired state. They combined their method of the quantum state preparation and measurement with feedback. As in the first experiment, they prepared a trial wavefunction Ψ_t with the help of the object wave (Fig. 15.4) and reconstructed it using the reference wave. In the next step, they compared the reconstructed trial wavefunction Ψ_t with the "target" wavefunction Ψ_g. The difference between the two functions was calculated and used to readjust the programmable pulse shaper (with independent control of the amplitude and phase) to change the optical pulse to reduce the difference. The authors used a simple feedback algorithm, in which the relative phase of each projection of the total wavefunction onto the eigenstates of the atomic Hamiltonian was changed by an amount equal to the difference between the measured and target phases. The feedback process continued until the target function was obtained. Remarkably enough, the experiment

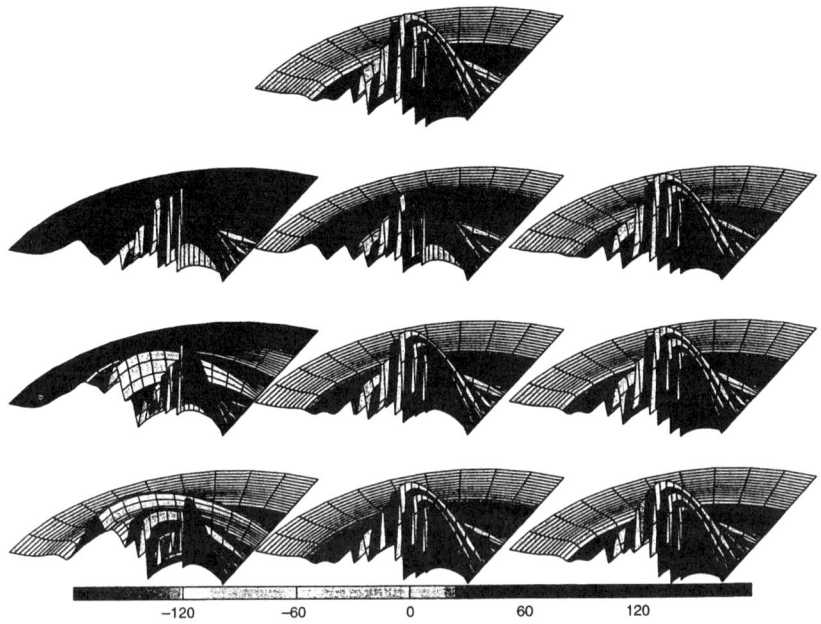

FIGURE 15.4. Two time-delayed pulses interfere, generating in each Rydberg state the total population, which is subsequently measured.

achieved the desired results within two iterations. The wavefunctions shaped in such a process are shown in Figure 15.5 (reprinted from the paper by Weinacht et al.[19]).

The top figure represents the target wavefunction. Each column in the figure corresponds to a subsequent iteration, and each row corresponds to a different delay time τ between the pulses.

15.4   Semiclassic Reconstruction of a Quantum Surface

In principle, it would be straightforward to generalize the simple geometry considered by Kolobov–Sokolov so as to make a squeezed state holographic representation for a fixed curved surface. For example, one could imagine introducing at each point of a smooth curved surface a strong nonlinear medium that produces parametrically downconverted photons. If the surface was irradiated with a single coherent pump beam, then the squeezed state photons produced at different points of the surface will be coherent and will produce an interference pattern on the detector plane that contains information about the shape of the surface. In principle, at least, the quantum state of the light falling on a detector plane can be determined using multiport homodyning techniques.[10] As in ordinary holography, one can think of the squeezed state hologram as arising from the superposition of many

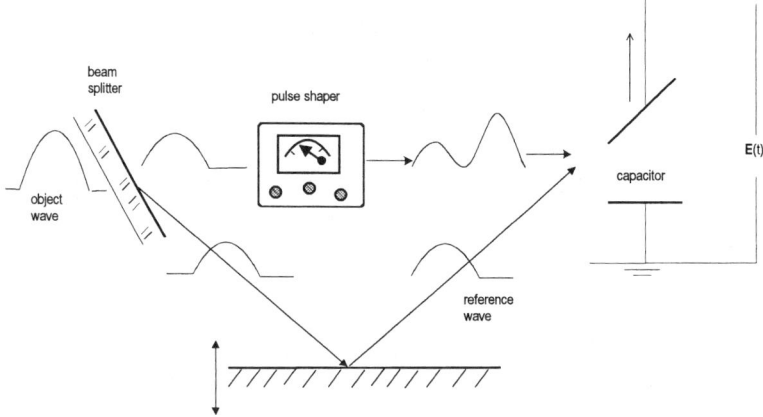

FIGURE 15.5. Reconstructed atomic Rydberg function. The top figure is the target wave-function. The last column shows the reconstructed wavefunctions for three different time delays. Reprinted with permission from Nature ("Controlling the shape of a wavefunction," by T.C. Weinacht, J. Ahn, and P.H. Bucksbaum, vol. 397, p. 234, 1999). Copyright ©1999 Macmillan Magazines Ltd.

elementary diffraction patterns; the elementary pattern in the quantum case being the spatial pattern of intensity fluctuations given in (6). If the parameter β in (6) is chosen so that the photon counting statistics are sub-Poissonian, then the surface reconstruction can proceed more or less as in classic holography. A perhaps more interesting problem of surface reconstruction arises, though, in the case of strong phase squeezing. In this case, we wish to draw attention to the possibility of using the quantum theory of inverse scattering in its reincarnation as a theory of adaptive optics[20] to reconstruct the surface.

The basic idea is that instead of a fixed classic surface, we consider a deformable classic surface whose shape is chosen to just compensate for the phase shift $\phi(\sigma, t)$ of the coherent pump wave incident on the surface at position s on the surface. We proceed by writing equations that describe the interplay between small deformations in the shape of a surface and changes in the intensity of light on a set of holograms. Defining an equivalent optical path length $\Delta l = \phi(\sigma, t)/k_1$, the first equation supposes that we have a control system that adjusts the absolute position $X(\sigma, t)$ of the surface with sufficient accuracy so that the intensity of light on the hologram at time t and position x is a linear function of the error $e(\sigma, t) = \Delta X(\sigma, t) + \Delta l(\sigma, t)$; i.e.,

$$I(x, t) = I_0(x) + \int_S B(x, \sigma)e(\sigma, t)d^2\sigma, \tag{14}$$

where $I(x, t)$ is the recorded intensity of light on the hologram at time t and position x, and $I_0(x)$ is the recorded intensity on the hologram when the surface is illuminated with a "standard" beam with no imposed variations in phase with respect to position or time. In writing equation (14), we are neglecting the light propagation time between the surface and the hologram.

The second equation relates the deformation of the surface at point σ, $\Delta X(\sigma, t)$, produced by the feedback control to the intensity of light on the hologram:

$$\Delta X(\sigma, t) = \int d\Omega \int_{-\infty}^{t} A(\sigma, x, t')I(x, t), \tag{15}$$

where the integral over $d\Omega$ means sampling the light intensity of the phase squeezed light at a sufficiently large number of points on the hologram as is required to determine the parameters $t_1, t_2, \cdots t_N$, which define the shape of the surface. When photon counting noise is neglected, i.e., when we regard the hologram as a classic object, then (14) and (15) have the solution (in vector shorthand)

$$e = [1 - \hat{A}\hat{B}]^{-1}\phi. \tag{16}$$

Equation (16) shows that when the negative feedback is strong, the error $e(\sigma, t)$ can be reduced to a small fraction of the perturbation in path length. That is, the position of the surface just tracks locally the change in optical path length. Upon reflection, one soon realizes that this just the classic principle of least action! What is perhaps most remarkable about this setup, though, is that the problem of changing the shape of the surface to compensate for changes in the optical path of the illuminating beam becomes equivalent to solving the multichannel Schroedinger equation when the effects of photon noise on the hologram are taken into account. This situation is qualitatively different from the classic case because the negative feedback in (16) will amplify the photon noise. Therefore, unless the photon noise is suppressed, there is now a limit to how strong the negative feedback can be made.

It is not hard to show that in the presence of photon noise, the two-point correlation function for the errors averaged over a time long compared with the characteristic time for photon number fluctuations has the form (again, in operator notation):

$$< e_1 e_2 >= [1 - \hat{A}_1\hat{B}_1]^{-1}[1 - \hat{A}_2\hat{B}_2]^{-1}\{U_{12} + \hat{A}_1\hat{A}_2\delta_{12}I_0\}, \tag{17}$$

where U_{12} is the average $< \phi_1\phi_2 >$ over the same time and $\delta_{12} = \delta(\sigma_1 - \sigma_2)\delta(t_1 - t_2)$. In contrast with the classic case, an optimal choice for the feedback matrix $\hat{A}$ is somewhat arbitrary. However, following Dyson, we take as the criterion for optimizing the feedback system that a quadratic function of the feedback errors should be minimized, and then it can be shown[19] that the optimal feedback matrix A(s,x,t') can be expressed in the form:

$$\hat{A} = \hat{K}\hat{B}^T I_0^{-1}, \tag{18}$$

where $\hat{K}(\sigma_1, \sigma_2, t_1 - t_2)$ is a causal matrix satisfying the nonlinear operator equation

$$\hat{K} + \hat{K}^T + \hat{K}(\hat{B}^T I_0^{-1})\hat{K}^T = 0. \tag{19}$$

This equation is a matrix generalization of the Gelfand–Levitan equation, and it is essentially equivalent to the multichannel inverse scattering theory of Newton and Jost.[21] In our quantum optical realization of their theory, the discrete channel labels $\alpha = 1, \ldots, N$ represent a set of illuminations of the surface.

In the case of a Riemann surface, for example, the channel indices can be taken to index the 2-g independent harmonic 1-forms on the surface, whose support coincides with the support of harmonic functions f_j such that $\omega_j = df_j$. Physically, this corresponds to choosing a canonical set of illuminations such that the ω_j describe the polarizations of the oscillating electric currents induced on the surface by the illuminating beams. The necessity for having 2-g illuminating beams in order to fully reconstruct the surface can be understood from the fact that g represents the number of topological handles on the surface.[22] Obviously, at least two directions of illumination are needed for illumination of each of these handles.

The similarity of the feedback control equations (14) and (15) to the inverse scattering equations for multichannel quantum mechanics means that our scheme for using the intensity pattern of a phase-squeezed hologram to control the shape of a surface provides a kind of quasi-classic analog for quantum mechanics. Of course, such a feedback system would not be actually quantum mechanical in the sense that at any given time the surface has a well-defined shape. However, by correlating the well-defined asymptotic phases derived from a sequence of phase-squeezed holograms with the shape and position of the surface, one can hope to construct over a period of time what is in effect a "wavefunction" for a deformable surface.

15.5   Entangled States and Quantum Geometry

Because the wavefunction construction described in the last section requires numerical computations, it is limited by the storage capacity of the computer. Thus, despite the use of nonclassic states of light, one is still bound by the well-known limitations of using a classic system to simulate a quantum mechanical system. In order to go beyond the limitations of using a classic system to represent a quantum surface, one must evidently replace representing the shape of a surface by a product wavefunction for single mode states by a quantum superposition of such representations. That is, to approach the situation of a truly quantum mechanical surface described by a quantum wavefunction, one must use a quantum superposition of squeezed state representations.

Of course, we don't really expect that it will be possible to produce a quantum superposition state for a real physical surface, as such a feat is likely to prove experimentally elusive. Indeed, to date, no one has succeeded in producing a superposition of distinguishable quantum states for any macroscopic object. However, a fascinating question for the future is whether it may instead be possible to create a quantum superposition state for a virtual surface. One approach to the practical realization of such a possibility would be to use the entangled pair of photons created by parametric downconversion to construct entangled holographic representations for a surface. In particular, by using a sequence of pump pulses and both of the downconverted photons in Fig. 15.1, it ought to be possible to record a quantum superposition of two squeezed state holograms of the Kolobov–Sokolov type.

Unfortunately, direct production of a quantum superposition of more than two squeezed state holograms would require using multiphoton entangled states, which have yet to appear in the laboratory. Although such quantum states may become available in the future, for the present, one will have to be content with making use of ensembles of two photon holograms. It may be noted in this connection that the progress that has been made over the past few years in creating intense sources of entangled photon pairs[23] represents a significant step toward the practical realization of ensembles of two-photon holograms. Even though different two-photon holograms in such an ensemble would not be coherent, one could follow what was done in the time domain to sculpt Rydberg atom wavefunctions[5, 19] and make use of second-order correlations between the two-photon holograms to reconstruct a virtual quantum surface.

In the low-intensity limit, where only the photon states $|0\rangle$ and $|1\rangle$ contribute for each independent transverse spatial mode, the most general quantum wavefunction that we would want to associate with a virtual quantum surface will have the general form

$$\Psi = \sum_{\{\sigma_j\}} a(n_1, n_2, \cdots n_N) \prod_{j=1}^{N} |n_j\rangle, \tag{20}$$

where $n_j = 0, 1$ is the photon number in transverse spatial mode j and N is the number of independent spatial modes. As discussed in Sections 15.2 and 15.3, when homodyned with a reference wavefunction, this wavefunction will produce a distinctive spatial and temporal pattern of photon counting correlations on a detector plane that can be used to deduce the complex amplitudes $a(n_1, n_2, \cdots n_N)$. By comparing the pattern of photon correlations expected for some target set of amplitudes $a(\{n_j\})$ with the pattern inferred from a set of local photodetector measurements, one may hope to follow the same type of strategy that was used by Weinacht et. al. in the case of a Rydberg atom[19] to reconstruct the wavefunction for a target quantum surface. More specifically, one may hope to continuously adjust the amplitudes $a(n1, n2, \cdots n_N)$ by using the type of feedback scheme represented by 14 and 16 of Section 15.4 to continuously adjust the path lengths $\Delta l(\sigma, t)$ of the coherent pump waves used to produce the multiphoton entangled state (20). Of course, because available electro-optical phase modulators or electromechanical systems for deforming a surface are classic systems, the physical surface being used as a template will not be in a pure quantum state. However, it seems to us not unreasonable to identify a wavefunction (20) for a fixed classic template surface with a wavefunction $\Psi(\tau_1, \tau_2, \cdots \tau_N)$ for a fluctuating virtual quantum surface centered on a classic template, where surface parameters τ_i are quantized.

Describing the geometry of the virtual quantum surface associated with the wavefunction (20) is an intriguing mathematical problem. The deformations of a Riemann surface with g handles are described by g complex numbers. Interpreting these g complex numbers as Cayley–Klein parameters suggests[24] that the smooth deformations of a Riemann surface can be identified with the configuration space of g spin 1/2 quantum spins, i.e., g qubits. More generally, one may use the Wigner–Weyl quantization method for a smooth curved surface[25] to identify the appropriate quantum configuration space. Rather amusingly, it can be shown[25] that this quantization procedure brings us back to the Fock space quantization for holographic representations discussed in Section 15.2.

We have pointed out two ways in which nonclassic states of light may be used to represent the quantum fluctuations of a deformable surface. These constructions build on:

(1) The use of coherent pump waves to produce phase squeezed states
(2) The development of homodyning techniques for measuring quantum wavefunctions
(3) The use of quantum inverse scattering theory to relate asymptotic phases to deformations of a surface.

In the case of a Riemann surface, these constructions apparently provide a means to construct any desired wavefunction for g qubits. Thus, our quantum optical construction of the wavefunction for a deformable surface may provide a means for carrying out quantum computations.

15.6   Conclusion

It is worth noting in this connection that the dynamical behavior of our feedback system relating the amplitudes of detector photocurrents to the deformations of a surface is entirely determined by the path length pair correlation matrix U. This is a consequence of the linearity of the fundamental equations (14) and (15) and exactly mirrors the fundamental theorem of quantum computation,[26] which states that any quantum computation can be effected by a quantum logic gate that consists of a unitary operator acting on an arbitrary choice of two input variables.

References

[1] J.C. Solem and G. Chapline, "X-ray biomicroholography," *Opt. Eng.* **23**, 193 (1984).
[2] M. Tegze, G. Faigel, S. Marchesini, M. Belakhovsky, and A.I. Chumakov, "Three Dimensional Imaging of Atoms with Isotropic 0.5 Å Resolution," *Phys. Rev. Lett.* **82**, 4847 (1999).
[3] P.A.M. Dirac, *Quantum Mechanics* (Cambridge University Press, Cambridge, U.K., 1932).
[4] C. Leichtle, W.P. Schleich, I. Averbukh, and M. Shapiro, "Quantum State Holography," *Phys. Rev Lett.* **80**, 1418 (1998).
[5] T.C. Weinacht, J. Ahn, and P.H. Bucksbaum, "Measurement of the Amplitude and Phase of a Sculpted Rydberg Wave Packet," *Phys. Rev. Lett.* **80**, 5508 (1998).

[6] Z.Y. Ou, C.K. Hong, and L. Mandel, "Coherence properties of squeezed light and the degree of squeezing," *J. Opt. Soc. Am. B* **4**, 1574 (1987).

[7] D.F. Walls and G.J. Milburn, *Quantum Optics* (Springer-Verlag, New York, 1994).

[8] M.I. Kolobov and V. Sokolov, "Spatially multimode squeezed states of light and quantum noise of optical images," *Izv. Aca. Nauk SSSR* **69**, 1097 (1989).

[9] A. Granik and J. Caulfield, "Quantum Holography," *Holography*, SPIE Series **8**, 33 (1991).

[10] N.G. Walker and J.E. Carroll, "Multiport homodyne detection near the quantum noise limit," *Opt. Quantum Electron.* **18**, 353 (1986).

[11] J.W. Noh, A. Fougeres, and L. Mandel, "Measurement of the quantum phase by photon counting," *Phys. Rev. Lett.* **67**, 1426 (1991).

[12] D.T. Smithey, M. Beck, and M. Raymer, "Measurement of the Wigner distribution and the density matrix of a light mode using optical homodyne tomograpy: Application to squeezed states and the vacuum," *Phys. Rev. Lett.* **70**, 1244 (1993).

[13] S. Schiller, G. Breitenbach, S.F. Pereira, T. Muller, and J. Mlynek, "Quantum Statistics of the Squeezed Vacuum by Measurement of the Density Matrix in the Number State Representation," *Phys. Rev. Lett.* **77**, 2933 (1996).

[14] G. Breitenbach, S. Schiller, and J. Mlynek, "Measurement of the quantum states of squeezed light," *Nature* **387**, 471 (1997).

[15] T.J. Dunn, I.A. Walmsley, and S. Mukamel, "Experimental Determination of the Quantum-Mechanical State of a Molecular Vibrational Model Using Florescence Tomography," *Phys. Rev. Lett.* **74**, 884 (1995).

[16] D. Liebfried, D.M. Meekhof, B.E. King, C. Monroe, W.M. Itano, and D.J. Wineland, "Experimental Determination of the Motional Quantum State of a Trapped Atom," *Phys. Rev. Lett.* **77**, 4281 (1996).

[17] C. Kurtsiefer, T. Pfau, and J. Mlynek, "Measurement of the Wigner function of an ensemble of helium atoms," *Nature* **386**, 150 (1997).

[18] M. Freyberger P. Bardroff, C. Leichtle, G. Schrade, and W. Schleich, "The art of measuring quantum states," *Physics World*, November, 41 (1997).

[19] T.C. Weinacht, J. Ahn, and P.H. Bucksbaum, "Controlling the shape of a quantum wavefunction," *Nature*, **397**, 397 (1999).

[20] F.J. Dyson, "Photon noise and atmospheric noise in active optical systems." *J. Opt. Soc. Am.* **65**, 551 (1975).

[21] R.G. Newton and R. Jost, "The construction of potentials from the S-matrix for systems of differential equations," *Nuovo Cimento* **1**, 590 (1955).

[22] H.M. Farkas and I. Kra, *Riemann Surfaces* (Springer-Verlag, New York, 1992).

[23] P.G. Kwiat, K. Mattle, H. Wienfurter, A. Zeilinger, A.V. Sergienko and Y. Shih, "New high-intensity source of polarization-entangled photon pairs," *Phys. Rev. Lett.* **75**, 4337 (1995).

[24] G. Chapline, "Is theoretical physics the same thing as mathematics," *Physics Reports*, **315**, 95 (1999).

[25] G. Chapline and A. Granik, "Moyal quantization, holography, and the quantum geometry of surfaces," *Chaos, Solitons, and Fractals*, **10**, 590 (1999).

[26] D. Deutsch, A. Barenco, and A. Ekert, "Universality of quantum computation," *Proc. R. Soc.* (*London*) *A* **449**, 669 (1995); S. Lloyd, "Almost Any Quantum Logic Gate is Universal," *Phys. Rev. Lett.* **75**, 346 (1995).

Index